The Challenge of Linear Time

Leiden Series in Comparative Historiography

Editors

Axel Schneider
Susanne Weigelin-Schwiedrzik

VOLUME 7

The titles published in this series are listed at brill.com/lsch

The Challenge of Linear Time

Nationhood and the Politics of History in East Asia

Edited By

Viren Murthy and Axel Schneider

BRILL

LEIDEN • BOSTON
2014

Library of Congress Cataloging-in-Publication Data

The challenge of linear time : nationhood and the politics of history in East Asia / edited by Viren Murthy and Axel Schneider.
pages cm. — (Leiden series in comparative historiography, ISSN 1574-4493 ; volume 7)
Includes bibliographical references.
ISBN 978-90-04-26013-9 (hardback : acid-free paper) — ISBN 978-90-04-26014-6 (e-book)
1. China—Historiography. 2. Japan—Historiography. 3. Historiography—Political aspects—China. 4. Historiography—Political aspects—Japan. 5. Time—Political aspects—China. 6. Time—Political aspects—Japan. 7. Nationalism—China. 8. Nationalism—Japan. 9. China—Intellectual life—20th century. 10. Japan—Intellectual life—20th century I. Murthy, Viren. II. Schneider, Axel.

DS734.7.C38 2014
951.0072—dc23

2013028013

This publication has been typeset in the multilingual "Brill" typeface. With over 5,100 characters covering Latin, IPA, Greek, and Cyrillic, this typeface is especially suitable for use in the humanities. For more information, please see www.brill.com/brill-typeface.

ISSN 1574-4493
ISBN 978-90-04-26013-9 (hardback)
ISBN 978-90-04-26014-6 (e-book)

This book is printed on acid-free paper.

FSC www.fsc.org MIX Paper from responsible sources FSC® C109576

Printed by Printforce, the Netherlands

CONTENTS

RECOLLECTION OF THE PAST AND THE POPULARIZATION OF HISTORY

HISTORY AND THE DEFINITION OF SPATIAL, CULTURAL AND TEMPORAL BOUNDARIES

LIST OF CONTRIBUTORS

SUN GE (PhD. 2003, Tokyo Metropolitan University) is Professor of literature and intellectual history at the Chinese Academy of Social Sciences. She is a public intellectual in Japan and China and has published numerous books on Chinese and Japanese intellectual history. She is currently working on a book on the famous Japanese intellectual, Maruyama Masao.

TZE-KI HON (Ph.D. 1992, University of Chicago) is Professor of History at State University of New York at Geneseo. He has published monographs, edited volumes and many articles on late imperial and modern China, including *The Yijing and Chinese Politics* (SUNY Press, 2005), *Revolution as Restoration* (Brill, 2013), the edited volumes *The Politics of Historical Production in Late Qing and Republican China* (Brill, 2007), and *Beyond the May Fourth Paradigm* (Lexington, 2008).

YA-PEI KUO (Ph.D. 2002, University of Wisconsin, Madison) is currently a Research Fellow at the Käte Hamburger Kolleg at Ruhr-Universität Bochum, Germany. She worked as an assistant professor at Tufts University, US in 2003–2008, and was a Research Fellow at International Institute for Asian Studies, Leiden University, the Netherlands, in 2009–2010. Her forthcoming book, *Debating "Culture" in Interwar China* (Routledge), analyzes the debate on China's national identity in the 1920s and early 1930s. She has also published on the changing meaning of Confucius cult in the late Qing, and Protestant missionaries' Chinese writings in the 19th century.

HAIYAN LEE (Ph.D. 2002, Cornell University) is Associate Professor of Chinese and comparative literature at Stanford University. She is the author of *Revolution of the Heart: A Genealogy of Love in China, 1900–1950* (2007), winner of the 2009 Joseph Levenson Prize from the Association for Asian Studies for the best English-language book on post-1900 China.

LONG-HSIN LIU 劉龍心, (Ph.D., National Chengchi University) is Associate Professor of history at Soochow University in Taipei, received her Ph.D. in history from National Chengchi University, Taiwan. She works on historiography and on modern Chinese intellectual and cultural history. Her most recent publications are "*Academy and Institution: The Disciplinary*

Process and the Foundation of Modern Chinese Historiography" (2002, and revised to Chinese simplified version in 2007), and "Historical Lessons and the History of Knowledge in the Late Qing Examination System," in Brian Moloughney and Peter Zarrow ed., *Transforming History: The Making of a Modern Academic Discipline in Twentieth-Century China* (2011). She is currently working on a new book-length project on modern Chinese history and historiography within the context of "knowledge transformation" and national identity.

Viren Murthy (Ph.D. 2007, University of Chicago) is Assistant Professor in Transnational Asian History at the University of Wisconsin-Madison. He specializes in Chinese and Japanese intellectual history and is especially interested in the critique of capitalist modernity and imagining Asian identity.

Takahiro Nakajima (Ph.D. 2009, University of Tokyo) is Associate Professor of Chinese philosophy at the University of Tokyo. His publications include *The Philosophy of Evil: Imaginations in Chinese Philosophy* (Tokyo: Chikuma Shobō, 2012), *Praxis of Co-existence: State and Religion* (Tokyo: University of Tokyo Press, 2011), *The Zhuangzi*, (Tokyo: Iwanami Shoten, 2009), *Philosophy in Humanities* (Tokyo: Iwanami Shoten, 2009), *The Reverberation of Chinese Philosophy: Language and Politics*, (Tokyo: University of Tokyo Press, 2007). He is now interested in the phenomenon on Confucian revival in East Asia.

Naoki Sakai (Ph. D. 1982, University of Chicago) is Goldwin Smith Professor of Asian Studies at Cornell University. He has published in a number of languages in the fields of comparative literature, intellectual history, translation studies, the studies of racism and nationalism, and the histories of semiotic and literary multitude—speech, writing, corporeal expressions, calligraphic regimes, and phonographic traditions. His publications include: *Translation and Subjectivity* (in English, Japanese, Korean, German forthcoming); *Voices of the Past* (in English, Japanese & Korean); *The Stillbirth of the Japanese as a Language and as an Ethnos* (Japanese and Korean); *Hope and the Constitution* (in Japanese; Korean forthcoming). He edited a number of volumes including: *Knowledge and System under Total War: 1935–1955,* Tokyo, Iwanami Shoten, 2002; *Trans-Pacific Imagination* (with Hyon Joo Yoo), Singapore & London, World Scientific Publishing Company, 2012; *Translation, Biopolitics, Colonial Difference* (with Jon Solomon) Vol. 4, *Traces: A Multilingual Series of Cultural Theory and Translation.* Hong Kong: Hong Kong University Press, 2006; *Globalization Studies—From Total War System to Globalization* (with Yasushi Yamanouchi) Tokyo, Heibonsha, 2003; Specters of the West and the Politics of Translation (with Yukiko Hanawa) Vol. 1, *Traces: A Multilingual Series of Cultural Theory and Translation.* Ithaca: Traces, Inc., 2001; *Deconstructing Nationality* (with Brett de Bary and Iyotani Toshio) Tokyo: Kashiwa Shobô, 1996 (English translation thereof, *Deconstructing Nationality.* Ithaca: East Asia Program, Cornell University, 2005). Sakai is the founding senior editor of *Traces: A Multilingual Series of Cultural Theory and Translation.*

Axel Schneider (Ph.D. 1994, Bochum University) is Professor of Modern Sinology at the University of Göttingen. He specializes in modern Chinese intellectual history, especially the history of historical writing and historical thinking.

Susanne Weigelin-Schwiedrzik (Ph.D. 1982, Ruhr University Bochum) is a professor of Chinese Studies and Vice Rector for Research and Career Development at the University of Vienna. She has published on 20th century Chinese history and historiography and is currently completing a book on East Asia in the 19th and 20th centuries. She has also published articles on memory issues related to the Great Famine and the Cultural Revolution.

INTRODUCTION

Viren Murthy and Axel Schneider

The papers collected in this volume congeal around a debate about the ways and extent of the dominance of linear time and progressive history and the concomitant delineation of the nation in Chinese and Japanese historiography. During the late nineteenth and early twentieth centuries, as China and Japan entered the global capitalist system of nation-states, the Chinese and Japanese regimes implemented a number of reforms, which resulted in transformations that affected everyday experience. In the face of imperialism and the perceived threat of being split up, the Meiji and late Qing governments radically reoriented policies in order to become wealthy and powerful in the global arena. They encouraged business, focused on Western style education and attempted to extend state control to local areas through legal reforms. The effects of these transformations cannot be fully discussed here, but for our purposes, increased commodification and international trade, the acceptance of international law in the form of signing treaties, along with the circulation of new ideas due to educational reform are especially important. International trade and diplomatic relations required that Japanese and Chinese rethink their notions of time and space and new conceptions of time found expression in Japanese and Chinese society.[1] People not only began to experience time and space in new ways, but elites also were increasingly exposed to Western theories of history and concepts of nationhood, which became dominant. These changes contributed to the production of new types of historical consciousness and collective identity.

Recently, a number of authors have charted the shift in East Asia from traditional modes of understanding time and history to more linear models. For example, focusing respectively on Japan and China, Stephan Tanaka and Prasenjit Duara both argue that beginning in the late nineteenth century, intellectuals began to consider history as linear so as to legitimate the modern nation-state.[2] Faced with the threat of Western

[1] Tanaka 2004, Kwong 2001, Wang Fansen 2011.
[2] Tanaka 2004, Duara 1995.

imperialism, intellectuals felt it imperative to develop a strong nation-state and historians wrote history as part of a project to create the subjective conditions for a wealthy and powerful country. In this context, history was seen as race track in which nations competed to achieve their goal, namely a strong nation-state, as quickly as possible.[3] Through reading such histories, people were to be transformed from passive subjects of a monarchy to active national citizens. As this mode of narrating history became dominant, other modes of storytelling and history writing were suppressed.

The essays in this volume develop the idea that the dominance of linear temporality and progressive history, but at the same time, the authors also complicates the above narrative in a number of ways. By focusing on how intellectuals inscribed previous modes of writing narratives in order to create history, the authors in this volume illuminate the different choices made by intellectuals in China and Japan during this period of immense transformation and the often unpredictable theoretical results that emerged from their conceptual maneuvers. In both China and Japan, before the nineteenth century, scholars envisioned the past by invoking temporal narratives. For example, classical historiographical works about Chinese pasts refer to the moral and political authority of the Three Dynasties and thus often told a story of cyclical change or of decline.[4] The two oldest classic historical works in Japan, the *Ancient Records* (*Kojiki*) and the *Chronicles of Japan* (*Nihon shoki*), both embedded narratives about temporal development in a world emanating from gods and goddesses and were bereft of a simple notion of moral progress. It is the role these configurations, now turned into "traditions" by the arrival of Western "modernity," played in China and Japan in the adoption and transformation of linear models of time and history that forms our horizon of inquiry.

The essays in this volume examine how early twentieth-century Chinese and Japanese intellectuals and officials rejected previous temporalities and resources from the past, or incorporated such resources into linear narratives, and/or drew on them to resist linear forms of historical writing and historical representation. Through a multiplicity of strategies, the authors of this volume illuminate not only the complexity of the histories of Japan's and China's respective incorporation into the global capitalist

[3] Tang 1996, Yü, 1994, Schneider 1997.
[4] Sato 2007, Schwartz 1996, Yü 1999, and Chevrier 1987.

system of nation-states, but also the contradictory nature of modernity itself. Japan and China were of course latecomers to modernity and as a consequence, Chinese and Japanese intellectuals not only attempted to catch up and compete with their Western counterparts, but to varying degrees questioned the linear modernist framework within which history was written. Thus, in analyzing how some intellectuals adopted and others questioned the linear narrative of history, it is clear that a framework larger than modernity and the nation-state is required. Although the essays in the volume do not provide a single, unified framework for making sense of the various narratives of history and time that emerged in the twentieth century, several of them propose theoretical strategies for building such a framework.

We have organized the papers around the following themes: Time, History, and Moral Responsibility; The Burden of the Past and the Hope for a Better Future; Recollection of the Past and the Popularization of History; and History and the Definition of Spatial, Cultural, and Temporal Boundaries.

Naoki Sakai opens the first section by examining the historical writing of Ienaga Saburō and Maruyama Masao, intellectuals who would become famous in postwar Japan. Although today Ienaga and Maruyama are well known for having been promoters of democracy, Sakai analyzes the intellectual histories they wrote earlier, during the interwar period, and the way in which they each drew on the Kyoto School philosophers Tanabe Hajime and Miki Kiyoshi. Sakai reminds us that despite their reputation as liberals, both Ienaga and Maruyama wrote histories that were imbricated in discourses about civilization and were intimately connected to the Japanese project of creating a multiethnic empire in East Asia. By underscoring how the philosophies of Tanabe Hajime and Miki Kiyoshi informed the respective histories of Ienaga and Maruyama, Sakai poses fundamental questions about the paradigm of historicism present in these two authors, a paradigm which includes problems of dividing geographical space and periodizing linear history.

Viren Murthy tackles the problem of linear time and history from a different angle by discussing the late Qing intellectual Tan Sitong's famous work *A Study of Cosmic Love* (*Renxue*) and his contemporary Zhang Taiyan's critique of this work. In the late nineteenth century, both Tan and Zhang developed visions of evolution but at the same time drew on categories from premodern Chinese thought. Tan's *Study* drew on the concept of humanity prevalent in Confucianism and synthesized the ontological reading of this concept during the Song and Ming periods with modern ideas about science and progress. Zhang Taiyan countered this idea of progress by developing a theory of evolution that combined concepts from Daoism, *The Book of Changes*, and modern biology. Such an eclectic fusion of ideas can be understood in terms of salient notions of hybridity, but Murthy suggests that one should conceptualize the theories of Tan and Zhang in relation to the epistemological transformations occurring in modern capitalism.

Through his analysis of Liu Yizheng, a historian writing in republican China, Axel Schneider addresses the intersection of morality and history. Schneider shows how Liu, who started out as a modernizer translating Japanese textbooks that were applying the new linear narrative of history to East Asia, developed an alternative to progressive models of history. He stresses Liu's emphasis on Confucian ethics as part of his critique of linear models of history and the concomitant suppression of the role of ethics and human agency. Against much of the historiography that stresses the centrality of the May Fourth movement and its iconoclasm, Schneider shows that Liu related to older notions of time and history, put them to work in a new context, and used them to argue for a view of history and historiography centered on ethics while at the same time adopting new methods of research introduced by modern Western historical science.

Sun Ge combines Chinese and Japanese contexts and examines the epistemological presuppositions of writing history through reading the works of the postwar Japanese sinologist and literary critic Takeuchi Yoshimi. She invokes Takeuchi's works as a retort to what she sees as the contemporary Chinese attempt to stuff reality into ready-made theoretical frameworks such as evolutionary or progressive versions of history. Sun explores this opposition between ossified theoretical frameworks and an emphasis on history by looking at a debate between the Marxist Toyama Shigeki and Takeuchi Yoshimi about the commemoration of the Meiji Restoration in postwar Japan. She underscores that Takeuchi counterposed Toyama's concept of social responsibility to his own ideal of scholarly responsibility. Through this, Sun touches on the rift between history writing and making history, which continues to haunt intellectuals committed to radical social transformation.

Moving to the next theme, "The Burden of the Past and the Hope for a Better Future," Takahiro Nakajima's essay also focuses on the work of Takeuchi Yoshimi during the postwar period but brings out both his concept of morality and his eschatological view of time. Nakajima turns our attention to Takeuchi's methods to convey the experience of war to a generation of Japanese who had not experienced it. Invoking an obscure passage in a review that Takeuchi wrote of his friend Takeda Taijun's book on Sima Qian, Nakajima teases out Takeuchi's notion of eschatological time, which he compares to Walter Benjamin's idea of history. Eschatological time, like Benjamin's messianic time, refers to an era of human liberation. In Nakajima's view, Takeuchi believed that coming to grips with the wartime past was a precondition for affirming this eschatological future.

Susanne Weigelin-Schwiedrzik continues this theme of imaging a different future in the context of the 1970s in China by focusing on the Campaign to Criticize Lin Biao and Confucius. She contends that Chinese history, far from being seen as simply linear, has been perceived by many as characterized by cyclical returns of tropes and rhetorical practices in different contexts. Hence the possibility of a different future for China will depend on breaking free from practices that have haunted China from ancient times.

In the next section, "Recollection of the Past through Different Media," Long-hsin Liu examines popular representations of history during the 1930s and wartime China. She shows that attempts by Chinese historians to write a popular form of history met with two challenges. First, how to write a popular history that would not immediately give rise to the suspicion on the side of the Guomindang government that these historians were taking sides with the Communists? Second, how to combine new contents with old forms so as to attract the largely rural, uneducated audience and yet convey a new, modern message to them? Because of the nationalist mood during wartime China, the historians of the Popular Reading Publishing House ultimately failed to avoid the suspicion of being leftist and never managed to bridge the gap between intellectual writers and the people.

Haiyan Lee discusses the representation of the past in museums during the 1950s in China. She analyses the notions of time, race, class, and nation in the exhibition "The Rent Collection Courtyard," which was displayed in a museum in Sichuan. The representation of history in this exhibition constructed the landlords as class enemies of progress. Drawing on the work of Etienne Balibar, Lee suggests that this representation is a type of class-racism. In other words, she contends that in the historical narrative of Communist China, class often functioned like race to designate the inimical Other, which negatively defined one's own identity.

Tze-ki Hon and Ya-pei Kuo bring this volume to a close with their essays in the section on "History and the Definition of Spatial, Cultural, and Temporal Boundaries." Hon argues against the widespread interpretation of the scholars known as the National Essence group as regressive or reactionary. Although they looked to Chinese antiquity in their search for China's "national essence," they reinterpreted the Three Dynasties in such a way that they were able to justify a pluralistic view of society and politics. Their aim was to ground their vision of racial (anti-Manchu) and political (anti-monarchical) revolution in history, without, however, depriving the successors of the traditional gentry elite of their role as sociopolitical leaders, because it was still in the classics—interpreted by the elite—that this vision was expressed most clearly.

Ya-pei Kuo returns us to Liu Yizheng and his famous text, *A History of Chinese Culture*, and explains Liu's use, during the 1910s and '20s, of the term "culture" to demarcate the difference between Chinese and non-Chinese. In stark contrast to the Liu Yizheng of the 1930s and '40s, in his *A History of Chinese Culture* Liu is fully situated within a modern, secular, and linear concept of history, emphasizing the civilizational role of the people while downplaying aristocratic and religious aspects of early Chinese history. However, unlike Hu Shi's objectivistic approach to history as science, Liu wanted to maintain a positive link to the past through an empathetic reading of history.

Both, Hon Tze-ki and Kuo Ya-pei again show how complicated and heterogeneous the transition from traditional historiography to modern history as an academic discipline was and how closely this process was intertwined with the task of nation-building and competing political visions.

References Cited

Chevrier, Yves (1987), "La Servante-Maîtresse: Condition de la référence à l'histoire dans l'espace intellectuel Chinois," in *Extrême-Orient, extrême-occident IX*, no. La référence à l'histoire (1987): 117–44.

Duara, Prasenjit (1995), *Rescuing History from the Nation: Question Narratives in Modern China and India.* Chicago: Chicago University Press.

Kwong, Luke S.K. (2001), "The Rise of the Linear Perspective on History and Time in Late Qing China 1860–1911," in *Past and Present* 173 (2001), 157–190.

Sato, Masayuki (2007), "The Archetype of History in the Confucian Ecumene," in *History & Theory* 46 (2007): 218–32.

Schneider, Axel (1997), *Wahrheit und Geschichte: Zwei chinesische Historiker auf der Suche nach einer modernen Identität für China.* Wiesbaden: Harrassowitz Verlag.

Schwartz, Benjamin I. (1996), "History in Chinese Culture—Some Comparative Reflections," in *History and Theory* Theme Issue 35 (1996): 23–33.

Tanaka, Stephan (2004), *New Times in Modern Japan.* Princeton N.J.: Princeton University Press.

Tang, Xiaobing (1996), *Global Space and the Nationalist Discourse of Modernity: the Historical Thinking of Liang Qichao*. Stanford, CA: Stanford University Press, 1996.

Wang, Fansen (2011), "The Impact of the Linear Model of History on Modern Chinese Historiography," in Moloughney and Zarrow eds. (2011), *Transforming History: The Making of a Modern Academic Discipline in Twentieth-Century China*. Hong Kong: Chinese University Press, 135–68.

Yü, Ying-shih (1994), "Changing Concepts of National History in Twentieth-Century China," in Lönnroth, Molin, Björk eds. (1994), *Conceptions of National History. Proceedings of Nobel Symposium 78*. Berlin/New York, 155–174.

—— (1999), "Überlegungen zum chinesischen Geschichtsdenken." in Rüsen, Jörn ed. (1999), *Westliches Geschichtsdenken, Eine interkulturelle Debatte*. Göttingen: Vandenhoek & Ruprecht, 237–268.

TIME, HISTORY, AND MORAL RESPONSIBILITY

CHAPTER ONE

NEGATIVITY AND HISTORICIST TIME: FACTICITY AND INTELLECTUAL HISTORY OF THE 1930S

Naoki Sakai

Allow me to begin with an open acknowledgment. Indeed, it is plausible to recognize something oriental, or Asiatic, in the thought of Japanese thinkers. Without much difficulty we may be able to uncover a few dozen if not a hundred, genealogical passages that trace back to some origins or features that are characteristically oriental or Asiatic, or to some institutionalized traditions such as Confucianism, Shintoism, and Buddhism—and even Hinduism and Islam—that were once believed to be authentic markers of the Orient or the non-West. The problem we would then confront is how to pin down the distinctive identities of these civilizational categories, the Orient, Asia, or the West, to which we customarily allocate a range of "religions." Today, we are increasingly insecure about the transferability of religious identities onto the cartographic configuration; we no longer believe that the world map can be a colored mosaic of religious regions. At primary and secondary schools, we used to be taught that Eastern Europe belongs to the Orthodox, Southern Europe to the Catholics, Northern Europe to the Protestants, North Africa and the Middle East to the Moslems, South Asia to the Hindus, Southeast Asia to the Theravada Buddhists and the Moslems, and Northeast Asia to a combination of Confucians and Mahayana Buddhists.[1] Being better informed on the complexity of the distribution of religions, we cast a skeptical look at such a world map today. This is partly because we are not as sure as before of what "religion" means. In addition, we are aware of a history in which the notion of "world religions" was constructed in reaction to the secularization of Christianity,[2] on the one hand, and to the globalization of colonial modernity on the other. But, most important of all, we are

[1] The term "Theravada Buddhism" itself was coined in Mahayana Buddhism, hence the very opposition of Theravada and Mahayana Buddhisms was promoted from the viewpoint of the historians who mostly conceived of the history of Buddhism from the modernizationist perspective.

[2] See Masuzawa 2006.

aware of the makeshift nature of civilizational categories themselves, so that we now know that they are even less historically enduring or persistent than the once-cherished identity of the "nation." In due course, only with a certain semantic frivolity do we use these civilizational categories; we have learned to refrain from expecting too much coherence or conceptual rigor in the terms Orient, Asia, Africa, or the West. We know of another set of categories that we have learned to use with careful but similar frivolity: the racial categories.

We are tired of reasoning based on the categories of race; we are wary of syllogisms such as this: "Blacks conduct themselves in this or that way. Of course, this particular individual is black. *Therefore*, he is bound to think or feel this or that way." In this respect, civilizational categories are not dissimilar to racial categories in their functioning in historical reasoning. It goes without saying that we do not divest semantic seriousness from civilizational categories because we are lighthearted about what they do in history. On the contrary, we stop ourselves from investing too great a desire for coherence in these terms—just as we caution ourselves about racial categories—precisely for their inordinate, aberrant, and disastrous consequences in history.

Despite the apparent prudence often granted to a genealogy of civilizational character, can we demand as an imperative for our academic argumentation that one ought to *resist* this civilizationist recognition or identity politics? Can we *hesitate*, legitimately and unwaveringly, before the civilizational categories are applied in the reading of Japanese texts? Can we put together a persuasive narrative without the endorsement of the civilizational categories beyond casual uses of them as geographic markers? How can we be *just* in refusing to proceed in the protocol of a civilizationist reasoning? What are we then to say by hesitating to endorse the categories of civilizational distinctions when confronted with the texts of Japanese thinkers of the 1930s and early 1940s? It goes without saying that the problematic that guides this entire chapter is my query as to how one can possibly write a history without slipping into the conventions of history writing based upon civilizational categories such as "the History of Western Philosophy" or "the History of Japanese Thought."

Let us note that these texts we are about to discuss are far from being innocent of either civilizationist or racial categories in themselves; how are we then to read them without committing to the preliminary incentives of civilizational categories, incentives which could easily lure us into reading them according to the categorization of the Occident and the Orient, of the West and the Rest? As a matter of fact, we can hardly argue that the categories of the Occident and the Orient, for instance, were imposed upon these Japanese texts against their immanent orientations, and hence were secondary with regard to their historicist projects. Without doubt the Japanese thinkers of the 1930s and early 1940s conceived of the historical existence of the human being starting with speculations on the categories of civilizational identities such as the Occident, Orient, Christianity, and Buddhism. Nevertheless, many of the leading intellectuals of the 1930s and early 1940s—particularly those philosophers and intellectual historians to be discussed in this chapter except for such ethnic nationalists as Watsuji Tetsurō—were hardly free of the lure of universalism that attempted to transcend civilizational particularities. However, it is also true that they were committed to the imperial nationalism of the Japanese empire, to an imperial nationalism that considered itself to be exceptional in its universalism. In order to unravel the interlacing of universalistic claims and national exceptionalism, therefore, I insist that we ought to *hesitate* before yielding to civilizational categories.

Intellectual History in the Multiethnic Empire

During the Asia-Pacific War, which dragged on for nearly fifteen years from 1931 until 1945, Japanese academia witnessed the publication of two outstanding works in intellectual history, Ienaga Saburō's *Development of the Logic of Negation in Japanese Intellectual History*[3] and Maruyama Masao's *Intellectual History of Tokugawa Japan* (*Nihon seiji shisō-shi kenkyū*).[4] After the demise of the Japanese empire in 1945, the two historians, Ienaga (1913–2002) and Maruyama (1914–1996), would be known, not only for their roles as the leading historians of Japanese thought and culture but also for their reputation as public intellectuals and their commitment to postwar democracy. Already in their wartime works, one can find many seminal insights and theoretical formulations to inspire their argument after the war, whereas, implicitly but incorrigibly, their scholarship after the defeat was no longer inspired by the ideal of a multiethnic empire. In due course a deep-seated dislocation can be detected in their postwar erudition; retrospectively it is evident that, after the loss of the empire, Ienaga and Maruyama stopped addressing large portions of their prewar audience whom their scholarly work used to eagerly

[3] Ienaga 1997: 5–78.
[4] Maruyama 1952, 1974.

embrace as "fellow countrymen." Immediately after the initial serial publication in the journal *Rekishigaku kenkyū* (Historical Study) in October, November, and December 1938 and prior to the publication as a monograph in Japanese in November 1940, *Development of the Logic of Negation in Japanese Intellectual History* was translated into Chinese and published in the Chinese journal *Beijing Modern Scientific Library Publications* in 1939.[5] At the time of its initial publication, Ienaga was obviously aware that his article was addressed to readers not only within the territory of the Japanese empire but also to some areas in China under Japanese control. The same can be said about the first and second parts of Maruyama's *Intellectual History of Tokugawa Japan.* As soon as the Japanese empire was lost, Ienaga's and Maruyama's mode of address seemed to undergo a drastic change. No longer was it, for either of them, a matter of primary concern whether their argument was of any relevance to readers on the Korean peninsula or in Taiwan, for instance. This aspect of their intellectual history, in which some fundamental dislocation took place, has been long overlooked, but how one speaks, whom one addresses, and with whom one discusses are not something accidental to the genre of intellectual history.

When Japanese intellectuals were engaged in debates concerning the East Asian Community in the 1930s, and later the Greater East Asian Co-prosperity Sphere in the early 1940s, they were aware that they had an audience in Korea, Taiwan, and even China. But, as the Allied Powers occupied Japan proper, with very few exceptions such as Takeuchi Yoshimi, they seemed to have given up any efforts to persuade those who had now dropped out of the national imaginary.

Maruyama Masao's life was closely interwoven with the lives of Koreans since not only was his father a noted journalist managing the *Keijō nippō* (Seoul Daily) in the 1920s but also he himself was assigned to a military station in Pyongyang as a draftee in the last phase of the war, prior to his transfer to Hiroshima. One would presume it more than plausible that, in his postwar work, he could have referred in one way or another to how he was involved in the ongoing campaign of the empire. Yet, he remained silent in his scholarship about his mid-life rupture induced by the loss of the empire (Maruyama was 31 years old when Japan surrendered) until

---

[5] *Rekishi kenkyū* (Historical research), vol. 8, no. 10, 11, and 12, 1938. The monograph was published by Kōbundō in Tokyo, entitled "Nihon shisōshi ni okeru hitei no ronri no hattatsu," in Chinese in *Beijing Jindai Kexue tushuguan kan* no.6 and no. 7, 1939, trans. By Zhang Wojun (張我軍).

his death in 1996, and it seems that he never openly acknowledged his implication in the ideological campaign of the multiethnic empire.

In contrast to Maruyama, the postwar scholarship of Ienaga Saburō was openly redemptive. He candidly wrote about his involvement in the war and said that he believed it to be a just war and that he contributed much through the production of academic knowledge to the legitimization of the Japanese state's policies towards East Asia and the Pacific (Ienaga, 1998). One can easily recognize in almost every publication and intellectual activity of Ienaga Saburō in the latter half of the twentieth century the persistent pledge that propelled his scholarship throughout the postwar period: repentance for and critical reflection on his involvement in wartime nationalism.

In English-speaking academia, both Maruyama Masao and Ienaga Saburō have been commemorated as testimonies to the conscience of Japanese democracy and to the presence in Japanese society of those who courageously resisted the Cold War alliance of the United States anticommunist conservatives and the former Japanese colonial bureaucracy and totalitarians under the increasingly global domination of the U.S. collective security system. They were often seen as symbols of liberal but anti-American democracy in Japan where the term "liberal democracy" had been appropriated—in the name of the conservative party that has been in power more or less continually since 1955 until September 2009, for instance—by pro-American conservatives with a direct inheritance from the prewar imperialist bureaucracy and big business. Not surprisingly there have been few occasions where the two intellectual historians' implication in Japanese imperial nationalism has been rigorously investigated.

However, the task I undertake in this chapter is not to assess the extent to which their historiography adhered to their commitment to wartime Japanese imperial nationalism. Rather, it is a preliminary groundwork for such an investigation. In other words, I am concerned with the general topics of historicism and the manufacture of subjectivity that continued to haunt many intellectuals from the 1920s well into the 1960s. What I pursue here is an outline of the configuration, or paradigm if not strictly in the Kuhnian sense of the word, of historicist problematics in which Maruyama Masao's *Intellectual History of Tokugawa Japan* and Ienaga Saburō's *Development of the Logic of Negation in Japanese Intellectual History* should be read. For this purpose, it is necessary to refer to Tanabe Hajime (1885–1962) and Miki Kiyoshi (1887–1945), philosophers representative of the 1930s, whose works both Maruyama and Ienaga were heavily indebted to.

The Paradigm of the 1930s and the Narrative Tropes of Periodization and Civilizational Cartographic

I would like to point out that the two methods of figuration guided historical investigations of Ienaga Saburô and Maruyama Masao. The first is the problematic of comparative evaluation, consisting of two separate pairings; the first is a geopolitical opposition between Japan and China, and the second a chronological comparison between two distinctive periods: the pre-Christian era (antiquity) and the Christian era (the medieval period) in European history. Evidently, because of these pairings introduced in their works, how to manipulate civilizational categories is of decisive significance in Ienaga's and Maruyama's historiographies. But, it is also important to note that, as I will elucidate in reference to Miki Kiyoshi's *Rekishi tetsugaku* (The Philosophy of History), civilizational categories carried different significance and referred to a different set of questions, depending upon how "the register of history" is taken as a point of focus.

The second problematic is that of the negativity in terms of which they articulated the issues of *subjectivity* to the formation of historicist consciousness. It is because of their commitment to the philosophical discussion of *subjectivity* in terms of negativity that their approaches to history can be qualified essentially as belonging to historicism. The Hegelian notion of *mediation* which Tanabe Hajime followed as the guiding thread in his social ontology in *Logic of Species*, therefore, played an essential role in both Ienaga's and Maruyama's texts.[6]

What is to be pursued in my reading of *Development of the Logic of Negation in Japanese Intellectual History* and *Intellectual History of Tokugawa Japan* is how the comparative method was used to construct the historical narrative of modernity during the 1930s, which would

[6] *The Logic of Species* (*Shu no ronri*) usually refers to the philosophical ontology of social beings that Tanabe Hajime attempted to construct mainly in a series of essays published from 1932 until 1946. The *Logic of Species* has two theoretical focuses: Hegelian dialectic reformulated from the view point of mathematical concept of discontinuity to comprehend social identities such as ethnicity, nationhood, individual, social class, and the state, and the schematism of the world in which a new national subject is transformed and manufactured. It is important to stress the universalistic tendency of the *Logic of Species* and the fact that it was appropriated by the Japanese state for the construction of the East Asian Co-prosperity Sphere. Ienaga Saburō was particularly attracted to the *Logic of Species*. Cf. Ienaga discussed how he was attracted to Tanabe's philosophy in his book entitled *Tanabe Hajime no shisô-shiteki kenkyû—sensô to tetsugaku-sha* (An intellectual history of Tanabe Hajime—war and philosopher), in Ienaga 1998: vol. 7.

continue to provide the framework for the discussion of modernity in postwar Japan.

The geopolitical figuration, which prepares the juxtaposition of Japan as the sphere for Japanese intellectual history, is straightforward. Both historiographic ventures clearly fall into the genre of intellectual history precisely because Japan is identified as a topos of intellectual production, and the positing of China enables this identification. Neither Ienaga nor Maruyama bothered to elucidate what is implied by China. Their positing of China is very much like the positing of the Rest in the Eurocentric identification of the West in the paradigm of the West and the Rest. What is ambiguously and unspecifically identified as China makes it possible to speak of Japanese thought as a unity with characteristics. Neither Ienaga nor Maruyama is particularly interested in the positive originality of the national tradition. In this respect, Japan is not the site of the cultural origin or of the primal beginning for the two intellectual historians. After all, they would never deny the derivative nature of Japanese Buddhism, the main topic for Ienaga, and of Japanese Confucianism for Maruyama. The focal point of their inquiry is how Buddhism and Confucianism transformed social reality and were transformed by social praxis in Japan. Already it is implicit that Japan is not an entity to be given but is instead manufactured through social praxis. Yet this topos is distinguished only insofar as another topos is juxtaposed with it. Japan emerges as a meaningful location from their comparative perspective only when China is posited as a representative figure of the Rest of Asia. It goes without saying that Ienaga and Maruyama are both engaged in the imperialist enterprise of demonstrating the exceptional character of Japan in Asia, of its leadership in the modernization of Asian people. And, in investigating how Buddhism and Confucianism were re-interpreted, re-constructed, and re-formed in Japan, both unambiguously refer to a certain historical narrative of Christianity. One can hardly fail to recognize that they are, wittingly or otherwise, committed to a certain exceptionalist logic similar in structure to that of Eurocentrism. Although the Christianity of late antiquity (the third and fourth centuries AD) is taken as the point of reference, what is at stake in their respective arguments is modernity that is believed to distinguish Japan from the Rest of Asia just as the West was then and is still now, to a large extent, believed to distinguish itself from the Rest of the World.

Comparative Periodization and Praxis in History

The debate over the essential difference between Greek antiquity and the Christian era started in 1930 when Tanabe Hajime published an essay entitled "Nishida sensei no oshie wo aogu" (Request for Professor Nishida's Guidance).[7] This scheme of periodization was initially concerned with the relationship between philosophy and praxis, and this debate itself was evidently a response to Marxism, in which the philosophy of praxis was the central issue. And, this issue of praxis was closely related to history as well.

In reference to a fairly well-established characterization of the two periods variously advocated by Neo-Kantians, Bergson, and so forth, Tanabe introduced two contrasting features of ancient Greek philosophy and the radical change introduced by Christianity. Classic philosophy is a sort of natural ontology that aspires to discover the harmonious order of the universe, and seeks a ruptureless accommodation of the past in the present. It is ultimately a form of aestheticism that finds its contentment in quietist meditation on the innate order of the universe. Eventually this natural ontology proved to be fatally inadequate when the problem of the individual presented itself as a contradiction to formal logic, the characteristic motor of Greek ontology, of the genus and the species. The individual thing cannot be understood either as the most specific in the hierarchical classification of the genus and the species. For as Tanabe Hajime insisted, there is no room for the individual in the natural ontology of ancient Greece.[8] As we know, Tanabe would further elaborate this insight into the fundamental contradiction of Christian universalism of the individual against Aristotelian logic in terms of his *Logic of Species* (*Shu no ronri* 種の論理).[9]

Tanabe tried to illustrate the historical significance of what his mentor, Nishida Kitarō, had been struggling with. Using the language of formal logic, Nishida was reaching the stage where he could demonstrate that what he termed *jikaku* or self-awareness—or apperception in Kantian terminology—cannot be theoretically determined within the scope of for-

7 Tanabe 1963a: vol. 4, 303–28.

8 See "Sonzairon no dai san dankai" 存在論の第三段階 (The Third Stage of Ontology) in Tanabe 1963a: vol. 6, 269–98.

9 I provide a rough outline of Tanabe's *Logic of Species* in Sakai 2000.

mal logic.[10] The individual was a scandal in the Greek ontology of nature, and the notion of spiritual personhood, for instance, was foreign to the Greeks of antiquity. By referring to this periodizing scheme in this way, Tanabe underlined two aspects of Nishida's logic of *basho* or *khōra*.

On the one hand, Nishida demonstrates that the agent of enunciation (the subject of enunciation) in the construction of a proposition can never be located at the site of the propositional subject (*shugo*; this is also "the subject of the enunciated"), and that what he calls "self-awareness" or the sense of "I" necessarily embodies a logical rupture between "I" as the propositional subject posited as the theme or subject matter of knowing (*shudai*) and the agent of predication which necessarily eludes thematization. Therefore, he argues that it is necessary to assume something like the Platonic *khōra*, which can accommodate the logical incommensurability between the propositional subject and predication but remains unaffected by the operation of predication. It is well known that Nishida called this *basho*, and that, just like the Platonic *khōra*, it makes the determination of predication in the propositional form possible, but it cannot be determined either as an object of inquiry (being posited as the *shudai*, the theme = subject, in the proposition) or as a function of the predicate. On the other hand, he follows Aristotelian logic to the point where so-called formal logic cannot deal with the being of the individual. Nishida expressed the very limit of formal logic with the famous formula "the individual infinitely transcends even the most specific of the species." What is customarily referred to as the auto-designation of the individual human being must be a contradiction, which cannot be accommodated in the Greek logic of the species and the genus. It can only be hinted at in the *predicative* direction of the proposition. The subject (*shutai*) of self-awareness irreducibly and essentially eludes identification. The subject (*shudai*) of praxis can be neither signified nor indicated by the subject (*shugo*) of the proposition. The human individual as an agent of praxis always escapes the logic of identity (*dōitsusei ronri*). What Nishida's inquiry of the individual thing disclosed was the

10 By "the individual," Nishida never yielded to the liberal logic of individualism. He always said "the individual thing," began with a reference to "the individual thing (*kobutu*)" within the scope of Aristotelian logic, but he did not endorse the classical treatment of the individual things in philosophy. Nor did he immediately equate the subject in the sense of propositional subject to the subject in the sense of the human agent of praxis.

very limit of formal logic, beyond which the founding economy of the species and the genus must change quantitatively and radically. This explains why, following Nishida's insight, Tanabe deliberately inherited the trinity of classical terminology: the species, the genus, and the individual, of formal logic. But he made them mean entirely different things in the *Logic of Species*. For Tanabe, Nishida's *Logic of Basho* marked an axiomatic event involving the entire field of philosophy. While Nishida pursued the problem of modern subjectivity, however, Tanabe suspected that his mentor could not get rid of the style of questioning, which was so indicative of the natural ontology of antiquity.

Following Nishida's analysis of "you" and "I" in "Watakushi to nanji" (I and Thou), Tanabe asserts that the individual can exist only in relation to another individual as a practical *subject* (*jissen no shutai*). Therefore, the individual cannot be comprehended in natural ontology where there is no conceptual capacity for creative individual beings. Only Christianity could deliver the teaching of negative reversal, of providing eternal life to those ready to sacrifice their lives. According to Christian teaching, the spiritual being is regained only in the negation of nature, and by way of rupture and transformation that posits the future as a negation of the past. Thus, Tanabe claimed, an ontology of the human spirituality replaced the natural ontology of ancient Greece.[11]

It goes without saying that pairing of the ancient Greek ontology vs. the Christian ontology of practical subjectivity is only schematic. But, to the extent that every measure of periodization functions as a figurative trope whereby to organize historical narrative, it is one of the most important periodizing devices.

Whereas Tanabe himself rarely referred to the dichotomy of Japan and China, Ienaga and Maruyama drew a parallel between this periodizing scheme and the geo-political opposition of China and Japan. Nevertheless, it is misleading to say this association of two dichotomous schemes establishes a marked contrast or division separating two territories. Although the pairing of Greek antiquity and the Christian era was aimed at the establishment of civilizational categories of the modernity of self-transforming Japan and the pre-modernity of stagnant China, it did not necessarily succeed in stabilizing them. Rather, I think, it allows us to see the dynamic nature of the relationship between the dominant and the dominated, much like the distinction of the West and the Rest, for such

11 Tanabe 1963a: vol. 6, 277.

a distinction has to be repeatedly established. Certainly, it is a type of mapping that projects two different domains, but it does not necessarily conjecture a one-to-one correspondence between cartographic representation (map) and physical reality (geographic space). Just as the West cannot be conceived of as an enclosed geographic space on the surface of the Earth, neither is Japan nor China an index of a geographic area. It introduces a certain hierarchy between Japan and China, a hierarchy closely associated with the notions of progress and modernization. But, there is virtually no guarantee that such unevenness persists all by itself.

We have so far followed a peculiar itinerary of historicist reasoning pertaining to civilizational categories. The dichotomy between Greek antiquity and the Christian era appeared, in the first instance, as a chronological one in terms of which one usually narrates the historical change of the Roman empire and establishes the two millennium of Christianity as the continuing tradition of the West. But, this very chronology begins to serve as something entirely different as soon as it is associated with a different pairing of Japan and China, and the emergence of Christian historicism.

We must now inquire into the ways periodization works with a view to negativity and subjectivity. For, history is a matter of the individual's praxis.

Three Registers of History and the Fact (jijitsu)

To further explore the implications of this periodizing trope, let me rely upon the theoretical formula that Miki Kiyoshi, the first translator of *The German Ideology* into Japanese,[12] introduced in his *Rekishi tetsugaku* (The Philosophy of History) (1932).[13] In terms of Miki's formula, we discern the three dimensions or registers in the very use of the periods that Tanabe referred to as Greek antiquity and the Christian era. Miki argues that the word "history" can be deployed with three different conceptual directives: (1) history in the register of *logos* (*logosu to shiteno rekishi*); (2) history in the register of being (*sonzai to shiteno rekishi*); and (3) history in the register of fact (*jijitsu to shiteno rekishi*).

First of all, history is invariably a story, and always tied to narration virtually in all the languages. Periodization in the register of *logos* refers

---

[12] Marx 1930.
[13] Miki 1967b.

to a narrative device or trope by which the historical narrative is organized since history in the register of *logos* emphasizes history as story or as a narration. Second, periodization marks the different types of social formation, polities, customs, economic conditions, dominant faith, and so forth. In short, periodization specifies the chronologically differentiated groups of referents to which the historical narrative is referred. Yet, we cannot overlook the Augustinian insight that one never has direct access to referents in the past or in the future, for things of the past must be remembered in the present and things of the future anticipated in the present. What comes to be at stake in history in the register of being is that the very mode of reference of the narrative of history-telling to its beings is also historically mutable. The way we are involved in history is also and necessarily historical.

It may appear that the register of *fact* seems very different from the registers of *logos* and being, for periodization does not seem to correspond to the past or the future there. I chose the term "fact" for *jijitsu*, but Miki's use of the term has little to do with fact in the sense of an observable empiricality.[14] In this word "fact," one may detect the resonances of past philosophical attempts such as Fichte's "*Tathandlung*" (act/fact) or Heidegger's "*Faktizität*" (facticity) in Miki's choice of the word *jijitsu*. What Miki stresses is the actional aspect of history, that is, history as an act or action (*kōi*).[15] Hence, history in the register of fact is confined to the present in which an action occurs. Yet, it is also important to note that an action is a process in which the future is anticipated and then articulated to the past.

Let me note two moments in history in the register of fact. The first moment is that of the instant or instantaneousness; a break in the continuity of time at which a decision or resolution takes place—it is an instant that introduces a rupture or *discontinuity* into the apparent duration of time. It implies the subjective (*shutai-teki*, as distinct from *shukanteki* or subjective in the epistemic sense) dimension of historicality, for there would not be praxis but for the instant of decision on the part of the agent who acts. Therefore, resolution constitutes an essential moment in praxis in a specific historical situation and the subject of praxis always exists in a situation.

14 Miki 1967b.
15 Miki 1967b: 26–37.

The other moment closely related to resolution is that of negation. The human being acts in order to bring about some change in this world. This is to say that the future anticipated in action must necessarily be related to the past through negation. Since the actualization of the future is nothing but the negation of the past, one would not intend to bring about a particular future unless one negates what actions of the past have created. The existential mode of hope, for example, is a persistent anticipation of the state of things other than what it is in the present. Hope is entirely immaterial unless the negation of what has been is projected. And action is not merely a state of anticipation; it is a coming of rupture by which the present state of things is denied.

Historically the human being exists in the situation, but the situation and the environment must be clearly distinguished from one another. An animal or plant lives *in accordance with* its proper environment, but it never exists in the situation because it does not exist in order to change the situation or go *against* it. It follows that the human being can be said to exist in the situation precisely because it negates its own situation. But for the moment of negativity, therefore, the human being cannot relate itself to the situation. The human being is a historical being because of the fundamental *fact* that "Man" is an animal not only of *logos* but also of *praxis*. And this negative relating or determining of the human being to or in terms of its ontological situatedness is the ground for its historicality. Thus Miki Kiyoshi argues that the *fact* determines itself, thereby denying or destroying itself.[16]

In this three-dimensional formula of historicality, it is not difficult to see that the successive articulations of the three different registers are organized by dialectic movement. The conceptual determination of the register of *logos* in history necessarily reveals its inadequacy; it is sustained by a not-yet-thematized excess which is not explicitly conceptualized in this register. It therefore leads to another dimension, the register of being, where history is distinguished from other kinds of narrative such as fable and myth. But, the conceptual determination of the register of being again proves itself inadequate precisely because it is impossible to find transhistorical criteria by means of which to determine the descriptive validity of a historical narrative to its referents. Then, finally Miki draws attention to the register of fact (*jijitsu*), historical ground of the human being, where history is discussed in view of historical praxis. When his

[16] This was already a formulation by Nishida Kitarō.

discussion reaches the register of fact, we can see that each register of history is a moment—in the strictly Hegelian sense of the word—of the historicality of the human being. Both Miki and Tanabe realized that historicality could not be adequately addressed without the logical procedure of dialectic.

Thus, it was impossible to think of *fact* independently of Hegelian *mediation* or "the self-othering of the subject in historical *praxis*." This explains why, perhaps implicitly against Heidegger whose intrinsic hostility to dialectic and negativity was already indicated in *Being and Time*, Miki and Tanabe—both of whom studied with Heidegger in Germany in the 1920's—had to reintroduce the dialectic into their investigation of *fact*.

After introducing these three directives of history, Miki claims that history in the register of fact is most fundamental. History in the register of *logos* depicts history in the register of being. History in the register of being must be represented in the present by means of memory or be hoped for by anticipation. Ultimately, both the histories in the registers of *logos* and being are based upon history in the register of fact. This is precisely the reason Miki Kiyoshi and Tanabe Hajime maintained that history comes from the future. In this respect, understandably, the historical is necessarily dependent upon the future-oriented facticity (*jijitsu-sei*) of subject (*shutai*),[17] and necessarily mediated by negativity.

Insofar as no one can escape the consequences of history, one who acts in history in order to bring about changes to the world is, without exception, also the one who has been made in history. Yet, one who acts in history for change is *passively* changed in the very action of bringing about changes to the world. This is because we are essentially and inevitably a *historical body*, immersed in the materiality. History in the register of fact necessarily discloses the materiality—or textuality—of historical action. The agent of history, therefore, changes *itself* in order to cause changes in the world. Without changing itself, it cannot act as an agent of praxis.[18] And this agent of change that is being changed in the course of action

[17] Miki 1967b: 48. Heideggerian "facticity" and Miki's *jijitsu-sei* must be carefully differentiated. We should keep in mind that Miki introduced Heidegger's Dasein analysis to Japan in the mid-1920s before the publication of *Being and Time* in German. He was well aware of the Heideggerian term of "facticity."

[18] Let me tentatively overlook the very important difference between praxis and practice. Nishida, Miki, and Tanabe all stressed the *poietic* nature of moral *praxis*. Moral action was never dissociated from the question of manufacture. This suggests the significance of Nishida's notion of "historical body."

is called "subject" (*shutai*). (The "*shutai*" in Japanese, a neologism for the term "subject," consists of two ideographic characters, *shu* and *tai*, etymologically implying "initiating or dominating" and "body or substance" respectively.) Subjectivity, therefore, always involves some *conversion* from the passive object of *poiesis* (making or manufacture) to the active subject of *poiesis*, which Nishida Kitarō called "*tsukurareta mono kara tsukuru mono e*" (from that which has been manufactured to that which manufactures). Let me note in passing that, in the discussion of historical praxis at the Kyoto School, the conceptual distinction between *praxis* in ethical action and *poiesis* in productive action is often sublated. Acting morally is a making of history.

Undoubtedly this conception of subject cannot be appreciated without reference to that of the subject in German Idealism, notably in Hegel. In due course, the term "dialectic" is frequently used in the writings of Tanabe Hajime and Miki Kiyoshi (from the late 1920s onwards), and this explains why, for instance, Tanabe characterized his *Logic of Species* in terms of the philosophy of absolute mediation (*zettai baikai no tetsugaku*). Yet, it is also significant that Tanabe and Miki attempted to add the conversionary rhetoric of Augustinian/Shinranian origins to their conception of subject.

In this respect, it is understandable that the historical is grounded in the future-oriented facticity (*jijitsu-sei*) of the subject, and necessarily mediated by negativity. As will be demonstrated, the concept of subject would play the central role in the historiographies of Ienaga Saburō and Maruyama Masao, historiographies that attempted to invoke the future to come in the late 1930s and early 1940s.

The Facticity (jijitsu-sei) *of Historical Subject: Conversion and Revolution*

Now we can find it plausible to see that the periodizing scheme, antiquity vs. the Christian era, is not only deployed in the register of *logos*—the organizing trope of historical narrative—and in the register of being—the evolutionary stages of social, cultural, and religious formations—but also in the register of fact where the periodizing scheme no longer serves to arrange events in a chronological sequence. It now means something more than the classification of the incidents of the past. Instead, the act of periodization itself imitates a temporal rupture through which the subject is brought forth in conversion. This is to say that the subject is the very splitting by which, as Gilles Deleuze brilliantly described in his thesis on Leibniz,[19] a new dimension of temporality is generated.

Here, it is possible to discern two contexts, once again. First, rupture means a sudden renewal or transformation of the subject. Or, to put it more specifically, a human being is entitled to be qualified as "subject" exactly because it distances itself from itself and denies its old self, thereby manufacturing a new one. Accordingly, on the one hand, the subject exists in the process of revolution with respect to the social conditions under which it has lived up to that moment, while, on the other hand, it exists in the process of conversion with respect to the integrity of its interiority. Of course, these two aspects are not mutually exclusive even though I described the ecstatic existence of the subject as if revolution and conversion were two distinguishable processes. They are interwoven, and, depending upon which perspective one wants to focus on, the use of the term "subject" (*shutai*) vacillates between the subject's social intervention and spiritual rejuvenation.

Consequently, Tanabe summarizes the contrasting periods in the history of Western philosophy as follows: the period of antiquity is best represented by Greek ontology. The Greeks contemplated the cosmos as the work of art whose ultimate *telos* could be found in the harmonious unity of the whole. Therefore, human action as conceived by the Greeks never implicated the transformation of reality through negation and destruction. In short, they could not comprehend a historical praxis that creates a new reality by negating and destroying the old one. For, in the final analysis, their ontology sought the harmonious and unified creation typical of the artistic *poiesis*.[20] After all, their action of creating objects never went beyond the purview of a speculative appreciation of art works. In their ontology of aesthetic speculation, the Greeks never allowed for "the presence of 'I' as the agent of action that is opposed to beings." In Greek ontology, there is no room for *the personality of freedom* that insists upon the priority of subjectivity over the harmonious unity of beings. Only when the freedom of an individual both as a contradiction and as a discontinuity disrupts the harmonious and continuous unity of the universe can "individuals" resistant and opposed to the whole emerge. With the introduction of the personality of freedom, the teleology of the universe could no longer be found in immediate reality; it had to be sought only

---

19 Deleuze 1988b.
20 Tanabe 1963b.

after the negation of guilt and evil, i.e., the negation of negation. What came into being with the demise of the Greek ontology of aesthetic speculation, Tanabe argues, is the religious redemption of *death and resurrection*.[21] The religious redemption of death and resurrection made it possible to reestablish the teleology of the world through the negation of negation. History as the existential dimension of subjective facticity was now brought into existence for the first time. Only with Christianity did history become possible. Tanabe's *Logic of Species* is decidedly historicist because the religious redemption of death and resurrection is the driving force throughout his philosophical argumentation. Even towards the end of the Asia-Pacific War, he never abandoned the rhetoric of redemption in reference to the City of God.[22]

Apparently both Tanabe Hajime and Miki Kiyoshi constructed their discussion of subjectivity on the basis of Heidegger's existential analysis of Dasein. Yet, it is important to note that they were quite hostile towards Heidegger by the early 1930s, and wary of him, particularly after the news of his participation in the Nazi Party was widely publicized in Japan.[23] Of crucial importance is how they became critical of Heidegger's politics. We know that both Tanabe and Miki had begun systematic readings of early Heidegger in the 1920s, focusing on his early works of phenomenological hermeneutics which culminated in *Being and Time*, and a rereading of Kant in his *Kant and the Problem of Metaphysics*. They were also involved in struggles against what were publicly regarded as "fascistic tendencies" in the early 1930s. In their critical reading of Heidegger, which would result in *the Logic of Species* by Tanabe, and *the Philosophy of History, the Philosophy of Technology* (*Gijutsu tetsugaku*), and *the Logic of Imagination* (*Kōsōryoku no ronri*) by Miki, they sought to find ways to struggle against

---

21 Tanabe 1963b: 25.

22 In an infamous lecture "*Shisei*" (Death and Life) which Tanabe delivered on May 19, 1943, to an audience including many drafted students, he continued to refer to *kuni* (the state) and *kami* (God) in the sense of the Augustinian City of God (Tanabe 1964b: 243–62).

23 The articles were published in the Tokyo Asahi shimbun on October 4, 5, and 6, 1933, entitled "Haidegga no daigaku ron" ハイデガーの大学論 (Heidegger on the University), "Haidegga no shisō hihan" ハイデガーの思想批判 (Critique of Heidegger's Thought), and "*Kokka no tetsugaku to seiji no ninshiki*" 国家の哲学と政治の認識 (State Philosophy and Political Cognition), respectively. But, at the time of the publication of this article "*Kiki no tetsugaku ka tetsukaku no kiki ka* 危機の哲学か哲学の危機か (The philosophy of crisis or the crisis of philosophy)" (Tanabe 1964a), Tanabe had not had access to Heidegger's Address. So, one can hardly evade the impression that, as a critique of Heidegger's politics, it was rather premature.

current fascism in Japan and elsewhere.[24] Of course, there were other contemporary works, including Tosaka Jun's (1936) *The Japanese Ideology* (*Nihon ideorogii-ron*), which were deliberately written against the global trends of fascism.

Until recently, however, the two distinct questions of Japanese fascism and of Japanese colonialism/imperialism have been obfuscated almost by design in the reading of the intellectual history of the 1930s in Japan, as well as in North America. Japanese intellectuals' commitment to the anti-fascist struggle is one thing, but their participation in Japanese imperial nationalism is another. I would never presume to believe that, as types of social formations and as sets of collective fantasies, fascism and imperial nationalism are independent of one another, but it is our task to comprehend how these were or are articulated to one another under specific historical circumstances. Therefore, it is important to keep in mind that an intellectual's commitment to the anti-fascist struggle does not necessarily free him or her from involvement in imperial nationalism. Nevertheless, since Japan's defeat, these two dimensions have scarcely been discerned in the retrospective evaluation of the intellectual and academic achievements of the 1930s. Clearly the regional hegemony of the United States required a systematic repression of the issues of imperial nationalism because the United States policy makers had pursued the strategy of appropriating the legacies of the Greater East Asian Co-prosperity Sphere into the American hegemony in East Asia since the late 1940s. They knew that a rigorous critique of Japanese imperialism could easily boomerang.

Accordingly, I want to emphasize that Tanabe and Miki were seriously committed to anti-fascist efforts, but far less reluctant to involve themselves in Japanese imperial nationalism. On some occasions, a number of leading intellectuals of the times, such as Miki Kiyoshi, Tanabe Hajime, Takata Yasuma, and Shinmei Masamichi, advocated the universalistic principles for a multiethnic nation as if the universalism they stood for had been sufficient to dispel fascistic tendencies. They were engaged in a theoretical investigation of the mutability of ethnicity and nationality. They all denounced racism by condemning the idea of the immutable identity of nation or ethnos (*minzoku*), but they could not escape the reductionism typical of imperial nationalism that reduces fascism merely

---

[24] Miki Kiyoshi, *Gijutsu tetsugaku* 技術哲学 (Philosophy of Technology) (originally published in 1941), in Miki 1967a: vol. 7, 195–330, Miki Kiyoshi, *Kōsōryoku no ronri* 構想力の論理 (Logic of the Power of Imagination), in Miki 1967b: vol. 8, 1–287.

to the fetishization of ethnic identity. They were by no means ethnic nationalists; they were far removed from the allegations of ethnic particularism often attributed to fascism, and their argument scarcely appeared "particularistic" at all even though their universalism definitely could not be free of many traits of colonial racism. Of course, it is hard to discover a great number of signs of petty particularism in Japanese colonial policies and nationalism in both Japan proper and its annexed territories. Racial discrimination against the colonized could be found everywhere in institutionalized forms as well as the everyday conduct of ordinary citizens. Just as in any colonial regime, the particularistic justification of their discriminatory practices was found in abundance. Yet, it is misleading to reduce the failure of Tanabe Hajime's and Miki Kiyoshi's thought to this type of ethnic particularism precisely because it would be too lenient; it would then overlook their more serious responsibility for their commitments to universal principles, thanks to which they were unambiguously implicated in Japanese imperial nationalism. In other words, we must never lose sight of the potentiality of racist and colonial violence not only in their particularistic indulgence in their ethnic identity but also in their universalism of integrationist nationalism. Here, I must repeat Étienne Balibar's extraordinarily important reminder that, just like nationalism and sexism, racism is not an opposite of universalism, but itself universalistic or a form of universalism.[25] It is clear that I have introduced the term "imperial nationalism" partly because I wanted to highlight the *imbrication* of nationalism and racism (and sexism) as universalisms.

Thanks to the oversimplified equation of fascism and imperialism in the academic study of modern Japanese intellectual history, little investigation has been done into how nationalism is imbricated with racism. Nationalism and racism are both universalisms—see my analysis of the imbrications of nationalism and sexism elsewhere[26]—but they cannot be synthesized. Being articulated to one another, nationalism and racism necessarily produce excesses, so that they cannot form a new coherence. They only serve to supplement one another.

What we must scrutinize, therefore, is Tanabe's or Miki's universalism and their commitment to imperial nationalism, just as today we cannot

[25] Balibar, 1997: 337–53. The original presentation that was given in English is reproduced as "Racism as Universalism" in Balibar 1994: 191–204. But, Balibar had already discussed racism as universalism on a number of occasions prior to this, including *Race, Nation, Class*.

[26] Sakai 2007.

afford to overlook the affinities between the American universalistic rhetoric of national integration and the exceptionalist superiority manifest in American practices of global police violence.

The World Schema and Discontinuity

The central topic in the philosophical discussion of *minzoku-sei* (ethnicity/nationality) during the 1930s was image and imagination. And the topic of image was talked about in terms of the problematic of the world as a *schema*. The concept of schema derives from the Greek term *skhêma*, meaning shape, figure, image, design, outline, or plan, and it occupies a middle or mediatory position that is not exclusively either in the order of idea or in the order of the material or sensational. It is well known that Immanuel Kant introduced the concept of schema in modern philosophy in order to elucidate the problem of temporality.

In an irretrievable way, Martin Heidegger introduced the concept of the world to philosophy in the 1920s. In essence, he characterized the word "world" in terms of the Kantian "schema," thanks to which Dasein is able to encounter beings and make sense of them in a variety of modes. Therefore, according to Heidegger, beings that Dasein encounters are "beings-in-the-world."

Tanabe Hajime tried to elaborate on this concept of the world as schema, and his series of essays under the title of the *Logic of Species* (*Shu no ronri*) between 1932 and 1946 is nothing but an accumulation of his further philosophical attempts to conceptualize "the schema 'world'" (*zushiki "sekai"*). Heidegger's viewpoint concerning the world clearly shows that Dasein transcends itself, constantly projecting itself towards the future. For the very transcendental character of Dasein, its existence of the exiting out of itself, can be analyzed in terms of "existentiality." However, Tanabe insists that Heidegger somewhat fails to fully appreciate the mutual contamination of space and time in the schema of the world.

In his elucidation of the schema of the world, Tanabe followed Heidegger's steps set out in *Kant and the Problems of Metaphysics*: in an incessant transition of "the present," Tanabe argued that the present of auto-affection is opposed to another "present" only insofar as it is externalized by an image, and that the "present" should be able to be attributed to many "here's." Yet, in view of Tanabe's reading of Heidegger with respect to the topic of image, we cannot evade a question: has Heidegger already elucidated Dasein's temporality as distinct from "time" in the binary of time and space in *Being and Time*?[27] Dasein is not a Being-in-the-world in the sense that a man is in a house. Being-in-the-world is prior to the state of being located in relation to some "beings-at-hand."

One may argue that the discussion of image and productive imagination, which Heidegger would further pursue in *Kant and the Problem of Metaphysics*, is already anticipated when he introduces the topic of the horizonal schema in *Being and Time*. Let us recall how the ecstatic unity—Dasein's existence—of temporality is elucidated in terms of horizonal schemata. No doubt, the horizon is a phenomenological term, and is beholden to Heidegger's enterprise of phenomenological hermeneutics.

In order for the world to be, there must be the existential-temporal condition, according to which the unity of temporality is a unity of Dasein's getting out of itself and, accordingly, the existential unity of temporality is accompanied by a horizon; temporality finds its unity in the ecstatic structure of a figure and a horizon; Dasein *ex-ists* and gets out of itself. But, of course, the existential-temporality of Dasein or Dasein's ecstasy does not simply mean that it gets carried away or is out of itself. It is a temporality anterior to the time of past-present-future. It gets out of itself toward somewhere; Dasein is always and already oriented in its getting out of itself. And, this 'somewhere' in the direction of which Dasein gets out of itself is framed, and this framing is nothing but the horizonal schema. In anticipation of Heidegger's interpretive strategy in *Kant and the Problem of Metaphysics*, one might say that, while, for the very reason of the finitude of man, the "human being"—a term Heidegger deliberately evades but neither Tanabe nor Miki does—cannot give itself an object of its receptivity, it is nonetheless able to spontaneously form the horizon for such a given; it is capable of producing a horizon as something "imaginary"—imaginary in the sense of being re-presented or pre-presented by imagination—in order to receive and apprehend something as an empirically given in the perception. The finitude of the human being does not allow the human mind to create an object of its receptivity, but the human mind is capable of creating an image *in anticipation of* that object. In order to apprehend something as a being in the world, the human mind must be able to spontaneously project an outline, an image as a horizon. The horizon, insofar as it is thus seen in view of the pure concept of understanding, is precisely a schema that bridges the two

[27] See the section entitled "The Temporality of the Spatiality that is Characteristic of Dasein," in Heidegger 1962: 418–21.

heterogeneous faculties, sensibility and understanding, of the human mind. "In so far as Dasein temporalizes itself, a world *is* too. In temporalizing itself with regard to its Being as temporality, Dasein *is* essentially 'in a world,' by reason of the ecstatico-horizonal constitution of that temporality."[28]

Apart from his overall objection to Heidegger's de-emphasis of the epistemological character of Kant's *Critique of Pure Reason*, two issues seem to be at stake in Tanabe's appropriation of Heidegger. The first is the problem of *discontinuity* in the ecstatic-horizonal constitution of temporality, and the second is Tanabe's accentuation of social antagonism in the historicality of the human being. In this respect, one cannot help hearing a certain resonance between Tanabe's conception of the world schema and Nishida Kitarō's critique of Henri Bergson.

> Life in the true sense of the word cannot be a continual inner development like what Bergson calls "creative evolution"; it must be a continuity in discontinuity. It should mean the life that is born through dying. Vital impetus [*élan vital*] is discontinuous. Death in the true sense is not implied in Bergson's life. Consequently his philosophy fails to elucidate the ground for spatial determination.[29]

As far as Tanabe Hajime could see, Heidegger regarded the ecstatico-horizonal constitution of the world fundamentally as being continuous, devoid of ruptures, incommensurabilities and contingencies. One might say that, essentially, Heidegger's transcendence is "smooth." Dasein transcends itself but remains integral in itself through self-transcendence. Since time as pure auto-affection and the transcendental apperception of "I think" are the same for Heidegger, time lies within pure apperception "as the ground for the possibility of selfhood".[30] Accordingly, the critical limit of the being-in-the-world is disclosed primarily in the death of selfhood. *My* own death is not in the time of beings-in-the-world; it is beyond the temporal structure of having been, in the present, and the future. Yet, for Heidegger, the other's death belongs to the worldly time of ontic beings. Since the selfhood of the "I" is already given in the very transcendence of time as auto-affection, its disclosure to itself is assumed to be without "discontinuity" in its ecstatic constitution of temporality.

---

28 See the section entitled "The Temporal Problem of the Transcendence of the World," in ibid. 416–17 (italics in the original).

29 "Watakushi to nanji" 私と汝 (I and Thou) in Nishida 1975: vol. 6, 356; I have adopted "vital impetus" for *seimei no hiyaku*, which refers to Bergson's *élan vital*.

30 Heidegger 1990: 131.

Accordingly, the sociality of "addressing" between you and me, which is theoretically construed in terms of "continuity in discontinuity," is not an essential moment in Heidegger's elucidation of the present in the existential temporality of Dasein.[31]

Tanabe was more concerned with the *discontinuity* internal to the world.[32] Following Nishida Kitarō, he asserted the fact (*jijitsu*) of history by saying that it determines itself, thereby the present determining itself for erecting *continuity in discontinuity*,[33] and attempted to modify the Heideggerian conception of the world in such a way that it could be the very schema for the fundamental *indeterminacy* of human sociality. What is of crucial importance is that there is no room in Heidegger's temporality for sociality for the other ego (*taga*), against or in response to which the integrity of the ego (*ga*) is asserted. For Tanabe, who followed Nishida, the I (*watakushi*) can be said to exist only insofar as it is addressing or being addressed by the you (*nanji*); I am I only insofar as I am open to you. The dialogic relating—and eventually dialectic one—is not merely a reflection of the system of personal shifters (I, you, we, and so forth) but it is ontological. The other ego is not entirely negligible for Heidegger, but it can be approached only as derivative in the domain of beings-at-hand. Tanabe conceives of the world in terms of the discontinuity of the self and the other, which justifies him discussing Hegelian mediation in the interaction of time and space in the transcendental power of imagination. Thereupon, Tanabe introduced the schema of the world instead of the

31 The year 1932 was very significant insofar as Nishida Kitarō, Miki Kiyoshi and Tanabe Hajime developed the thought of praxis. Nishida's essays to be contained *Mu no jikaku-teki gentei* 無の自覚的限定 (Apperception in Terms of Mu), Miki's *Rekishi tetsugaku* 歴史哲学 (Philosophy of History) and *Kiki-ishiki no tetsugaku-teki kaimei* 危機意識の哲学的解明 (The Philosophical Elucidation of Crisis-Consciousness), and Tanabe's *Zushiki jikan kara zushiki sekai e* 図式時間から図式世界へ (From the Schema "Time" to the Schema "World") all appeared in 1932.

32 Tanabe 1963b: 18.

33 See Nishida Kitarō, "Jiai to ta'ai oyobi benshōhō" 自愛と他愛及び弁証法 (Love of the Self, Love of the Other, and Dialectic), in *Mu no jikaku-teki gentei* 無の自覚的限定 (Apperception in Terms of Mu) in Nishida 1975: vol. 6. "It should be the case that time ought to be examined from the stand point of *mu* (nothingness) determining itself. Insofar as time is comprehended from the viewpoint of the present determining itself, the present cannot be shifted to another present just as a point in space generates another point. The transition of one present to another cannot be grasped in terms of continuity. The present vanishes instantaneously at each moment, and thereby a new present is begun. This is to say that the transition of the present must be thought according to the sense that one lives by dying. In other words, the present is continuity in discontinuity" (Nishida 1975: vol. 6, 264).

schema of time, and attempted to situate the radical mutual indeterminacy of time and space at the heart of the schema "world."

Consequently the determination of the present is construed as mediation, that is, self-othering of the subject in social praxis, self-othering of the self of the present and the other self of the old present. The formula of the series of "now's," once rejected by both Heidegger and Tanabe, returns in this context. The successive sequence of "now's" is no longer an incessant passing of the coming-forth or the image-forming of time. It is the fact (*jijitsu*) of self-othering transcendence, for "transcendence does not mean the transcending of the self continuously into what is posited as its own origin"; "transcendence must mean the transcending of the self into what is other than itself, into what is discontinuous with its origin".[34] It is an othering of the old now into the new now, of the old self into the new self.

But, is it possible to comprehend a sequence of "now's" without at the same time going through many different "here's"? The temporal distinction of one "now" from another "now" cannot be made without going through many "here's." In order to identify itself, the temporal element must be *mediated* by the spatial element; by the same token, the spatial element must be *mediated* by the temporal element in order to identify itself in space. And the network of mutual negation between temporal and spatial elements is nothing but "the world." The entire rhetoric of the negation of negation, which Heidegger deliberately rejected, also returns with Tanabe's logic of mediation.

Tanabe Hajime redefined the schema of the world in a manner somewhat analogous to the mathematical reformulation in which time as the fourth dimension of space was combined with the other three dimensions of space (*Minkowski spacetime*) instead of being granted an autonomous, orthogonal axis, in terms of Hermann Minkowski's formula explaining Einstein's theory of special relativity. In this way he could illuminate the element of indeterminacy to the relationship between time and space. While Heidegger's conception of the world in terms of Kantian schematism still retained Kant's adherence to Newtonian physics, Tanabe proposed revising Kant's schema of time so as to accommodate the new vision of the universe in the twentieth century.[35]

---

34 Tanabe 1963b: 18.
35 Tanabe 1963b: 22.

Besides its implications in the natural sciences, what is decisive in this revision of the schema "world" is that antagonism is introduced into the world as a fundamental feature. The self-transcendent (ecstatic) structure of Dasein is no longer seen merely as moving out of oneself: it is viewed as *an othering of one's own past*, so that Dasein confronts its own past spatially as its other. Since Dasein's free project (*geworfener Entwurf*) towards the future is mediated by the thrownness of its past, transcendence must have as its essential moment Dasein's past as its other. In other words, the existential structure of the human being is revealed essentially as a form of social conflict between the new ego to emerge and the old ego to be negated and destroyed.[36] "The old ego exists inside my ego and is opposed to *me*. Furthermore, in order for the old ego to be opposed to *me* as my old *me*, the old ego should be inside and outside *me* at the same time."[37] The thrownness of the past—or having-been[38]—cannot simply be abolished; it must be preserved for the project of freedom to sublate the thrownness of the past. In the future, the past will have been preserved in order for it to be overcome. Only insofar as there is the moment of antagonism or conflict can the futurity of the free project be synthesized with the thrownness of Dasein. And, this synthesis of the past in the ecstatic existence of Dasein in the future is a dialectic sublation in which the moment of antagonism is absolutely indispensable.

This splitting is exactly what Miki would have referred to as the "subject of the fact (*jijitsu*)." And the individual subject is the very site where social conflict in general takes place. Therefore, even conflicts between social groups are ultimately a matter of negation in the subject. The individual is in fact a site of social antagonism, a locale where discontinuity between social formations—let us tentatively call this formation "a species in Tanabe's sense"—actualizes. Tanabe argues,

---

36 Clearly I am deviating from Heideggerian terminology. Heidegger would never equate Dasein to some being such as the human being. Neither does Tanabe use such terms as *ningen* (the human being) without a certain precaution.

37 Tanabe 1963b: 13.

38 Heidegger does not confuse the past and the modality of having-been, but the distinction between the past and the having-been acquires significance from the viewpoint of anticipatory resoluteness (cf. chap. 3, 'Dasein's Authentic Potentiality-for-being-a-whole, and Temporality as the Ontological Meaning of Care" and chap. 4, "Temporality and Everydayness," in *Being and Time*). Anticipatory resoluteness is a very significant topic for Tanabe too, but since he does not ponder much over the authenticity of Dasein, the past and the having-been are not sharply distinguished.

> The negating subject in the action of self-negation must not merely be the general concept of a negating agent but a species with particular and positive content. Yet, the other species that negates a species should not exist externally outside that species which is being negated: the negating species must erupt from within the negated species, and its negative conflict must still be contained within the overall unity of the whole species in such a way that the negating species may return and relate itself internally to the whole. When two species are rendered continuous, one species does not exist extrinsic to the other but should penetrate the other in a manner comparable to the structure of Bergsonian duration. So, the unity of mutual penetration in duration is nothing but the structure of self-negativity for the species.[39]

Let me concentrate on Tanabe's remark that the antagonism between two social formations that happens within a social formation may be comparable to Bergsonian duration I doubt that negativity can be attributed to duration. Henri Bergson distinguished "qualitative multiplicity" from "quantitative multiplicity," and ascribed to "qualitative multiplicity" an incommensurate difference and "mutually penetrating" heterogeneity which cannot be reduced to the difference among species, to specific difference of the Aristotelian logic. Let me appeal to a Deleuzian notation momentarily. Duration is a difference that changes in kind in the process of differen*t*iation; it is a difference without being countable or numerical. It is actualized without being separable from the movement of its actualization, a difference that differentiates itself, or a heterogeneity which is actualized through differen*t*iation.[40] This is exactly what Tanabe calls "the absolute negativity" (*zettai hitei-tai*), the negativity of mediation that creates *continuity in discontinuity*.

With the new conception of the world as a schema, Tanabe claims that social antagonism and religious conversion can be seen simultaneously in the subject that is a heterogeneity actualizing itself through differentiation.

For both Tanabe and Miki, the problem of conversion remained crucial throughout their academic and political careers; they always responded passionately to the texts of Uchimura Kanzō, Saint Augustine, Blaise Pascal, Karl Barth, Shinran, and Søren Kierkegaard. But this does not necessarily mean that they were religious thinkers. After all, the concept of religion

---

[39] "Ronri no shakai sonzai-ron-teki kōzō" 論理の社会存在論的構造 (Socio-ontological Structure of Logic), in Tanabe 1963a: vol. 6, 321.

[40] Deleuze 1988a: 37–49.

has never been more unreliable than today. On the contrary, the question of faith was never divorced from that of revolutionary pathos, and from the disruptive transformation of social formation and personality.

The historicality of the human being can be sought not in the transcendental-horizonal existence of temporality but in the absolute negativity actualized in the reciprocal indeterminacy of time and space inherent in the schema "world." Tanabe Hajime argued that the historicality of the world could not be explored without reference to the originary accident or contingency (*Urzufall*), whose manifestation is the fact of history as "the ground of Being."[41]

Negation and the Logic of Social Transformation

The essay entitled "The Logic of Negation in Japanese Intellectual History" was originally published in 1938. As mentioned above, it was translated into Chinese and published in Beijing immediately following the initial serial publication. It undoubtedly suggests that Ienaga's argument was perceived to be a significant contribution to the then ongoing discussion about the East Asian Community. Without doubt, Ienaga Saburō was a passionate reader of Tanabe Hajime's philosophy and attempted to write an intellectual history following Tanabe's philosophical insights.

In this essay, the periodizing scheme of antiquity and the Christian era is directly applied to Japanese national history. Ienaga argues that the development in Japan of Buddhism differed markedly from that in China; Japanese Buddhism from the early days manifested an inclination towards a sharp distinction between the ideal and the existent worlds.

Prior to the importation of Buddhism, the ancient Japanese also tended to view this world and the one beyond in a continuous manner, just as the Greeks saw the ultimate orderliness of the world in the harmony of the universe. Surprisingly enough, there is a parallel development between the history of Western philosophy and Japanese intellectual history. Just as Christianity was a foreign religion implanted from the Orient into the Occident, Buddhism was evidently a foreign import from the Asian continent. But both Christianity and Buddhism became integral parts of the respective traditions in the West and Japan.

---

[41] Tanabe 1963b: 24, 387.

For the first time, Buddhism introduced into Japan the negation of the existing reality. For Prince Shōtoku (574–622), for example, the adherence to Buddhism could not be accomplished unless one comprehended the logic of negation, whereby this world was negated as a delusion and, through this negation, the followers of the Buddhist faith would seek a world of absolute positivity in the other world. On returning from the other world, then, the existing reality would not be totally rejected simply because of its delusive appearance. Instead, the followers of Buddhism would be able to aspire to reform reality from within, and to improve the state through moral cultivation. Yet, the developments in Buddhism in Japan during the early medieval period indicated obvious symptoms of regression in the logic of negation. Many longed for the other world, the Pure Land, during the Heian period, but rather than attempting to enter the absolute "other" world through the negation of what existed here and now, aristocrats of the time such as Fujiwara Michinaga (966–1028) wanted to create the illusion of the "other" world in this world. For them, the other world was not in the beyond, but could rather be in the world of their subjective interiority. Therefore, in the final analysis, they could never believe that the Pure Land existed in what one referred to as "the City of God;"[42] it was never posited on the side of the truly transcendent world.

Shinran (1173–1263) of the thirteenth century finally achieved an extreme form of the logic of negation. Ienaga seems to imply that in the figure of Shinran one can see the subjectivity of conversion and revolution, where the radical rupture of history was actualized. Shinran's Shin Buddhism marked the beginning of a truly historical era, but in Shinran as a person one can see the conversion being actualized through differen*t*iation.

Ienaga argues that Shinran destroyed any possibility of actualizing the other world in this world, and introduced the most rigorous distinction between the other world and this one. Accordingly, he denied the legitimacy of religious rites and symbolism whose authority was derived from the presence of the other world in this one. He encouraged Viren to

[42] Throughout Tanabe Hajime's career, the City of God played the role of the central theoretical trope in his social ontology. What he defines as the genus in the trilogy of the individual (*ko*), the species (*shu*), and the genus (*rui*) is a direct inference from the notion of the City of God. Under the overwhelming impact of Tanabe Hajime's philosophy, Ienaga Saburō incorporated some of Tanabe's formulations into his historiographic narrative about medieval Japan.

confront the miseries of this world, and thereby solicited them to aspire towards the Pure Land beyond this world.

Of course, Ienaga's argument about Shin Buddhism touches upon some of the issues I have discussed with regard to negativity and subjectivity. As Miki Kiyoshi argued in his unfinished manuscript on Shinran, Shinran's teaching was the most "historicist" among the leading figures of Buddhism. The radical implication of Shin Buddhism can be found in the relevance of Shinran's teaching to the problem of subjectivity. "Although essential differences can be found between Pure Land Buddhism and Christianity, I cannot deny that Shinran's history is similar to Christian eschatology," for "to be aware that one lives in the era of *mappō* [the last and most decadent era in Buddhist chronology] means that one reaches the transcendent (*chōetsu-teki naru mono*) *subjectively* (*shutai-teki ni*) through the acknowledgement of one's own sinful existence."[43] Like Miki, Ienaga argued that there must be a correspondence between the history of Western philosophy and the history of Japanese thought; but Ienaga as a historian was not particularly concerned with the empirical validity of such a historiographic proposal. His civilizational comparison between the West and Japan is not persuasive, except in the context of Japanese imperial nationalism, in the register either of *logos* or being; his historiography is oriented more toward history in the register of fact.

The Disintegration of the Feudal World-Image and Totalitarian Sovereignty

Maruyama Masao published *Intellectual History of Tokugawa Japan* (*Nihon seiji shisō-shi kenkyū*) in three separate essays in 1940, 1941, and 1944 in the *Journal of Jurisprudence Study* (*Kokkagakkai zasshi*). The third part that Maruyama said he finished just before being drafted into the Japanese army sounds markedly different in tone to the previous parts, and focuses more on how Japan's modernization was obstructed by many factors in the last decades of the Tokugawa rule in the early nineteenth century, but the first and second parts are devoted to a passionate demonstration of the early developments of Japan's modernity in the Kogaku school of Confucianism and National Studies in the seventeenth and eighteenth centuries. One may speculate that, by the time he completed Part III,

43 "Shinran," in Miki 1968: vol. 18, 467–68, 458. An explanation of *mappō* by Sakai (emphasis by Miki).

the prospect of Japan's defeat was imminent for many Japanese intellectuals including Maruyama himself.

The comparative orientation of Maruyama's historiography is manifest from the outset. He begins the entire book with a citation of Hegel's famous observation about the stagnant character of Chinese society, and tries, at least in Parts I and II, to present Japanese intellectual history as a counter-argument against Hegel's characterization of Asiatic societies. His argument in the first two parts is organized in such a way that, in contrast to the absence of negativity in the "Empire of Duration," the development of Japanese Confucianism and National Studies is nothing but an eloquent testimony to the dialectic movement of negativity. Ogyū Sorai becomes such an important figure precisely because he reads Ogyû's texts as a testimony to the presence of modern rationality, which disrupts the unified view of the universe characteristic of the premodern formation.

Unlike Ienaga's *Logic of Negation in Japanese Intellectual History*, however, Maruyama's *Intellectual History of Tokugawa Japan* does not directly appeal to the periodizing trope of pre-Christian antiquity and the Christian era as the principal narrative format. In the strict register of beings, what Maruyama is concerned with is not Christian eschatology but rather Japan's modernity in which capitalist developments should be accompanied by the forms of legitimacy particular to modern political agency or *sovereignty*. So, he shifts his attention to the seventeenth and eighteenth centuries when, in large cities in Japan, the signs of early capitalist development and the rise, though politically limited, of the bourgeois social class were observed. However, this does not mean that he is unconcerned with the periodizing trope of pre-Christian antiquity and the Christian era in the register of fact. As we will see when we return to this point with regard to "crisis," "historical consciousness," and "*shutaisei*" (subjectivity, or, perhaps "sovereignty" is more appropriate in this case), Maruyama attempted to show that the periodizing trope is operating powerfully—not in the register of *logos* or beings but in the register of fact—in Ogyū Sorai's logic of *seisaku* (invention or *poiesis*).

Intellectual History of Tokugawa Japan incorporates two grand narrative schemes of modernization: the first is the evolution of capitalism, which gradually rendered the premodern principles of social organization and the feudalistic forms of political legitimacy irrelevant, and the second is the transformation of the "perspectival structure" (*Aspektstructur*) that can be portrayed in terms of the disintegration of a harmonious and all-inclusive cosmology.[44] In feudal society, it was guaranteed that every domain of experience was ultimately intelligible in one totalizing thought system and the fragmentation of such a cosmology into essentially three mutually independent spheres of epistemology, ethics, and aesthetics is the most essential characteristic of modernity.

To highlight the historicist orientation of Maruyama's historiography, let me introduce two important points of reference, to clearly present some of the problematics he struggled with. One is Franz Borkenau's *From the Feudalistic World-image to the Bourgeois World-image* (*Der Übergang vom feudalen zum bürgerlichen Weltbild*),[45] which seems to have provided him with a general framework for historical explanation. *Intellectual History of Tokugawa Japan* has often been read in reference to Karl Mannheim, but at crucial points I find Maruyama's approach to have more in common with Borkenau than with Mannheim. I cannot identify the second context in terms of specific texts, but I am convinced that Maruyama was inspired by Carl Schmitt on a number of issues including the state of exception and sovereignty.

Following the historical insights deriving from Max Weber's studies on Protestantism and capitalism, Borkenau attempted to offer a historical analysis as to how the collapse of the medieval world-image led to the darkest period—Lucien Goldmann's (1964) *The Hidden God* would portray this early modern period—in the history of "Western Europe" while opening up possibilities leading to a new world-image that would eventually become that of the global modernity. In the feudalistic world-image in which every part was organically integrated into the totality of the universe, Borkenau argues, it was impossible to conceive of the natural law as it operates in modern natural science. The natural law and the theory

[44] In the English translation by Mikiso Hane (Princeton and Tokyo: Princeton University Press and Tokyo University Press, 1974), *Aspektstructur* is rendered as "frames of reference." Hane's translation facilitates readers' immediate comprehension, so I have objection to it. My choice of "perspectival structure" is based on a different consideration in that it indicates methodological concerns in Maruyama's approach, particularly in reference to Franz Borkenau's concept of the "world-image." He wanted to read Japanese Confucian texts of the seventeenth and eighteenth centuries mainly because of the way the capitalist transformation of social reality led to the disqualification of certain forms of political authority. His theoretical justification for selecting some Confucian texts in order to discuss Japanese modernity is found in its "perspectival" function of the Confucian categories. Very much within the horizon of neo-Kantianism, Borkenau conceived of the "world-image" as a table—or today one might well say a paradigm—of categories through which people of the times made sense of beings in their sensible encounter.

[45] Borkenau 1934.

of the rational order of nature were put forth by Thomas Aquinas, whose theological works Borkenau takes as the representative embodiment of the feudalistic world-image, in an endeavor to define the natural and rational order in society. For Aquinas, there could not be two different orders, the order of natural phenomena and the order of social relations, because both his conception of natural laws and his theory concerning the existence of rational order in nature derive from his effort to define the natural and rational order in society. Social order was an integral part of the natural order, so that ultimately social law could only be understood as an extension of the overarching law of nature. As a result, the system of Thomas Aquinas' doctrines left out the possibility of modern mechanics and the entire modern natural sciences that were based upon the mechanistic vision of nature.[46] By tracing the internal disintegration of the medieval Catholic world-image, Borkenau identifies the historical conditions under which the mathematico-mechanistic world-image became possible.

Borkenau cautions, however, that the emergence of capitalist modernity must be clearly divorced from developmental teleology. Historical transformation is not a pre-established process toward a higher degree of rationality. Neither is it attributable to doctrinal genealogy. Francisco Soárez, perhaps the most sophisticated Spanish reactionary of the sixteenth and seventeenth centuries, is exemplary; he devoted his scholarly career to the preservation of the Thomist world-image, but we can recognize even in the midst of his conservative scholarship against Protestant trends of his times symptoms of the disintegration of the medieval world-image.[47] Borkenau justifies this assessment by means of the following methodological premise:

> Let us put aside the question of whether or not there are *a priori* categories that are universally valid for every human being. As far as we are concerned, of significance is only the reality that every experience is formed in terms of historically changing categories. It is certain that what experience is sought for, received, regarded as an evidence, counted as an index of empirical positivity, or discarded as being nonsensical is totally dictated by the dominant categories of the times. Therefore, our task is to reduce the history of

[46] I rely upon the Japanese translation particularly the first part of the book where Borkenau explains his methodological premises (Borkenau 1965: 14–63).

[47] Borkenau 1965: 45.

> knowledge put forth from the perspective of the Enlightenment to the history of categories.[48]

I am not particularly interested in discussing how much Maruyama Masao was influenced by Franz Borkenau, even though Maruyama said that he read Borkenau's *From the Feudal World-Image to the Bourgeois World-Image* in the prewar translation.[49] What is of decisive importance is that the two methodological directives put forth by Borkenau above also helped Maruyama envision the overall narrative of the Tokugawa history of Confucianism and National Studies. Undoubtedly, Maruyama saw Zhu Xi's philosophy as an equivalent to the medieval world-image of Thomas Aquinas.[50] Furthermore, what he calls the "perspectival structure" helps define his term "structure of thought" (*shi'i yōshiki*). It is not difficult to see what Maruyama attempted to accomplish was similar to Borkenau's history of categories. In reference to Confucianism, Maruyama argues, "They [= thinkers of the Tokugawa period] regarded the various Confucian categories as 'frame of reference' (Aspectstructur) and observed the feudal social order only through these frames of reference."[51] The prevailing tradition of the times "was to regard scholarly investigation as the acquisition of knowledge from the ancient classics rather than the examination of social reality itself." Mostly it was "unaware of the real gap between what was in the classics and existing society." "The connection between their cognitive intentions and the existing social system simply had to be

---

48 Ibid., 40.

49 The first half of *Der Übergang vom feudalen zum bürgerlichen Weltbild* was translated into Japanese by Niijima Shigeru and Yokokawa Jirō and published in 1935. I have no access to this translation, so I do not know how many chapters are included in this part. The second half was not available until 1965. In reference to this partial translation, it seems, Maruyama mentioned this book in the postscript to the 1952 edition of *Nihon seiji shisō-shi kenkyū* (Intellectual History of Tokugawa Japan). The postscript is not included in Hane's English translation. Although the second half was not available in Japanese in the 1930s and early '40s, this book drew the attention of not only such intellectual historians as Naramoto Tatsuo and Matsushima Ei'ichi, but also philosophers and historians of science—Taketani Mitsuo, Hara Mitsuo, and so forth.

50 On this point, Maruyama is astonishingly frank. He argues, "In Zhu Xi's mode of thought, the ethical norms are *naturalized* doubly. First, the ethical norms are grounded in the order of the universe (the Principle of Heaven). Second, the norms are believed to exist a priori and innately in human nature as man's *original* nature (*nature as it should be* or *hon'nen no sei*)" (emphasis in the original). And he referred to the seventeenth-century testimony of a former Catholic missionary, Fabian Fukuhashi, about the similarity between Christian theology and Song Rationalism (Maruyama 1952: 202; in English in Maruyama 1974: 198–99, but I have amended Hane's translation).

51 Maruyama 1974: 192.

established via an intermediary known as the 'Five Relationships'."[52] Although I can see a number of theoretical confusions in Maruyama's argument,[53] the exceptionally important status Maruyama grants to Ogyū Sorai was a result of this arrangement. Reactionary ideologue though Ogyū was towards capitalist modernity, his texts could be read as symptomatic of the disintegration of the Zhu Xi (Chu Hsi) mode of thought in which the laws of nature and the laws of moral actions are continuous.

Here, one must be aware that one factor clearly separates Maruyama from Borkenau: Maruyama implicitly introduces the developmental teleology of modernity, so that the disintegration of the Zhu Xi mode of thought would naturally lead to the fragmentation of a unified world-image and to the establishment of the three autonomous realms of epistemology, ethics, and aesthetics. For Maruyama, the trajectory of modernity is somewhat pre-ordained, and he follows somewhat blindly the Enlightenment teleology against Borkenau's warning.

Subsequently, the assessment of the collapse of the medieval world-image opens up two significant problems of modernity for Maruyama. In the first place, the disintegration of Zhu Xi's mode of thought is nothing but an intimation of the more radical disintegration of the universe essentially into the three domains of knowledge, praxis, and feeling. He first traces the collapse of continuity between the world of nature and the world of praxis in such figures as Yamaga Sokō and Itō Jinsai. Then, Maruyama wanted to see the completion of the disintegration of the Zhu Xi mode of thought in the scholarship of Motoori Norinaga, who declared the authenticity of the subjective interiority of aesthetic sentiment. Consequently, his intellectual history of Tokugawa thought was intended to delineate the fragmentation of the premodern world into the three

52 Ibid.

53 The confusing points are: (1) Maruyama does not discern the theoretical requirement consequent to the imposition of the perspectival structure from the contingent traits particular to Japanese Confucianism. Clearly, he should adopt the first stance—it does not matter what mode of thought is at issue. It is precisely the perspectival structure that forces the intellectual historian to view what was held to be "social reality" as a consequence of historically contingent categorization—but he describes Japanese Confucianism as if it had been restricted by its particular blindness about what constituted "social reality" then. This leads to the second. (2) He talks about "social reality" without qualification. But, what is "social reality"? Should it be the very problem of his entire work to find out how a certain configuration of categories allowed Japanese thinkers to conceive of a new type of experience as "social reality"? (3) Then, how should we view the ethical norms in relation to social reality? If the Five Relationships or the Five Constancies are not part of "social reality" in Japan—as Motoori Norinaga would later begin to claim in the late eighteenth century—what historical conditions would allow such disparity to become conceivable?

separate worlds of epistemology, of praxis, and of judgment. For Maruyama too, the absence of integral cosmology should be the fundamental marker of modernity.

Secondly, Maruyama posed the problem of sovereignty in view of subjectivity and negativity.

What emerges when the disintegration of continuity between the order of nature and the order of ethical action is accomplished is history in its essential sense. Just as Tanabe and Miki—and Ienaga following Tanabe—argued before him that there was no history before Christianity, Maruyama too argues that there was no history before Ogyū Sorai or before the disintegration of the Zhu Xi mode of thought. This means that no social, political, or cultural order exists beyond history; such an order does not exist by itself, but if it did exist, it must be invented or manufactured (*seisaku*), and reproduced incessantly. It follows that social order should be able to be changed or abolished since it is an invention.

> The rejection of these naturalistic assumptions meant that the sages had to be removed from the natural order and given the role of producers of order out of absolute disorder. Before the sages' invention there was nothing; after it, everything. An abyss separates the age that preceded and the age that followed the sage's invention. It follows that before the advent of the sages, society was without norms of any kind, in a state of nature like that conceived by Hobbes.[54]

Some one must manufacture the social, political, or cultural order, and reproduce it; some one should be able to change it or abolish it. Insofar as the order applies to the totality of the populace under rule, this "some one," who is the Sage King in Confucianism, is a "sovereign" in the sense in which Carl Schmitt (1968) discusses it. It is necessary to posit the authenticity of Manufacturer in the Sage King beyond the legitimacy of the existing polity; the Sage King must transcend the polity, because it is not the polity but its Manufacturer that endows the polity with the ultimate legitimacy. In this respect, the Sage King is the *source* of political legitimacy but the legitimacy of the Sage King itself is not grounded anywhere. In other words, according to Maruyama Masao's analysis, the Sage King is *groundless*, and his legitimacy is absolutely *contingent*. Only when

54 Maruyama 1974: 212, original in Maruyama 1952: 213–14.

his legitimacy is contingent, the Sage King can be said to be "an absolute subject" (*zettai-teki shutai*).[55]

At this level, however, the concept of the period in the register of being loses its significance. "[I]f, after the invention of the sages, the Way is cut off from the inventing agent [manufacturing subject] and justified in itself as an objectified Idea, this means after all a return to the concept of natural order. But this approach could not provide any *political decisions* able to deal with existing reality".[56] The subject that makes a decision in the midst of a "crisis" is an absolutist sovereign, by whom order must be manufactured instantaneously each moment anew. Order should not last because it would gain its independence from the subject as soon as it is externalized from its manufacturer. Just like the world according to Occasionalism, it is not enough to say that God creates the world once; the world must be continually, and in every moment, created by God. Likewise, the Sage King is an absolute sovereign in the sense that the order he dictates must be at every moment his invention. Therefore, it is pointless to apply the concept of the period.

"In theories of natural order, the Idea is realized immanently and continuously, but in Sorai's case, the realization of the Idea is *discontinuous*, in that institutions must be invented anew by the ruler of each new regime".[57] In this sense, of course, Maruyama could insist that history became possible with Ogyū Sorai. It is the sense of history in which the beings referred to by historical narrative do not last in and of themselves, and, in which, therefore, periodization is essentially cancelled. The Sage King does not exist in the past in the register either of *logos* or being. It is only in the register of fact that, after all, Maruyama's intellectual history can complete its argumentation. "By extending the analogy of the logic of *invention by the Early Kings* to all ages, the ascendancy of persons over ideas was firmly established for the first time. This made it possible for political rulers to engage in inventive activities, directed towards the future, in order to overcome the crisis confronting them."[58] Maruyama Masao thus argued for the absolutist sovereignty in historicist terms.

---

[55] Maruyama 1974: 212, original in Maruyama 1952: 213. I have amended the translation. It should read: "So long as the concept of the sage has the significance of a universal idea, the sage as a concrete person who manufactured rites, righteousness, laws, and mores would ultimately have to derive his authority from the idea that exists beyond him. Under such circumstances, the Sage King is not an absolute subject."

[56] Maruyama 1974: 212, my emphasis.

[57] Ibid., my emphasis.

[58] Ibid., italics in the original.

Notwithstanding the implications of emperor absolutism and decisionist politics,[59] one could recognize the logic of *eternal revolution*, which he later ascribed to the implementation of democracy in the post-Asia Pacific War period.

Maruyama's historicism, just like Ienaga Saburō's, must be located in the historicist paradigm of the 1930s, and we will see that much of their work for postwar Japanese democracy retained some of the seminal insights put forth by theoreticians of the 1930s.

The lengthy but overtly cursory reading above of the wartime discussions of history and intellectual historiography returns us to my initial remark about civilizational categories. I hope this helps to illuminate why I see a need to hesitate. We must *hesitate* before applying civilizational categories in our reading of the historical texts by Ienaga Saburō and Maruyama Masao, not because they managed to elude these categories and simplistic binaries of the West and the Rest, and of Japan and China, but instead because we must investigate in what contexts, and what register of history, they yielded to such binaries. Indeed, we need to hesitate so as to disentangle ourselves from such civilizationalism, but we also have to hesitate since we cannot afford not to know how civilizational categories are still alluring to us.

References Cited

Balibar, Étienne (1994), "Racism as Universalism." In *Masses, Classes, Ideas*. New York: Routledge, 191–204.

—— (1997), "Le racisme: encore un universalisme?" In *La crainte des masses*. Paris: Galilée.

Borkenau, Franz (1934), *From the Feudalistic World-image to the Bourgeois World-image* (*Der Übergang vom feudalen zum bürgerlichen Weltbild*). Paris: Alcan.

—— (1965), *Hōkenteki sekaizō kara shiminteki sekaizō e* 封建的世界像から市民的世界像へ (*From the Feudalistic World-image to the Bourgeois World-image*). Mizuta Hiroshi et al. trans. Tokyo: Misuzu shobō.

Deleuze, Gilles (1988a), *Bergsonism*. Trans. Hugh Tomlison and Barbara Habberjam. New York: Zone Books.

—— (1988b), *Le Pli*. Paris: Les Éditions de Minuit.

Goldmann, Lucien (1964), *The Hidden God: A Study of Tragic Vision in the Pensées of Pascal and the Tragedies of Racine*. Philip Thody trans. New York: Humanities Press.

Heidegger, Martin (1962), *Being and Time*. John Macquarrie and Edward Robinson trans. New York: Harper & Row.

---

[59] In reference to Maruyama's conceptual distinction between idea (*rinen*) and ideal (*risō*), I have argued Maruyama's commitment to Japan's emperor system: "The Problem of 'Japanese Thought': the Formation of 'Japan' and the Schema of Co-figuration," in Sakai 1997.

—— (1990), *Kant and the Problem of Metaphysics*. Richard Taft trans. Bloomington: Indiana University Press.

Ienaga Saburō (1997), *Nihon shisō-shi ni okeru hitei no ronri no hattatsu* 日本思想史に於ける否定の論理の発達 (The development of the logic of negation in the history of Japanese thought). In *Ienaga Saburō shū*, vol. 1. Tokyo: Iwanami shoten.

—— (1998 [1974]), *Tanabe Hajime no shisō-shi teki kenkyū* 田辺元の思想史的研究 (Intellectual historical study of Tanabe Hajime). In *Ienaga Saburō shū*, vol. 7. Tokyo: Iwanami shoten. (originally published by Hosei University Press in Tokyo, 1974).

Maruyama Masao (1952), *Nihon seiji shisō-shi kenkyū* 日本政治思想史研究 (Intellectual history of Tokugawa Japan). Tokyo: Tokyo University Press.

—— (1974), *Intellectual History of Tokugawa Japan*. Mikiso Hane trans. Tokyo and Princeton, NJ: Tokyo University Press & Princeton University Press.

Marx, Karl (1930), *Doic'che ideologi* (The German Ideology). Miki Kiyoshi trans. Tokyo: Iwanami shoten.

Masuzawa, Tomoko (2005), *The Invention of World Religions; Or, How European Universalism Was Preserved in the Language of Pluralism*. Chicago: University of Chicago Press.

Miki Kiyoshi (1967a), *Rekishi Tetsugaku* 歴史哲学 (Philosophy of History), in *Miki Kiyoshi zenshū*, vol. 7, 195–330, Tokyo: Iwanami shoten.

—— (1967b [1932]), *Kôsôryoko no ronri* 構想力の論理 (The logic of imagination), in *Miki Kiyoshi zenshu*, vol. 8, 1967: 1–287.

—— (1967c [1932]), *kiki-ishiki no tetsugaku-teki kaimei* 危機意識の哲学的解明 (The philosophical exposition of the crisis consciousness), in *Miki Kiyoshi zenshu*, vol. 5, 3–30.

—— (1968), *Shinran* 親鸞, unfinished manuscript, in *Miki Kiyoshi zenshû*, vol. 18. Tokyo: Iwanami Shoten.

Nishida Kitarō (1975), *Mu no jikaku-teki gentei* 無の自覚的限定 (Apperception in Terms of Mu), in *Nishida Kitarô zenshü*. Tokyo:Iwanami Shoten, 1975, vol. 6.

Sakai, Naoki (1997), *Translation and Subjectivity: On "Japan" and Cultural Nationalism*. Foreword by Meaghan Morris. Minneapolis: University of Minnesota Press.

—— (2000), "Subject and Substratum: On Japanese Imperial Nationalism." *Cultural Studies* 14, 3/4 (2000): 462–530.

—— (2007), *Nihon/Eizō/Beikoku—kyōkan no kyōdōtai to teikokuteki kokuminshugi* 日本／映像／米国ー共感の共同体と帝国的国民主義 (*Japan/image/the United States: The sympathy and imperial nationalism*). Tokyo: Seido-sha.

Schmitt, Carl (1968 [1929]), *Legalität und Legitimität* (*Legality and Legitimacy*). Berlin: Duncker & Humbolt. I rely upon the Japanese translation by Tanaka Hiroshi and Harada Takeo (Tokyo: Miraisha, 1983).

Tanabe Hajime (1963a), *Tanabe Hajime zenshū* 田辺元全集 (*Collected Writings of Tanabe Hajime*), 4 vols. Tokyo: Chikuma shobō.

—— (1963b [1932]), "Zushiki jikan kara zushiki sekai e" 図式時間から図式世界へ (From the schema "time" to the schema "world"), in *Tanabe Hajime zenshû*, vol. 6, 1–50.

—— (1963c [1936]), "Ronri no shakai-sonzairon-teki kôzô" 論理の社会ー存在論的構造 (The social ontological structure of logic), in *Tanabe Hajime zenshû*, vol. 6, 299–396.

—— (1964a), "Kiki no tetsugaku ka tetsukaku no kiki ka" 危機の哲学か危機の哲学か (The philosophy of crisis or the crisis of philosophy)", in *Tanabe Hajime zenshû*, vol. 8, Tokyo: Chikuma shobô, 1964: 1–9.

—— (1964b), "Shisei" 死生, in *Tanabe Hajime zenshū*, vol. 8, 243–262.

Tosaka Jun (1977), *Nihon ideorogii-ron* 日本イデオロギー論 (*On the Japanese Ideology*). Tokyo: Iwanami Shoten (originally Tokyo: Hakuyô-sha, 1936).

CHAPTER TWO

ONTOLOGICAL OPTIMISM, COSMOLOGICAL CONFUSION, AND UNSTABLE EVOLUTION: TAN SITONG'S *RENXUE* AND ZHANG TAIYAN'S RESPONSE

Viren Murthy

Introduction

Written in 1896, almost immediately after the Sino-Japanese War (1894–1895), *Renxue*, or *A Study of Cosmic Love*, was one of the first works to radically reinterpret classical Chinese texts in the framework of modern Western philosophy and science.[1] Its author, Tan Sitong (1865–1898), is famous for bravely facing execution after the failure of the Hundred Days Reform in 1898 and hence *Renxue* was widely read when Liang Qichao published it in 1899. Zhang Taiyan (1869–1936), one of the many intellectuals who read *Renxue* in manuscript form shortly after Tan wrote it, was unlike most other readers in that he was scathingly critical of the text. He shared Tan's ideals of self-sacrifice and in 1899, he praised Tan's martyrdom. But in the same year, Zhang wrote "On Bacteria," in which he lambasted Tan's optimistic worldview and developed his own pessimistic ontology of confusion.

While many scholars have interpreted Zhang and Tan's respective works, they have failed to consider that their works are particularly significant because they express a larger trend, an epistemological turn that gave birth to a new sense of time and history that would shape Chinese discourses in the years to come. We can understand the shift during the late Qing as a movement towards philosophy. In other words, if as Benjamin Elman claimed, the transition from Song Confucianism to Qing dynasty evidentiary scholarship was a conversion from philosophy to philology,[2] then as the Chinese faced the world of global capitalism during the late

[1] *Ren* is notoriously difficult to translate; it is usually rendered "benevolence." However, in Tan's text, I render the term "cosmic love" since I believe Tan invokes it as an ontological ground rather than merely a moral maxim.

[2] Elman 1984.

nineteenth and early twentieth centuries, intellectuals either switched "back" from philology to philosophy or, more often, they reincorporated philology into a philosophical framework. This is why tropes related to Song-Ming Confucianism recur in late Qing writings. This new discursive framework entails a dialectic between abstract time and evolutionary history, which operates in different degrees in Tan and Zhang's respective writings.

The practice of philosophy and the dialectic of time and history point to deeper structural changes. Following Georg Lukács' analysis of German idealism, I contend that Tan and Zhang's philosophical concepts, such as equality, abstract time, and evolution are intimately associated with the logic of global capitalism. However, when Tan and Zhang bring modern philosophy into dialogue with classical Chinese thought, they produce philosophies that both affirm the modern world and attempt to resolve some of its contradictions by positing spaces outside of history and time.

Like the German idealists, Tan posits an ontological source with two sides: a mental side, which he sometimes calls cosmic love or mental power, and a more concrete side, which he labels ether or energy. The fundamental goal of his philosophy is to let this energy or mental power flow unobstructed and thus negate the temporal self, which hinders this flow. Tan draws on Buddhism and Daoism, interpreted through the lens of Song and Ming Confucianism, to present ways for humans to transcend time. By combining Song and Ming Confucianism and ideas from modern science, Tan attempts to transcend the finite subject, while preserving ethical and political meaning in an evolutionary vision of history. Zhang, on the other hand, posits confusion as the source of his ontology, and so is much more skeptical of letting things go. Drawing on Xun Zi, he stresses artifice and emphasizes the legal apparatuses.

Reification and the Antinomies of Capitalist Modernity

One can understand Zhang and Tan as expressing aspects of China's process of entering the global capitalist system of nation-states. This process is sometimes described as two-sided: an economic or capitalist dimension and a state-formation dimension. Following Giovanni Arrighi, Prasenjit Duara describes these two interrelated dimensions as "logics," both of which are required to maintain global capitalism: the logic of accumulation and the logic of territoriality.[3] Zhang and Tan each express these logics to some degree in their philosophies; what distinguishes them is a matter of emphasis.

Both these two logics imply tendencies linked to a more fundamental process. This process can be connected to rationalization, which we can understand in the context of practices associated with bureaucracy, the modern state, and capitalism. As Georg Lukács explains:

> Bureaucracy implies the adjustment of one's way of life, mode of work and hence of consciousness, to the general socio-economic premises of the capitalist economy, similar to that which we have observed in the case of the worker in particular business concerns. The formal standardization of justice, the state, the civil service etc., signifies objectively and factually a comparable reduction of all social functions to their elements, a comparable search for the rational formal laws of these carefully segregated partial systems.[4]

The scope of both capital and the bureaucracy is not only national, but global—for example, with respect to the scale of international laws, treaties, and of course imperialism. Toward the end of the nineteenth century, Chinese intellectuals and officials generally recognized a crisis associated with imperialism and nation-building and, since the Self-strengthening Movement in 1865, there was an organized attempt to develop modern bureaucracy and technology. Despite the subsequent denigration of the movement after China's loss in the Sino-Japanese War, the Self-strengthening Movement was fairly successful in attaining its goals of industrializing specific enterprises.[5] By the time of the Self-strengthening Movement, China was already responding to the pressures of the global capitalist world and for our purposes, it is significant that many late Qing intellectuals, including Tan Sitong, were educated in areas associated with the industrialization initiated by the Self-strengthening Movement.

As Chinese officials embarked on a project to develop modern bureaucracies, industries, and scientific knowledge, they initiated several conceptual transformations associated with what Lukács calls reification (*Verdinglichung*). Put simply, reification implies the emergence of a "world of complete things and relations between things," which stand against the subject. This idea of a world of things implies not only a new type of discretely divisible spatiality, but also a new view of time as a series of points.

---

3 Arrighi 1994, Duara 2003: 11.
4 Lukács 1971: 98, translation amended.
5 Elman 2004.

As people conceive of a world within a continuous stream of "now" points, they have a new secular sense of historical continuity, but they tend to efface the historical specificity of this particular mode of temporal continuity. The world presents itself as continuous and transhistorical.

In modern capitalist society time becomes an independent variable,[6] that is, a variable separate from change. Speed and efficiency are crucially important in modern society and concepts of this sort presuppose time as an independent variable. As Aristotle pointed out, when we say that something happens quickly, we imply that a change has taken place in a short time, and thus time must be independent from being, which changes.[7] While this vision of time can be read in Aristotle, we must emphasize that it becomes widely institutionalized and universal in capitalism.

Abstract time sets the frame for historical time, which refers to the movement of capital to greater levels of productivity.[8] This movement represents a novel social dynamic that propels firms to increase their rate of production just to stay in business. However, more important for our purposes, the dynamic of capitalism causes societies as well as the globe to transform constantly. Proponents of evolutionary history misrecognize the above-mentioned dynamic, which is specific to capitalism, as a transhistorical process. In other words, they posit a social dynamic of progress from primitive to modern societies, and usually conceive of this movement in terms of increased efficiency. Evolutionary historians combine both the above conceptions of time since they presuppose a subject that moves through time and whose speed can be measured through abstract or chronological time.

In this context, individual subjects appear confronted by external forces; they feel like objects rather than subjects of history. Reification makes it such that history seems freed from previous eschatological narratives and thus opens a space for action. However, at the same time, history appears as a force unrelated to one's own action and thus stands against human agency. As Immanuel Kant's famous antinomy captured lucidly, humans are at once free and at the same time determined by the forces associated with the world of things, the dynamic of history. This vision implies a

[6] Postone 1993: 201.

[7] "All changes may be faster or slower, but not so time; for fast and slow are defined by time, 'faster' being more change in less time and 'slower' being less change in more time" (Aristotle, 1968: 385).

[8] Postone 1993: 293.

dichotomy between the mind and the material, which late Qing texts both mark and attempt to overcome, often by drawing on older narratives.

The antinomy of freedom and necessity implies another contradiction in the world of things, namely a conflict between system and chaos. The modern impetus towards rational control is countered by the structural impossibility of this task. Individuals attempt to rationally order their lives, but the result of such a pursuit is often unintended chaos. At a larger level, consider a number firms each pursuing its individual interests, producing monetary crises. Perhaps the context most readily visible to late Qing intellectuals was the international world, where nation-states each seem to compete rationally, but their rational actions reproduce the chaotic world of global capitalism and imperialism.

When placed within this contradictory dynamic, individual subjects experience a split between control and chaos, between the task of exhaustive rational calculation and the impossibility of completing such a task. Since Kant, philosophers have struggled to reconcile the formal determinations of the subject with the chaotic giveness of concrete life. In this paradigm, where the subject is associated with the formal side, the side of control, the philosophical task is to overcome irrationality or chaos. Lukács describes this project as "grasping and holding on to the irrational character of the actual contents of the concepts," and striving "to go beyond this and erect a system".[9] With this gesture the German idealists attempted to posit a realm that goes beyond distinctions between subject-object/control-chaos and create an ideal unity.

If we can understand the development of philosophy in Germany as an attempt to erect a system, then Schopenhauer, Nietzsche, and Heidegger each attempted to destabilize this system by pointing to drives, chaos, forgetting, and blind will. Late Qing intellectuals generally harbor both of these tendencies. This is in part because they entered the global system on the fringes and hence could constantly see the problems, chaos, and inequalities of their world. Thus they conceived the world as chaos and posited its negation or transformation. But, at the same time, they generally accepted the project of developing a nation-state as part of an evolutionary scheme. Thus Tan Sitong expresses tensions between a view that negates time, a theory of human evolution, and the goal political action in the temporal world.

[9] Lukács 1971: 117.

Chinese intellectuals in the twentieth century constantly drew on traditional cosmology to develop a theory that dealt with some of the abovementioned antinomies. Tan anticipates the work of many other Chinese intellectuals, including Zhang Taiyan, by developing a new theoretical mélange of traditional Chinese cosmology and modern science, through which he posits an ontological source that overcomes conventional dichotomies between subject and object. Then from this relatively transparent ontology, Tan developed certain political conclusions. Zhang, on the other hand, explained the emergence of subject and object by positing a self-concealing ontological source, "sexual germs of confusion," which will lead to different political strategies. However, despite their differences, both Tan and Zhang's respective politics can be understood in relation to the goal of consolidating state power in a hostile world.

Tan Sitong's Life and Career

Tan Sitong was born in 1865 in Hunan, three years before Zhang Taiyan, and his early life was in many ways typical of that of reformers at the time. Tan was precocious and began studying the classics at the age of five, with great success. While studying for the examinations with various tutors, Tan learned a great deal about Song Confucians such as Zhang Zai and the early Qing thinkers such as Wang Fuzhi and Huang Zongxi. However, he would eventually fail the imperial examination six times. This is partially reflective of the crisis of the examination system during the late Qing. Due to the increase in population, the number of qualified candidates greatly surpassed the number of positions available in government. As a result, many talented people eager to help China at a time of war and crisis would seek other institutional means to achieve this goal.

Repeated failure in the examinations may have caused Tan and other late Qing intellectuals to disdain the eight-legged essay and many tenets of traditional Confucian culture. In this space outside of the examination system, intellectuals used their traditional knowledge in different ways. They learnt about both traditional and Western ideas in a new context informed by changes associated with China's response to the global world of modernity. As Elman notes, "By going outside the orthodox curriculum of the civil service examination, the newly educated in science, mathematics, and engineering inhabited the unprecedented institutional venues of arsenals, shipyards, and industrial factories that promoted non-degree-oriented engineering, mathematical, and science studies."[10]

Moreover, as part of the attempt to catch up with the technological and military superiority of the West, officials initiated translation projects at shipyards and arsenals. The Fuzhou Shipyard and the Jiangnan Arsenal were especially important for late-Qing intellectuals' education. The Jiangnan Arsenal was a major shipbuilding site in Shanghai and was arguably more advanced in shipbuilding technology than its Japanese counterparts.[11] The arsenal's organizers and advisors believed that China could create a new industry through manufacturing machines, training "machine workers" and engineers, and translating Western works about science.[12] Perhaps the person most important for the translation project was the Protestant missionary John Fryer (1839–1928), who came to China in 1861.

Shortly after the Sino-French War (1883–1885), Tan became extremely concerned about China's place in the modern world; however, until the early 1890s he still knew little about Western philosophy and scientific thought. In 1893, he traveled to Shanghai, met John Fryer, and bought many of the translations done at the Jiangnan Arsenal.[13] Fryer had translated a work on psychology by Henry Wood (1834–1909) entitled *A Method of Avoiding Illness by Controlling the Mind* (*Zhixin mian bingfa*) into Chinese; Tan read the book and claimed that if one combined the knowledge in it with traditional Chinese thought, one could save China. This was a work that underscored the importance of a mental power that transcended the ordinary world.

Given that the translation projects were intimately linked to industrialization, scholars interested in learning Western thought and science would often come into contact with the processes associated with the modern economy. Moreover, as Chan Sin-wai notes, in Shanghai, Tan was able to see firsthand many of the economic changes taking place in China.[14] These observations increased his admiration for the West and formed a context for his philosophical work.

Tan's interest in Western thought and culture naturally increased after China's defeat in the Sino-Japanese War. China's loss made a major

---

10 Elman 1999: 11.
11 Elman 2004: 295, Meng 1999: 16–17.
12 Elman 2004: 292.
13 Elman 2004: 305.
14 Chan 1985: 60.

impression on the minds of late-Qing intellectuals. In particular, because China lost the war to a former tributary state and conceded another one of its tributary states, namely Korea, Chinese intellectuals and officials felt a supreme sense of despair. For example, the governor of Tan's home province, Hunan, exclaimed: "Our country can no longer survive in the world."[15] As intellectuals reflected on the causes for the defeat, they generally blamed it on the failure of the Self-strengthening Movement. They also became more critical of traditional thought.

The jolt from the Sino-Japanese war elicited from Tan a dual response to Chinese traditional thought. On the one hand, he came to stress the universal values of Western science and become critical of classical Chinese thought, including Buddhism.[16] He was also assailed religion in general and advocated "transforming temples into schools." However, on the other hand, he extolled aspects of the classical thought he believed were compatible with Western science, such as Wang Fuzhi's materialism.

By 1896, the year he composed his *A Study of Cosmic Love*, Tan had metamorphosed into an avid believer of Buddhism, which he would separate from ordinary religion. Tan's turn to Buddhism may have been influenced by personal experiences. In 1896, Tan became an Expectant Prefect of Jiangsu in line with his father's wishes. The official corruption he personally witnessed likely played a role in his transformation into a Buddhist.[17] Shortly after serving in his new position, Tan went to study Buddhism with the renowned Buddhist scholar and teacher Yang Wenhui. He was also introduced to another Buddhist scholar, Xia Zengyou, whom Liang Qichao praised as the "forerunner of the intellectual revolution of the Qing" and the "first to realize the usefulness of the Weishi (Yogacara) School" in the modern era.[18] Tan came to know Xia through Liang Qichao and the three of them studied Buddhism and Western thought together.

Judging from Tan's life experience, he used Buddhism to address two aspects of the modern condition. On the one hand, Buddhism meshes with an imperative toward science and rationalization and evolution, which undermines previous religious modes of thought. But on the other

---

[15] Wong 1989: 8.

[16] Tan wrote in his "Notes": "The low Buddhists, in an effort to assert the so-called voidness, deny the existence of the world, comparing it with the unreality of sound and light. By so doing, not only do they cut themselves off from the world, but also exhibit a lack of understanding of sound and light. For heaven and earth are not unreal; neither are sound and light" (*Tan Sitong quanji*, vol. 1: 130–31; cited from Chan 1985: 63).

[17] Chan 1985: 69.

[18] Chan 1985: 71.

hand, it responds to an increasing sense of human finitude and points to overcoming the temporal world and any form of thought. In Tan's *A Study of Cosmic Love*, he reworks classical ontological systems so that they are compatible with the metaphysical assumptions of science and attempts to overcome the fissures associated with modernity.

Tan's Renxue

Tan's magnum opus *A Study of Cosmic Love* (*Renxue*)[19] begins by establishing an ontological source that can be considered a substance.[20] Simply put, Tan interprets the moral concept of "benevolence" (*ren*, 仁) as the cosmological ground of the universe. However, this substance is extremely difficult to pin down and is peculiarly hybrid.

Tan begins his treatise by describing *ren* as both ideal and material. Jiang Guanghui suggests that Tan continues a Chinese tradition of thought in which there is no clear division between the mind and body or the material and the psychic.[21] In Tan's text, *ren* has an ontological significance and is associated with material substances such as ether.

Tan's vision brings together unique aspects of premodern Chinese and modern Western patterns of thought. To some extent, we can see Tan Sitong developing an idea that Hajime Nakatani considers prevalent during the post-Han period, namely that moral characteristics are identical to states of *qi* or bodily characteristics.[22] For example, in his *Treatise on Personalities* (*Renwuzhi* 人物志), Liu Shao (189–244 AD) defined *ren* in the following manner: "Thus one whose bones are upright but resilient is called broad minded and strong willed. Being broad minded and strong willed is the essence of *ren*".[23] This passage associates *ren* with a certain physical disposition and behind this physical disposition is a configuration of *qi*. In Nakatani's words: "A person as conceived by medieval authors was then but a passing phase in the consecutive stages leading from *qi* endowment to its physiological, psychological, ethical and communal realizations."[24]

---

19 Tan 1984, 2002.

20 Because Tan gives the term *ren* a cosmological significance at times, I translate it as "cosmic love".

21 Jiang 1997: 366.

22 Nakatani 2006.

23 Cited in Nakatani 2006: 547.

24 Ibid.

Tan follows this pattern and connects *ren* to ether, which as Ingo Schafer has pointed out, in Tan's thinking has much in common with the traditional concept of *qi*.[25] Like earlier Chinese philosophers' conception of *qi*, ether is imperceptible and yet penetrates the world.

> The phenomenal world, the world of the void, and the world of sentient beings are permeated by something extremely vast and minute, the cohesive, penetrative, and connective power which embraces all things. Its form eludes the eyes; its sound, the ears; its taste, the mouth; and its smell, the nose. For want of a better term, let it be called "ether."[26]

Even if we limit our gaze to the Chinese context, the above passages of course draws on more than the medieval connection between *qi* and *ren*. Tan expands the medical metaphor associating the flow of *qi*, health, and *ren* to envelop the whole world. In his exploration of the influence of Song-Ming Confucianism on Tan's concept of *ren* as the world, Shimada Kenji draws our attention to the philosophies of Cheng Mingdao (1032–1085) and Wang Yangming (1472–1529). Cheng Mingdao invoked Daoist and Buddhist conceptions of the unity between humans and the cosmos. We see this vision in a number of classical sources. For example, Zhuang Zi wrote, "Heaven and earth were born together with me," and the Buddhist Zeng Bi wrote "the myriad things and I are one body."[27] However, Cheng fused this concept of the cosmos with the principles of Confucian morality. Tan developed the Song-Ming Confucian synthesis of Daoism-Buddhism and Confucianism, in which the ontological source takes on the characteristics of Confucian morality, and *ren* in particular. Human beings realize this ethical perspective when they allow the cosmos to flow without obstruction and identify their own bodies with the cosmos.

Tan invokes the identity between bodily states and ethical stances to explain moral and religious concepts around the world.

> When it [ether] reveals itself in application (*yong*), Confucius called it "*ren*," "the ultimate source," and "nature"; Mo Zi calls it "love without discrimination"; the Buddha calls it "the ocean of nature" and "compassion"; Jesus calls it "soul," "you shall love your neighbor as yourself," and "love your enemies like friends"; and natural scientists call it "centripetal force" and "gravitational force"—all refer to this thing.[28]

---

25 Schafer 2001: 267.
26 Tan 1984: 67, Tan 2002: 12.
27 Cited in Shimada 1958: 12.
28 Tan 1984: 67; Tan 2002: 12.

This passage again repeats the pattern that Nakatani identifies in medieval Chinese texts; something material, ether, serves as the source for moral concepts, such as "love without discrimination" or "compassion." But this passage also shows that Tan brings the theory of identity that we find in medieval texts into a space where materiality and ideality have taken on a new significance.

Tan describes the physical dimension of the things in terms of modern scientific concepts such as gravity, which acts on all bodies equally. From this perspective, we might say that the duality between the material and Tan's cosmological source can be compared to Spinoza's description of God or Substance having the attributes of thought and extension, which are identical to one another. According to Spinoza as well, every idea can also be understood as some type of physicality or extension. In fact, he even claims that the mind is the idea of the body.

However, while Spinoza avoids calling substance ethical, Tan constantly uses metaphors of the all-pervading substance of ether to derive ethical meanings, such as loving one's neighbor as one's self and loving one's enemies like one's friends. While previous thinkers may have used *qi* to establish hierarchical relations, all of the above-mentioned concepts express something similar to the standpoint of exchange value, according to which any thing can be compared to any other. From this arises the ability to stand in for the other. While in Tan's text this exchangeability or equality is attached to a moral maxim, such as "love thy enemy," at the same time, it represents a physical force such as gravity.

While in the modern worldview a force such as gravity is devoid of moral content, Tan tries to infuse physicality with the feeling of love and morality. However, unlike medieval authors, Tan faces a world in which physicality appears to resist the imposition of moral norms and human feelings. Thus Tan makes the identity of subject and object or mind and original body or ether into a goal for the reflexive subject to pursue. In other words, the identity between mind and body is Tan's presupposition but not quite a description of the way things are in the temporal world. In fact, as we shall see, the unity between the physical and the mental is actually a means to overcome time and finitude; realizing this unity then becomes the true precondition for moral action.

Time and Transcendence

Tan's key concept of ether has material and non-material sides, both of which transcend time. Since the perception of time requires change, at the level of unchanging ether there is no time and no death. One of the key goals in Tan's *A Study of Cosmic Love* is to unite the self, which arises from the false use of concepts and distinctions, with ether and cosmic love. Tan discusses at length how this process involves overcoming time and death and he develops two narratives of overcoming the temporal. On the one hand, he points to a type of annihilation of time, duration, and being, which we could say is his attempt to revive classical Chinese schemes in a modern world. This amounts to an annihilation of abstract time and a negation of subject and object. But this framework does not allow him to develop a theory of political action specific to his context and thus he develops an alternative narrative of evolutionary temporality. He thus attempts to transcend change while remaining within time. This move corresponds to a utopian misrecognition of historical time within abstract time; one evolves to a point at which one is in the world but no longer changed by it.

Tan's anxiety about time expresses its increasing importance in a global capitalist world, which he saw firsthand in the social and economic transformations in late nineteenth-century Shanghai. Moreover, his discussion represents the way time takes on a two-fold character in the modern capitalist world. That is, on the one hand, time is an independent variable,[29] which can be separated from events, and, on the other, time refers to a process of history moving to increasingly high levels of productivity, which Tan grasps as evolution.

In his *A Study of Cosmic Love*, Tan makes a number of statements that express the abstract and fleeting nature of time. For example, in section sixteen he asks:

> How can we know that the present (今日) exists?[30] We know it exists by comparing the present to the past and the future. But how can we know that the present exists when it has already passed and what is to come has

29 Postone 1993: 201.

30 The term *jinri* is usually translated "today" and Chan Sin-wai follows this convention. However, it does not make sense to say that "by the time we know that today exists, it has already passed." Moreover, Tan contrasts *jinri*, not with *zuori* and *mingri*, yesterday and tomorrow, but with past and future (*guoqu* and *weilai*). Since *jinri* was also used to talk more generally about the present, I translate as "present."

> not yet come? By the time we know that the present exists, then surely it is a present just past.[31]

Thus Tan describes a fundamental *aporia* within abstract time itself, namely as time becomes an independent variable—a series of now-points—it becomes increasingly difficult to grasp the now. That is, it becomes impossible to locate the existence of the now, without giving it duration. If the now has duration, then it is no longer time but in time; that is, it ceases to be an abstract measure, but becomes the measured. Moreover, if the now has duration, it would thus traverse more than one now-point. Thus time splits into two nonexistent spaces, the no-longer and the not-yet.[32]

As Tan notes, Buddhists had dealt with this problem, but he stresses the acuteness of the human anxiety caused by the experience of time. "The past lingers in our memory, but the future comes in a continuous stream. Throughout our lives we receive this flow without rest and in the end we cannot take it any more. Is this not lamentable?"[33] In Tan's view there is a disjuncture between our experience of the past as lingering and an overwhelming future. The lamentable nature of this experience is produced by a rapid change in time, which increases one's awareness of human finitude or death. This is a future-oriented outlook, in which people feel anxiety because they are caught within the force of the past and are unable to make sense of what comes into the present from the future.

Tan deals with the problem of anxiety related to the experience of time on two levels. He invokes traditional Buddhist and Daoist resolutions to finitude and then incorporates these themes into an evolutionary narrative of the nation-state. First, Tan asserts that the human anxiety related to the experience of finitude results from ignorance:

> Those who love life and hate death may be said to be confused (*huo*) and without understanding. For they are ignorant of the fact that all are by nature unborn and undestroyed. Their ignorance gives rise to delusion; hence even if they know what is righteous, they are overwhelmed by fear of death and they refrain from doing it.... Seeing that life is so transient, and that every-

---

[31] Tan 1984: 102; Tan 2002: 51.

[32] In Ousia and Grammé, Jacques Derrida claims that this *aporia* is the fundamental *aporia* of time throughout history, Derrida 1985.

[33] Tan 1984: 104, trans. amended; Tan 2002: 51.

> thing around them is so distressing, they just opt for anything that pleases them. This being the case, how can all under heaven still be ordered?[34]

Here Tan resolves the problem of finitude by negating change. He links this extremely abstract solution to the problem of human finitude to concrete human action and claims that some people become overwhelmed by the ephemeral nature of existence and opt for hedonism because they are lost in their finite selves and ignorant of their true eternal nature. These are the ethical consequences of the loss of meaning in a world changing aimlessly.

This response to the problem of human finitude is similar to Buddhist and Daoist responses in that Tan alludes to the illusory nature of the self by pointing to a primordial ontological source outside of time. Although the goal of Buddhists and Daoists was often personal enlightenment, the context of the late Qing required active political intervention and by connecting this hedonism to the disorder of all-under-heaven, Tan clearly refers to the political consequences of the ignorance of the true self. He notes that ignorance about one's self could lead to cowardice and an inability to act righteously.

Tan contrasts this cowardice and hedonism with an active interpretation of Buddhism, which he explicitly distinguishes from Daoism:

> For how can Buddhism be like this [i.e., inactive]? Inactivity in Buddhism is such that good actions are brought forth, resulting in the salvation of sentient beings. To put it more philosophically, activity is inactivity, and inactivity is activity, which means that these two relative terms are actually unnecessary. That is why all who have a good understanding of Buddhism never fail to be vigorously aroused and courageously powerful.[35]

Here Tan tries to put what Lukács calls the contemplative view, which transcends matter, into action. Tan could link Buddhism with "good actions" by invoking his ontology of *ren*. The ontological source of both the universe and the self is cosmic love, which goes beyond everyday experience and self-interest. To some extent, Tan develops the Buddhist link between wisdom and compassion. According to Mahayana Buddhism, the realization that the self does not exist entails a compassion for the rest of the world.

There is of course a gap between the realization of nothingness or even *ren*, and the concrete world in which one acts; Tan bridged this gap with

---

[34] Tan 1984: 92; Tan 2002: 40.
[35] Tan 1984: 116; Tan 2002: 64.

the politics of the nation-state. Although Tan discussed the emptiness of the self at a universal level and spoke of "saving the world" (救世), he also mobilized this discourse for the purpose of the nation-state. At this point, he invoked the temporality of history, periodized in terms of the old, the present, and the new:

> The two continents of Europe and America became prosperous by being fond of what is new; the Japanese learned from them and changed their dress, food, and habits. The three continents of Asia, Africa, and Australia fell through attachment to what is old. We Chinese often resort to citing precedents in the ancient systems. When death is staring us in the face, we still dote on the savage and uncivilized age as if the present does not exist.[36]

In English one cannot quite grasp the difference between the two ways Tan discusses time, since the terms *jinri* and *jin* both can be translated as either "the present" or "now." However, the meaning of this "present" changes depending on the terms against which it is contrasted and the general framework of temporality evoked. As we have seen, in section sixteen of *Renxue*, Tan contrasts the word *jinri* with the "past" (*guoqu*, 过去) and the "future" (*weilai*, 未来). But in discussing the fate of continents, he uses *jin* to refer to the present age as opposed to the old (*gu* 古). While Tan argues against the existence of the present as a now-point by stressing the way it splits into past and future, he also underscores the importance of recognizing the present age as opposed to the old. Tan asserts that the new is better than the old by listing a number of characters that have the component *gu* (古) "old" in them and pointing out that these characters all have negative meanings.[37] Tan connects this second temporal distinction between the contemporary and the old to an evolutionary perspective, which evokes the second notion of time we mentioned earlier.

As Michel-Rolph Trouillot notes, the view of evolutionary time as a linear progression has implications for a theory of actors as well. "For as soon as one draws a single line that ties past, present, and future, and yet insists on their distinctiveness, one must inevitably place actors along that line."[38] When one emphasizes distinctiveness, one already links time

---

[36] Tan 2002: 59, cf. Tan 1984: 112. My translation.

[37] Tan's list is long, but here are a few: 苦 *ku*, bitter; 枯 *ku*, decay; 辜 *gu*, sin; 痼 *gu*, serious illness; 故 *gu*, death; 沽 *gu*, wine from a shop, which Confucius would not drink (Tan 1984: 112, Tan 2002: 59).

[38] Trouillot 2002: 225.

to space; that is, distinctness in this context implies occupying different places on a line. But the concepts of time we mentioned above entail different modes of "distinction." In the case of mere abstract time as an independent variable, although we have a line, it is blind and the various points are not meaningful; every point on this line is qualitatively and quantitatively identical. Notice that when Tan is arguing in a Buddhist mode, he specifically attacks the distinctness of past and present, claiming that it is impossible to separate dimensions of time; in short, he tries to negate or overcome time.

However, when discussing the future of the Chinese nation, he distinguishes between old and new and places actors along an evolutionary time line. This notion of time requires more then a mere line; it requires subjects, which unlike identical now-points, can be new or old, advanced or underdeveloped. Tan discusses how nations and continents that stress the new became prosperous, advancing along the line, but those that stressed the old—Asia, Africa, and Australia—lost their independence because of attachment to the old. In other words, Tan links the new to a more prosperous sovereign nation, while he identifies the old with a loss of sovereignty or a loss of the nation.

These two notions of time are inextricably related; specifically, time as an independent variable is in some sense the measure of evolutionary time and this makes possible the category of speed or efficiency, which is so important in the modern world. Marx notes that in capitalist society, "Time is everything, man is nothing; he is at most the embodiment of time (*die Verkörperung der Zeit*)."[39] Moreover "the pendulum of the hour has become the exact measure of the performance (*der Leistung*) of two workers, just as it is for the speed of two locomotives."[40] In global political economy, as nations compete for capital and resources, be they symbolic or material, they appear similar to the two workers Marx mentions. Once Tan has set the trajectory of nations in terms of civilization, he can compare the progress of two nations in terms of how far and how fast they have moved towards this goal, just as one compares two locomotives.[41]

---

39 Cited in Lukács 1971: 89, translation amended.

40 Ibid.

41 Later Hu Shi would express this view clearly: "When we look at culture from an historical perspective, we see each nation traveling on the same original path—the difference is that some travel the path with more ease, some faster and some slower. Hence the speed that nations travel the path is different and some arrive earlier than others" (Hu 1996: 179, see also Zhang Rulun 2001: 359).

Tan brings progressive time-history into the picture both to give hope and to re-create a type of anxiety after the above-mentioned Buddhist/Daoist-type resolution. We saw earlier that Tan develops the Song-Ming Confucian discourse of principle and draws on earlier attempts to identify the moral and material worlds. However, in his discussion of evolutionary time, he constantly stresses the goal of overcoming time through negating the material dimensions of the world. Tan believes that the ends of evolution are limitless and claims that mental power, which is another gloss on the term *ren*, could eventually overcome the physical world. "In addition, we might be able to fully study eugenics [进种之学, literally, "the study of improving races"] so that the coming generations will be better than the previous ones, and when this goes on for many generations, a different kind of man will then be created, who uses only intelligence, but no strength, who has a soul but no body."[42] This part of Tan's thinking is evidently influenced by ideas of modern science.

Tan was also influenced by an apparent trend of the times, namely an increasing control over nature by machines. He links machines to the overcoming of the physical dimension of life and, in particular, to overcoming time, which he claimed was a goal of the ancient Chinese sage kings. Hence it is paradoxically because the Westerners have been able to save time with machines that they could emulate the Three Dynasties:

> ... within a very short span of time, the civilization of the Western countries emulated that of the three dynasties. They relied on nothing other than time-saving, so that they never fell short of time, and this is like putting into one man all the energies of a few dozen men. It is written in the *Great Learning*: "One must do things quickly." Only machines can do this.[43]

With these words Tan provides an answer to human finitude different from the Buddhist negation of the subject; in this case, we see the expansion of the power of the subject so that it can increasingly overcome duration. Instead of overcoming the distinction between life and death, here Tan invokes a "prolongation of life," which would occur because less time in one's life would be wasted. Hence by looking at machines, Tan constructs a vision of hope within the ordinary conception of time. In this passage, Tan also expresses the movement of capitalist society in ideological form. As Postone notes, "capital unfolds historically in such a way that the level of productivity becomes less and less dependent on the direct labor of

[42] Tan 1984: 213; Tan 2002: 158.
[43] Tan 1984: 132–33; Tan 2002: 80.

workers."[44] This perhaps corresponds to ideals of machine-oriented production so prevalent during the Self-strengthening Movement.

Tan reads this idea of time-saving back into the ancient past. Like most late Qing intellectuals, Tan understands the world using the categories associated with the nation-state, and so he measures China's decline not only with respect to its past, but in relation to the West, which is advancing. He perceives a time lag between China and the West and sees these different nation-states or regions as involved in different but related historical processes. It is as if Tan describes Hegel's idea of the trajectory of Spirit, which moves from China to the West, but in Tan's text, we could say that Spirit begins in the Three Dynasties, which has significance as a moral and political ideal and adds legitimacy to China as a nation-state. China's progress to mechanized production would simply involve becoming more truly itself by realizing the ideals of the Three Dynasties. Tan believed that although the trend of history was towards greater mechanization and time-saving, in China, "not only is time not saved... [t]he civil service examination is burdened with too many subjects; the promotion of persons is hampered by the seniority system.... Alas when time is wasted, it will lead to the loss of the country as well as to the extinction of the race."[45] With the above argument, Tan connects abstract time, namely time one can save through acceleration, with historical and evolutionary time. If individuals and the nation do not move to further mechanization and overcome the limitations of the physical, time is lost, and the people of the nation will decline.

Tan explained this history as decline or devolution vividly:

> During this long period of darkness, we have been through thousands of disasters under the thick curtain, learning of no new principles and seeing no new systems. And so after two thousand years, we have fallen from the culture of the Three Dynasties to that of the barbarians of today. In another two thousand years' time, we will degenerate from the present barbarian culture to that of apes, orangutans, dogs, pigs, frogs, and clams. By then, the cycle of reproduction will come to an end, and what is left will be a barren land, like a desert uninhabited by any human beings or living things.[46]

Notice that here Tan retains the ancient ideal of the Three Dynasties, and when this ideal combines with Darwinism, rather than a theory of

44 Postone 1993: 296.
45 Tan 1984: 133; Tan 2002: 80.
46 Tan 1984: 167; Tan 2002: 117.

evolution, it produces a theory of degeneration.[47] In the overarching framework of progressive history, only some nations move forward, while others may move backward. Hence, Tan claims that during the Three Dynasties, China had ideas associated with democracy, but gradually declined, while Western countries began to implement these principles.

In order to connect his narrative of decline/evolution with modern political goals, he infuses the Three Dynasties with the ideals of democracy. Tan presents a historical narrative based on a version of the Confucian ideal of the people as root (*minben*):

> When people first emerged in the world, there were no such distinctions as that between ruler and subject; everyone was just "people." As the people could not govern themselves and did not have the time to do so, they therefore collectively raised one person from the multitude to be the ruler. As the ruler was said to be "collectively raised," it was not the ruler who selected the people, but the people who selected him. As the ruler was said to be "collectively raised," this station was neither above nor below the people. As the ruler was said to have been "collectively raised," the people existed before the ruler, and as such, the ruler was the branch, and the people, the root (本).[48]

Here Tan draws on Mencius' philosophy of the people as root and Huang Zongxi's development of this theory in the late Ming dynasty in order to highlight the egalitarian and democratic nature of the human condition. However, the evolutionary narrative is still at work since he claims that the need for a ruler arose from the imperatives of time. In Tan's view, the first ruler was elected democratically, and Mencius and Zhuang Zi theorized this practice. Mencius discussed the democratic element of Confucian thought, while Zhuang Zi developed a critique of the rulers. In this way, Tan argues that the basic teachings of Daoism emerged from Confucianism.[49] Despite their differences, Mencius and Daoism have a common presupposition, a point that is developed by Ming neo-Confucians (especially of the Wang Yangming school), namely that human nature is spontaneously good.

Of course, as Tan expounds this egalitarian reading of Confucianism, he must deal with the aspect of Confucianism that he openly criticizes, namely the five relationships, which were explicitly hierarchical. He does

---

[47] This theory had its precursor in the work of the late Ming/early Qing dynasty philosopher Wang Fuzhi.

[48] Tan 1984: 154; Tan 2002: 102.

[49] Tan 1984: 146; Tan 2002: 95.

this through a narrative of intellectual decline. In Tan's view, Confucian philosophy was hijacked by Xun Zi, who called himself a Confucian, but stressed the hierarchical divisions between rulers and people and consequently undermined the democratic dimension of Confucianism. In the 1920s, Liang Qichao noted that Tan's critique of Xun Zi was part of a general movement by reformers to attack Xun Zi.[50] In line with this movement, Tan notes that when common people practice the philosophy of Xun Zi they "openly debase themselves" to obtain high positions and "call it 'loyalty' when they help the ruler oppress the people." Rulers use Xun Zi's philosophy to make themselves superior to the people and keep them ignorant.[51] According to Tan, from the time of the Qin emperor, rulers and people began to practice the philosophy of Xun Zi, and thus he concluded that

> government over the last two thousand years has been government in the style of the Qin emperor; all those who governed were great robbers. The teaching of the last two thousand years has been the teaching of Xun Zi; all those who following it were conformist hypocrites. Only great robbers were good at taking full advantage of conformist hypocrites and only conformist hypocrites were skilled in pleasing great robbers. They made use of each other in the name of Confucius. If we blame Confucius because of the wrongdoings of those great robbers and conformist hypocrites, how can we understand Confucius?[52]

Tan clearly describes a narrative of decline that we have seen him invoke elsewhere as the regression of the civilization of the Three Dynasties to the barbarism of today.

Like many other reformers of the time, Tan thinks of political evolution as a return to Mencius, who symbolically supported reformist plans for implementing local autonomy by empowering the local gentry. Tan's faith

[50] Zhu 1996: 333.

[51] Tan 1984: 147; Tan 2002: 95. The famous intellectual historian Li Zehou thus echoes Tan Sitong when he contends that Mencius drew on aspects of "primitive democracy" in ancient Chinese shamanism. See Li 1985.

[52] Tan 1984: 149; Tan 2002: 96. In this narrative, Confucianism becomes not just the symbol of the nation-state, but the hope for a new world. However, as Tan notes, Confucianism was misinterpreted by Xun Zi, whose philosophy dominated after the Qin. Moreover, again in line with the New Text School, Tan claims that Liu Xin, the Han dynasty scholar, implemented the philosophy of Xun Zi and forged key texts of the classics, which were originally written by Confucius. In fact, Tan contends that after the decline of Confucianism and the establishment of the Qin empire, there were basically no books worth reading except for the *Posthumous Works of Wang Chuanshan* and Huang Zongxi's *A Plan for the Prince*, since both of these works "obliquely complain about the separation between the ruler and the people".

in the local gentry is related to his optimistic ontology, which stresses that problems both physical and political emerge through blockage. Thus the gentry were to keep channels flowing smoothly between the ruler and the people. We shall now turn to Zhang Taiyan, who develops a different political theory based on a pessimistic ontology.

Zhang Taiyan's Critique of Tan's Thought and His Ontology of Confusion

Zhang offered a theory of evolution to counter many elements of Tan's philosophy and posited an alternative vision of political reform. Before looking at Zhang's early work and his response to Tan Sitong, we must note how his style differs from that of Tan. In particular, while in 1896 Tan completed a fairly coherent philosophical treatise, Zhang's early work consisted of several articles published in various journals and newspapers on a vast variety of scholarly and political themes. In what follows, I will focus on Zhang's early ideas of evolution, which are expressed primarily in his essay "On Bacteria," published in 1899. Among Zhang's voluminous collection of early essays, "On Bacteria" is particularly significant both because it was well received by scholars such as Yan Fu and because it introduces ideas that Zhang would develop later. Specifically, it shows Zhang's hesitance to overcoming the moment of temporality or chaos and confusion, which we could also associate with the political world. Hence, in Zhang's view, evolution does not eventually transcend confusion; rather, it stems from the temporal world and inflects it differently.

Zhang Taiyan's Early Involvement in Politics

Zhang Taiyan was born in 1868 in Yuhang district, Zhejiang. At the age of 16, he went to sit for the imperial examinations, but could not take them due to a fit of epilepsy. He would never go to take the examinations again and like Tan, failure to succeed in the examinations may have contributed to his disdain for official life. Zhang would continue to learn the classics and in 1890 he joined the Gujing Academy of the Classics in order to study with famous Old Text School scholars, such as Yu Yue and Sun Yirang. The academy provided Zhang with a space to pursue scholarship in an environment somewhat separate from the imperatives of imperial politics. Zhang Taiyan and the Old Text School scholars are famous for stressing the autonomy of scholarship from the agenda of the palace. However, if the space of the Gujing Academy was removed from imperial politics, this distance also allowed Zhang to think about the emerging politics of the nation in a way that did not completely conform to the mainstream. Unlike Tan Sitong and Yan Fu, Zhang did not receive his early intellectual training by studying in or frequenting the shipyards and arsenals and was relatively removed from the world of politics until he was in his late 20s.

In 1895, after the Sino-Japanese War, Kang Youwei organized a "Self-Strengthening Society," which was something like a modern political party.[53] Zhang sent sixteen yuan to join the society and eventually was invited by Liang Qichao and Zhang's in-law Wang Kangnian to join their newly founded paper, *The Times* (*Shiwubao*). A few months later, Zhang resigned from the Gujing Academy to join the paper and become a political activist. Although reformers like Kang Youwei praised Zhang's early writings in *The Times*, due to Zhang's resistance to arguments advocating Confucianism as a religion, on April 16, 1897, Zhang left *The Times* to work in the *Statecraft Journal* (*Jing shibao*) and the *Substantial Learning Paper* (*Shixue bao*).[54]

The *Statecraft Journal* was run by Song Shu (1862–1910), a prominent Buddhist and also a relative of one of Zhang's former teachers, Sun Yirang. Zhang thought highly of Song's work and Song is probably the first person who really taught him about Buddhism. However, in his chronological autobiography, Zhang makes the following comment about Song Shu, Tan Sitong and Buddhism:

> In the spring of the twenty-thirdyear of the Guang Xu reign [1897] I was in Shanghai. Liang Qichao and his group were calling for a "Confucian religion," to which I was greatly opposed.... At this time, Song Shu, also known as Song Pingzi, came to Shanghai and we had a talk full of mutual sympathy. Song showed me a copy of Tan Sitong's *Renxue*. Tan had indiscriminately gathered together any- and everybody's ideas. I couldn't understand it, nor was I the least bit interested. Song asked "Have you read the Buddhist classics?" I responded: "At the prompting of Xia Zengyou, I have cursorily glanced

53 Liang Qichao claims that Tan Sitong also wanted to join this party, but when he went to Beijing to meet Kang Youwei, Kang had already left for Guangdong. However, there are those who dispute Liang's narrative and claim that Tan had no interest in joining.

54 Zhang began to quarrel with the editors of the *Statecraft Journal* because they thought he was too zealous a reformer and his words were too direct. Hence Zhang shifted to *The Substantial Learning Paper*, where he would also leave for the same reason. In the *Substantial Learning Paper* he concentrated on a critique of Confucio-centrism and while in *Statecraft Paper* he critiqued the reform project in general and grappled with concepts in Western political theory such as equality and democracy. Zhang would republish many of the essays he wrote for these two journals in *A Book of Urgency*.

> through such sutras as the Fahua, Huayan, and Nirvana sutras, but without any depth of understanding." Song said: "You should try to read the Sanlun" [The Three Treatises: namely, the Zhonglun or Treatise on the Middle, the Bailun or the Treatise on the Hundred Verses, and the *Shiermenlun* or the Treatise on the Twelve Gates]. So, I tried to read through them but found I didn't like them at all. It was at this time that I had become enamored of Xun Zi's notion of *rushu* [emphasis on ritual and politics and on classical philological research], so I found the expedient of an abstract theory highly displeasing. I would occasionally read the *Da sheng qi xin lun* (Treatise on the Rise of Faith in the Greater Vehicle) for spiritual nourishment, but I did not enjoy it.[55]

Although Zhang is now famous as a promoter of Buddhist politics, in his early works Zhang saw Buddhism as opposed to the scientific method and explicitly criticized it. These two points, namely the emphasis on science and praise for Xun Zi, are a recurrent theme in Zhang's early writings.

Like Tan, Zhang was interested in Western thought. When he left the *Substantial Learning Paper*, Zhang went on to found "The Translation Society." Although he only worked in this society for a brief time, his interest in translation continued and in 1898, he rejoined Wang Kangnian to work on the *Flourishing Talks Paper* (*Changyan bao*). In this paper he edited a translation of Spencer's work on evolution, which would clearly influence his early views on history and politics.[56]

On September 21, 1898, the Empress Dowager Ci Xi staged a coup d'état which marked the failure of the Hundred Days Reform and ended any immediate hopes of political change. The Qing government executed six reformers, including Tan Sitong. Zhang fled to Taiwan, where he both wrote for the *Taiwan Nichi Nichi Newspaper* and contributed to Liang Qichao's journal, *The Pure Opinion Paper* (*Qingyi bao*). In one of its first editions, Liang published the whole of Tan's *An Exposition on Cosmic Love* in installments along with Liang's own preface introducing the importance of Tan's work.

In 1899, Zhang's views on the reformers were similar to his attitude towards Tan Sitong. He agreed with the reformers' cause and admired Tan Sitong's willingness for self-sacrifice, but he was critical of their theoretical and scholarly views. We can see this in some of his writings around this time, such as his long essay "On Bacteria" and his self-selected collection of essays, *A Book of Urgency*.

---

55 Cited in Shimada 1990: 51, translation amended.

56 Wong 1989: 13.

Zhang's Ontology of Confusion

In what is perhaps his most famous philosophical article of this period, "On Bacteria" (*Jun shuo*), Zhang critiques Tan's metaphysical worldview and constructs his own theory of evolution. We shall see that although Zhang's position differs from Tan's in important ways, like Tan, Zhang propels classical ideas into the framework of modern science to create a hybrid discourse framed by basic tropes related to modernity. While Tan's ontological source is an overarching totality, Zhang constructs his ontology on the minute actions of sexual bacteria and physical atoms. Moreover, these minutia produce confusion rather than enlightenment. In this way ontology is associated with confusion and thus posited against human order.

Zhang derives the title of the essay "On Bacteria" from a line in the second chapter of the *Zhuang Zi*, which Burton Watson translates as "music (*yue*) from empty holes, mushrooms springing from the dampness" (乐出虚, 蒸成菌). Watson seems to follow the commentaries of Guo Xiang[57] and Cheng Xuanying, both of whom interpret the characters *yue* and *jun* respectively as music and mushrooms. However, at the beginning of "On Bacteria," Zhang links these two concepts to modern medical science by reading the character 乐 as *le*—happiness or emotion—and glossing *jun* as bacteria or germs. Zhang writes:

> I once read Zhuang Zi's "Discussion on Equalizing Things" in which it is written "emotions come out of empty space and mushroom like bacteria are formed from the warm moisture" but we don't know the basis of his statement. The music of peoples hearts emerges from the void, but to say that this moisture can produce bacteria with form, is this not nonsense? Recently, I got Duncan J. Reid's book *A Discussion of Humans' Struggle with Small Bacteria* and know that the above words are not nonsense.[58]

Zhang then explains that Robert Koch discovered that tuberculosis was caused by small bacteria. But he does not simply use modern biology to interpret the *Zhuang Zi*. He also brings Robert Koch into the world of traditional Chinese medicine and philosophy. In Western medicine, bacteria can be used to explain sickness and a number of transformations;

---

[57] A Wei Jin period commentator on the *Zhuang Zi*.

[58] Zhang Taiyan 1982: 54. Duncan J. Reid was an English doctor who gave a lecture on the above-mentioned book in Shanghai in 1892. A translation of his book was subsequently published the same year in *Gezhi huibian*.

however, Zhang tries to make bacteria (*jun*) into a more fundamental category and interprets them as the root of human existence.[59]

He continues by connecting bacteria to the polysemic character in classical Chinese *gu* (蠱), which can mean insect, confusion, or affairs. Zhang explains that Koch

> of course discovered that cholera was produced by small bacteria and named the bacteria that causes tuberculosis, *tubercle bacillus*. These are all the same thing, namely bacteria. I will not mention cholera, but tuberculosis begins with extreme desire and lasciviousness. Excessive desire becomes a mycotoxin and crawling bacteria emerge therein. Is this not what is meant by "emotions come out of empty space and mushroom-like bacteria are formed from the warm moisture?" Although their contaminating nature is not related to enjoyment, these bacteria always begin with enjoyment. The famous doctor Yihe said "Women are objects of men's desire and when having sex at night, if one overindulges, one will get the sicknesses of internal heat and the *gu* [a legendary venomous insect] will arise."[60]

By associating various characters with one another, namely moving from emotion/bacteria (*jun*) to sexual desire and insect, Zhang links the basic element of his worldview to enjoyment. Zhang explains that sexual desire, emotion, and sickness can mutually produce one another. "Nonetheless, it is not only lasciviousness that can produce bacteria and the *gu*-insect. Once the bacteria and poison have been formed, they can control people's will and cause people to be lascivious. Besides lasciviousness, emotions such as joy, anger, sorrow and happiness all are influenced by this bacteria-poison."[61]

In addition to connecting insect and emotion, Zhang further moves the concept of *gu* to the ideational level by playing on the two senses of the word *gu*. In the *Book of Changes*, the character *gu* is glossed as both confusion (惑) and affair (事). Zhang brings these two meanings of the word together by invoking the interpretation of Wang Bi (also known as Fu Manrong) of the *gu* trigram, which links confusion with affair. Wang Bi writes: "*Gu* is confusion and chaos. The myriad affairs and things arise

---

[59] Zhang makes this link between the human existence and bacteria by citing an obscure passage from the *Huai Nanzi*: "I read Huai Nan Zi's chapter 'Di Xing Xun' in which he says 'The *baren* (person with thick hair produces) the dark person. The dark-skinned person produces *ruojun* (bacteria) and the ruojun (bacteria) produce the sage and the sage produces the common people. All *baren* are produced by the common people.' Huai Nanzi places the bacteria between the dark-skinned person and the sage; hence the *ruojun* (bacteria) are also people" (Zhang Taiyan 1982: 59).

[60] Zhang Taiyan 1982: 55.

[61] Ibid., 57.

from chaos. Hence confusion is affairs and things."[62] Through Wang Bi, Zhang interprets bacteria in relation to confusion (*gu*), which he in turn associates with the Buddhist idea that human beings owe their existence to ignorance or confusion.

Hou Wailu praises Zhang's essay "On Bacteria" for being materialist and resisting Tan's Confucian idealism.[63] However, as in Tan's case Zhang's ontology is two sided. Zhang links bacteria to confusion and uses Wang Bi to show that all things arise from confusion, hence putting into question the materiality of bacteria. Zhang develops this position further when he notes that human beings are not created by God but through confusion, which inhabits the sperm:

> People have the desire to mate and transfer this desire to their sperm. The sperm receives this and it feels the same way as the person in whose body the sperm resides. The sperm then admires the human form and hence becomes a fetus to imitate humans.... Sperm cells create human beings. Human beings begin with confusion and the sperm cells complete their project using confusion. It is not that "god" creates them; rather, things create themselves. Hence Zhuang Zi says "all take what they want for themselves, but who does the sounding?"[64]

Desire and confusion transcend the particular, but are always already embodied in individuated things. Unlike Tan's cosmic love, desire and confusion do not themselves become a metaphysical entity. In this passage Zhang uses Zhuang Zi to attack the Christian God, but he may also target the prime mover (cosmic love/ether) in Tan's framework. Zhang continues his critique by moving from discussing sperm, bacteria, and insects to discussing atoms (*adun*). On this point once again Zhang quotes Zhuang Zi:

> Small Knowledge asked "In the four directions and the various dimensions of the cosmos how do the myriad things emerge?" Tai Gongtiao answered, "Sometimes Yin and Yang harm each other and sometimes they complement each other. The four seasons mutually substitute, produce, and kill one another. Desire and hatred, acceptance and rejection arise in succession like the Qiao Bridge. The pairing of male and female becomes a regular occurrence. Danger and peace interchange. Calamity and good fortune arise together.

62 Ibid., 60. Wang Bi's original text can be found in the *Wenyuange siku quanshu* (Taipei: Taiwan shangwu yinshuguan), vol. 7: 677.

63 Hou 1978: 345.

64 Zhang Taiyan 1982: 60. The full passage from the *Zhuang Zi* is: "Blowing on the ten thousand things in a different way, so that each can be itself—all take what they want for themselves, but who does the sounding?" (second chapter of the *Zhuang Zi*).

> Speed and slowness complement one another. Congealment and dispersal complete each other. One can note down name and reality and write down the minute details. Order sets limits and the rhythm of things moves them. When something is exhausted it moves in the opposite direction. When one thing ends another begins. This is the way of things."[65]

From this passage, Zhang derives a theory of atoms as elements that repel and attract one another and he contends that it is through this attraction and repelling that things are produced. With this emphasis on movement, action, and repulsion, Zhang shifts from the realm of biology and medicine to that of physics. Here, instead of bacteria, he discusses atoms. Zhang will eventually stress the materiality of atoms to criticize Tan's attempt to connect ether with mental power. However, his conception of atoms also contains a mental element. He claims that if it is the case that the various atoms have the tendency to attract, repel, and resist, then "although air, *qi*, gold and metal are hard, they also have a small amount of consciousness."[66]

Zhang's conceptions of conscious and material atoms/bacteria perform three tasks: first, they serve to criticize Tan Sitong's theory of ether; second, they form the foundation for a theory of evolution linked to human agency; and third, because atoms are linked to bacteria and confusion, they form the basis for a Xunzian theory of human nature, which, as we shall see below, legitimates the use of external legal structures to attain order. This last gesture also goes against Tan's vehement critique of Xun Zi and his attempt to develop a political theory based on Mencius.

Zhang focuses his criticisms on Tan's attempt to gloss ether as "mental power." Tan had written, "The first meaning of cosmic love is interconnection. Ether, electricity, mental power, all refer to the means through which cosmic love interconnects."[67] "Buddhists call it the ocean of being."[68] But he also claims that "electricity must be the brain without form or substance." Zhang's response brings out the difficulty of positing mind-body identity in the modern world of science:

> Some people say that the "the sea of being is ether." But ether is moving light waves, which can pass through glass: but it moves quickly or slowly depending on the color of the light. Its substance is the atom; an inch of substance can be separated into 50,000,000 elements and each element is

65 Zhuang Zi 1999: 727–28, Zhang Taiyan 1982: 62.
66 Ibid.
67 Tan 1984: 61; Tan 2002: 6.
68 Tan 1984: 67; Tan 2002: 12.

> the size of an atom, which has a form one can measure. The movement of ether is smaller than this, but since it has speed, it cannot be without substance (*ti*).[69]

As we have seen, Tan's conception of ether/mind developed the neo-Confucian idea of the heart and mind as a foundation of ethics. Moreover, by making cosmic love transcend time, Tan attempts to resolve the problem of human finitude and create the conditions of sacrificial agency. But when he places this theory within the reified space-time of modern physics, it is difficult to think the identity of mind and body. Zhang stresses the material dimension of ether/atoms and, by placing the emphasis on individual atoms, he criticizes Tan's effort to link the individual to a timeless sea of being. Against Tan, Zhang connects abstract ideals to biological organs and desire:

> So, although the organs are temporal, the theories of universal love (*jian ai*) and mutual aid originate in the organs. Why? This is because all knowledge cannot be formed outside of the senses. Human desire is expressed in relation to sound, color, smell, taste, and touch; humanness (*ren*) and righteousness (*yi*) also arise from desire. A selfish king acts only for himself and is called a "greedy thief"; being good at dealing with people and learning from examples is called "humane and righteous." Hence the *Book of Changes* says "benefiting things is sufficient to be in accord with righteousness." This clearly shows that when there is no benefit (利) there is no righteousness. Once there is righteousness, then there are differences with respect to distant and close relations and there are differences between small and great acts of kindness.... By ruling with rituals and righteousness, one cultivates desires and allows their pursuit, ensuring that they are measured and within boundaries. [See Xun Zi, "On Ritual".]) This is what I mean by "outside of illusion there is no reality."[70]

Zhang makes his critique of Tan's abstract cosmology and democratic politics with reference to an empiricist epistemology of confusion. In Zhang's view, knowledge emerges from sense experience and thus he rejects metaphysical transcendence. Given that there is no reality outside of illusion (*wang*) and bodily desire, Zhang advocates Xun Zi's proposal to tame the passions through the use of ritual and law. Although during his *Minbao* period Zhang would vehemently criticize corporate bodies, in this early period, he stresses Xun Zi's theory of corporate bodies as an effective social order and also as a means of evolution.

---

69 Zhang Taiyan 1982: 69.
70 Zhang Taiyan 1982: 70–71.

Given that Zhang rejects transcendental principles, he requires a new way to deal with the problem of human finitude and at this point draws on the power of the corporate body/nation. In particular, his vision of history and progress must be different from Tan's reliance on transcendent principles. The principle of evolution can channel the anxiety created by human finitude into the nation-state. Tan both emphasizes the anxiety associated with the possible death of the nation and invokes other ideals of transcendence. As we have seen, Tan builds on both the Song dynasty conception of principle, the medieval idea of the identity of mind and body, and the conception of mind-control developed by John Fryer. The ideals that Tan invokes go beyond past, present, and future and he mobilizes these ideals to legitimate sacrifice for the nation. Zhang reacts against positing a transcendent principle to which individuals must be subservient, and throughout his essay "On Bacteria" he stresses Zhuang Zi's idea that things create themselves without an external creator or principle. Hence he must argue for the possibility of self-transcendence and nationalist-identification within a generally immanent framework. He advocates that people identify with the corporate body (*qun*), which transcends the individual.

> "If once a human body falls and dies, it does not awake even after one thousand years, who will dare to die?" Some think, because of this type of attitude, Chinese people are cowards and cannot compete with the countries based on the Buddhist and Christian religions.[71] However, although there is no idea of ghosts and spirits, the idea of generational continuity with ancestors alone is similar to other countries' idea of religion.[72]

The idea of generational continuity serves to ground an experience of transcendence without recourse to God. By making this gesture, Zhang invokes an ancient immanent view of the cosmos against Tan's tropes of transcendence. In this case, only the ancient conception of ancestor worship serves as a quasi-transcendental principle, which would legitimate sacrifice for the nation in times of war and crisis. Like his theory of evolution, Zhang's view of nationalism represents only transcendence within confused temporal existence.

Zhang stresses communal identity from an evolutionary perspective as well and in so doing departs somewhat from his materialist ontology. In fact, Zhang's theory of evolution gestures in the direction of transcending

---

71 Japan, the United States, and such countries.

72 Zhang Taiyan 1982: 79.

the material. First, let us examine how he explains evolution in the non-human realm. We have seen that Zhang claims that all things are conscious and he uses this idea to envision an evolution based on mental agency. Once again as in Tan Sitong's work, although one can find in Zhang's essay a mind-body identity that might be trace to premodern works, the mental side becomes a site of possibility, change, and in a word, evolution.

Zhang argues that there are two ways that things evolve. "Some think of using force to transform themselves. Some do not rely on force and only use their minds to transform themselves."[73] Zhang contends that certain birds constantly stretch their beaks, which causes the necks of future generations of birds to become longer. However, Zhang explains phenomena such as the camouflaging of bird eggs in relation to the birds' intent or will. "The Lu Chao bird fears that people seek its eggs and so its eggs are mostly not white.[74] This is extremely different from birds that build nests in dark places. Hence birds that make nests in straw and grass must lay eggs that are green like sprouts. Birds that make nests in the river must lay eggs that are dark green like reeds."[75] Note that unlike a neck, which can be transformed by physical stretching and exercise, Zhang contends that the color of the eggs must be influenced by some kind of mental power (*sili*).

Zhang opines that evolution from various other animals to human beings is also a result of this type of mental power. "The various things that slowly change and become humans do so by using thought to incur a self-transformation. If this is so, then all of the above examples are produced by so-called delusion and are what Fu Manrong calls confusion (*gu*) or what the *Huai Nanzi* calls bacteria (*jun*)."[76] The confused consciousness of these minute bacteria and atoms becomes the motor of evolution. In other words, evolution represents a possibility within confusion, rather than a movement beyond it.

As in Tan's case, the evolutionary framework does not imply actual progress and this opens a space for a peculiar type of human agency that propels evolution for humans.[77] Towards the end of Zhang's other famous essay on evolution, "On the Origins of Change," he explains that the key

[73] Zhang Taiyan 1982: 64–65.

[74] The Lu Chao bird's nest is in the open.

[75] Zhang Taiyan 1982: 65.

[76] Ibid.

[77] What we see here is again a hybrid formation or what Lydia Liu calls translingual practice Liu (1995). In other words, when placed in the late Qing context, the meaning of evolution changes.

to understanding human change lies in the *Book of Changes*, in particular in the interpretation of the hexagrams. By combining the *Book of Changes* with a notion of linear evolution, Zhang stresses the idea of reversal within a linear narrative of human development.

Wang Bi explains the idea of change expressed in the *Book of Changes* in the following manner: "The constraint appropriate to one moment of time can undergo a reversal and turn into an occasion to exert oneself, but the good fortune of one moment of time can undergo a reversal and turn into misfortune."[78] The *Book of Changes* clearly does not express an evolutionary conception of the time and hence "there is no fixed track for affairs to follow" (ibid.). By combining the possibility of regression adumbrated by the *Book of Changes* with an evolutionary view of time, Zhang, like Tan before him, creates a framework in which a linear track is more or less fixed, but the direction in which the train of history moves can change. In other words, although lower-level animals gradually evolved to become human beings, if humans are not diligent, they may regress.

Zhang is clearly concerned with the possible extinction of the race and the regression of humans into more primitive forms such as apes and expresses this concern most clearly in his essay "On the Origin of Change":

> When beings slowly increase their wisdom, they become intelligent and big. When they slowly decrease in wisdom, they become stupid and their form becomes like crippled turtles.... If we look carefully, we see that fish and shrimp are blind, but not because they never had eyes. They do not use their eyes. Whales have legs but do not use them to run. Rams have horns but do not use them to fight. The camel bird has wings but does not use them to fly. When these ... things cannot be used, after a long time, they will be discarded. So, when a person neglects his intelligence, it will whither and fade; he will become a type of ape. If a person is forced to enter a secluded cave, then one cuts his wisdom and his organs cannot link with his mind and so this person's wisdom will be discarded and he becomes like a type of ape.[79]

Unlike in Tan's analysis, speed is not a key factor in Zhang's vision, which is perhaps why, at this point, abstract time does not play a role. Increase and decrease of wisdom both happen slowly (浸). But the above passage also shows that, although both Tan and Zhang accept the goal of evolution, they conceive of a bi-directional movement. Later, during his *Minbao* period, Zhang asserts that the dialectic of confusion propels history in

[78] Wang 1994: 29.
[79] Zhang Taiyan 1982: 101.

both directions at once, but here evolution and regression are held up as possibilities, with human agency playing the decisive role in determining which way a species travels. Hence, Zhang Taiyan must answer the question: what can people do in order to evolve to a type of super-human-being rather than regress to apes?

Zhang connects biological evolution to political forms and claims that if people do not form corporate bodies (*qun*) and develop their intelligence, they can regress and become apes.[80] Zhang explains this using a biological metaphor, which has political implications. "When a given being knows the way of corporate bodies, even things minute like bees that make sounds as they fly but move forward and backward as a troop, cannot be hurt."[81] Zhang juxtaposes this ideal conception of the group with the human being who leaves society in order to get away from greed and corruption. Although in his *Minbao* period, Zhang will defend such a rejection of society, in his early work, Zhang contends that such a person would become inhuman. In this way, Zhang made sense of the world using an ontology of confusion and desire, in which group identity and the state would effectuate evolution. Thus just as he combines Zhuang Zi and modern biology to develop an ontology, he combines Xun Zi's political theory with a modern vision of the state to propel China onto the world stage as a power.

Conclusion

We have seen how the theories of Tan Sitong and Zhang Taiyan react to and express different dimensions of internationally circulating discourses related to global capitalist modernity. Tan draws on Buddhism to promote a philosophy of absolute equality, and posits a unity between a cosmic ontological source ("cosmic love or ether") and individual subjectivity. Based on his optimistic views of ontology, epistemology, and spontaneous action, he sought to endorse a version of democracy.

Zhang, on the other hand, counters Tan's assumption about ontology by positing a type of self-concealment at the origin of being. Given that what lies at the source of being is not interconnectedness but fragmented

---

[80] Moreover, in Zhang's view, this has already happened among people in certain regions of the world. For example, he claims that the aborigines in Australia are really a type of ape. See Zhang Taiyan 1982: 101.

[81] Zhang Taiyan 1982: 103.

sexual delusion, Zhang stresses artifice over chaotic nature. Tan has faith that evolution will eventually lead to a realm of freedom, while Zhang believes the cosmos is chaotic and thus needs to be brought under control. This has political implications: Zhang argues in favor of legal restrictions and against spontaneity and local autonomy. In this way, rather than the equality of interconnectedness, Zhang stresses equality and hierarchical divisions based on law, where officials and the state attempt to rationalize contingencies and confusion. In his 1908 essay "Should We Implement Representative Government?" Zhang will argue that to ultimately bring confusion and self-interest under control requires self-transcending officials. Zhang's emphasis on creating new officials detached from individual interest will lead him to draw on aspects of Confucianism and Buddhism to develop a type of person.

Zhang's concept of confusion will continue to influence his later writings. Specifically, Zhang's early writings on confusion inflect his interpretations of Buddhism and Daoism and move them in a direction different from his contemporaries, including Tan. When Zhang accepts Buddhism in his *Minbao* period, he will follow Tan in thinking of the negation of this world characterized by the antinomies of chaos and control; however, Zhang's emphasis on confusion will give him a critical perspective on a host of ideologies, such as evolution and science, both of which were intimately connected to China's state-building project.

References Cited

Aristotle (1968), *Physics*, in Philosophic Classics: Thales to Ockham, Walter Kaufmann ed. New Jersey: Prentice Hall Inc., 1968.

Arrighi, Giovanni (1994), The Long Twentieth Century: Money, Power and the Origins of Our Times. London: Verso.

Chan Sin-wai (1985), *Buddhism in Late Ch'ing Political Thought*. Hong Kong: Chinese University Press.

Derrida Jacques (1985) "Ousia and Grammé" in *Margins of Philosophy*, Alan Bass trans, Chicago: University of Chicago Press.

Duara, Prasenjit (2003), *Sovereignty and Authenticity: Manchukuo and the East Asian Modern China*. Lanham, MD: Rowman and Littlefield.

Egginton, William (2004), "Time the Memorious: Borges, Derrida and Heidegger on 'Vulgar' and 'Original' Time." Unpublished manuscript.

Elman, Benjamin (1984), *From Philosophy to Philology: Intellectual and Social Aspects of Change in Late Imperial China*. Cambridge, MA: Harvard University Council on East Asian Studies.

—— (2004), "Naval Warfare and the Refraction of China's Self-Strengthening Reforms into Scientific and Technological Failure, 1865–1895," in *Modern Asian Studies* 38, 2: 283–326.

Hou Wailu [ed.] (1978), *Zhongguo jindai zhexueshi* 中國近代哲學史 (History of Modern Chinese Philosophy). Beijing: Renmin chubanshe.
Hu Shi (1996), "Du Liang Shuming xiansheng de dongxi wenhua jiqi zhexue" 讀梁漱溟先生的東西文化及其哲學 (Reading Liang Shuming's "Eastern and Western Culture and Philosophy"), in *Hu Shi wencun*. Hefei: Huangshan shushe, vol. 2: 167–82.
Jiang Guanghui (1997), *Zouchu lixue* 走出理学 (Leaving behind the Learning of Principle). Shenyang: Liaoning jiaoyu chubanshe.
Li Zehou (1985), *Zhongguo gudai sixiang shilun* 中國古代思想史論 (On the history of ancient Chinese thought). Beijing: Renmin chubanshe.
Lydia Liu (1995) *Translingual Practice: Nation, Culture and Translated Modernity-China 1900–1937*, Stanford: Stanford University Press.
Meng Yue (1999), "Hybrid Science versus Modernity: The Practice of the Jiangnan Arsenal," in *East Asian Science, Technology, and Medicine* 16 (1999): 13–52.
Nakatani, Hajime (2006), "Empire of Fame: Writing and the Voice in Medieval China," in *Positions: East Asian Cultures Critique* 14, 3 (2006): 535–66.
Postone, Moishe (1993), *Time, Labor and Social Domination: A Reinterpretation of Marx's Critical Theory*. Cambridge. UK: Cambridge University Press.
Schafer, Ingo (2001), "Qi and Yitai," in Lackner, Amelung, and Kurtz eds., *New Terms for New Ideas: Western Knowledge and Lexical Change in Late Imperial China*. Leiden: Koninklijke Brill, 257–69.
Shimada Kenji (1958), "Chūgoku kinsei no shukan yuishinron: Banbutsu yitai no jin no shiso," 中国近世の主観唯心論: 万物一体の仁の思想 (Subjective Idealism in Early Modern China: The Philosophy of the Myriad Things as All Being on in Humanity), in *Tōhōgakuhō* (Kyoto) 28 (1958): 1–80.
—— (1990), *Pioneer of the Chinese Revolution: Zhang Binglin and Confucianism*. Joshua Fogel trans. Stanford, CA: Stanford University Press.
Tan Sitong (1984), *An Exposition of Benevolence: the Jen-hsueh of T'an Ssu-t'ung*. Chan Sin-wai trans. Hong Kong: Chinese University Press.
—— (2002), *Renxue*. Beijing: Huaxia chubanshe.
Trouillot, Michel-Rolph (2002), "Otherwise Modern", in Bruce M. Knauft, ed., *Critically Modern*. Bloomington: Indiana University Press, 2002.
Wang Bi (1994), *The Classic of the Changes: A New Translation of the I Ching as Interpreted by Wang Bi*. Richard John Lynn trans. New York: Columbia University Press.
Wong Young-tsu (1989), *Search for Modern Nationalism: Zhang Binglin and Revolutionary China*. Oxford: Oxford University Press.
Zhang Rulun (2001), *Xiandai Zhongguo sixiang yanjiu* 現代中國思想研究 (Research on Modern Chinese Thought). Shanghai: Renmin chubanshe.
Zhang Taiyan (1982), *Zhang Taiyan xuanji: zhushiben* 張太炎選集: 注釋本 (Selected Writings of Zhang Taiyan. Annotated Edition). Zhu Weizheng and Jiang Yihua, eds. Shanghai: Shanghai renmin chubanshe.
Zhu Weizheng (1996), *Qiusuo zhen wenming: wanqing xueshu gailun* 求索真文明: 晚清學術概論 (In Search of True Civilisation: Outline of Late Qing Scholarship). Shanghai: Shanghai guji chubanshe, 1996.
Zhuang Zi (1999), *Zhuangzi buzheng* 莊子補正 (Complement and Rectification for the *Zhuangzi*). Liu Wendian, ed. Hefei: Anhui daxue chubanshe.

CHAPTER THREE

NATION, HISTORY AND ETHICS: THE CHOICES OF POST-IMPERIAL HISTORIOGRAPHY IN CHINA[1]

Axel Schneider

The history of modern Chinese historical thought and writing is characterized by a close interaction between indigenous traditions and the modern Western post-Enlightenment views of history and historiography. In this process of interaction, a progressive, often also teleological view of history was dominant, in combination with modern, professional institutions and approaches to the writing of history aiming at understanding "what actually had happened". Although modern Chinese historiography is divided into many different academic and political camps, and is influenced in varying ways by pre-modern traditions, there were few attempts to resist these dominant trends. Liu Yizheng (柳詒徵 1880–1956) in his later years was one of those historians, who, after initially wholeheartedly adopting modern views of history, began to voice doubts about progress, causality in history and the objectivity of historical research. Situated within a dual horizon of modern Western theories of history and Chinese historiographical traditions, Liu struggled to find a consistent position. His attempts to overcome the contradiction between modern claims to objectivity and his profound ethical concerns, and how he first doubts and then discards notions of progress and causality help us to understand the complicated process of interaction.

A Dual Horizon

Modern Western theories of history are closely connected to the project of nation-building.[2] Nation-states needed a past supporting their claims

[1] I would like to thank Hon Tze-ki, Kuo Ya-pei, Luo Zhitian, Brian Moloughney, Viren Murthy, Wang Fansen, Susanne Weigelin-Schwiedrzik and Daniel Woolf for their comments on this paper. This article has earlier been published in Moloughney and Zarrow 2012, 271–302. It is reprinted here with kind permission of Chinese University Press.

[2] Kramer 1997: 520.

to legitimacy and loyalty. In order to foster national consciousness newly forming nation-states invested heavily in the establishment of a national educational system, including the development of the modern academic discipline of history. In Germany the Historical School of Law around Friedrich Karl von Savigny, the historiography of Leopold von Ranke, and of Heinrich von Treitschke are situated within this new approach to history focusing on the nation-state and with a concomitant concept of world history.[3] In this process of nation-building and the related development of modern historical research the concept of history underwent a fundamental shift.[4]

The pre-modern notion of history as one story among others—*history* hence always understood in the plural as *histories*—told by a writer in order to exemplify moral lessons taken from somewhere else, mostly from moral philosophy or cosmological speculations, was replaced during the Enlightenment by a notion of history as denoting singular facts. Time was quantified and transformed into a quasi-mechanical, abstract time of experience as opposed to the qualitative time of a normative order. In the West coherence and hence the meaning of history was conceptualized as being ascribed to historical facts by the historian from a position outside of history.

Before long this mechanical and rational notion of time and history underwent further change, leading to a concept of history in the singular. History became the sum of all individual (hi)stories encompassing all that ever happened; and what once was divine history (providence) now was desacralized and became universal history. History thus acquired the status of an actor, of a process external to human beings influencing if not determining their fate. Meaning, previously ascribed to history in a rational, constructive act (as typical for the Enlightenment), was now thought to be immanent in history. Although in this speculative version of history[5] reason manifested itself in history, it was not yet historicized and particularized.

Towards the middle of the nineteenth century this speculative concept of history was severely criticized by a burgeoning academic discipline of history. Be it the positivism and naturalism of a Comte or Buckle,[6] or

[3] Kelley 2003: chapter 7, Rüsen 1993, Iggers 1968.
[4] The following summary is based on Koselleck 1975, Rüsen 1993, Schnädelbach 1974.
[5] As e.g. the philosophy of history of Hegel. See Pöggeler 1973, 1995.
[6] Kelley 2003: chapter 9.

be it the idealistic, religious historicism (*Historismus*) of a Ranke,[7] both attacked the philosophy of history for its speculative violation of facts. Whereas positivists modeled history after the natural sciences, historicists stipulated that history is man-made and has to be understood as the expression of ideas. Positivistic history was explicitly universalistic in its approach to and understanding of history, i.e. positivists saw universal laws of nature at work in history, a view that later was combined with notions of linear time and concepts of biological evolution. Historicists, however, saw history as the unique expression of ideas and conceptualized every historical epoch as individual and ultimately incomparable. They condemned attempts to subjugate history under a speculative universal reason or under universal laws of progress, and demanded that the historian should abstract from his own personality. Only then would the historian be able to let the facts speak for themselves.

The decline of notions of history as the progress of reason, or whatever other absolute, contributed to the historicization of norms and values once conceived as both timeless and universal. As a result of this process, Western historiography and philosophizing about history in the modern period came to be characterized by a tension between, on the one hand, rationalistic or naturalistic concepts such as the notion of history as the progress of reason or as characterized by linear time, and, on the other, the breakdown of metaphysical or theological certainty caused by thorough historicization, here understood not as integration into and determination by a historical time line, but understood as thorough historical relativization.[8]

Whatever the vision of history was, the biggest change of all was the central position history as such now acquired. *History* became one of the main modes, if not *the* mode, of interpreting human life. Human existence in all its facets was increasingly seen as historical, albeit with different meanings. It is this feature of the modern period that marks it off from preceding periods, when history was nothing but a private pastime for personal edification far removed from the centers of political power or the temples of philosophical and cultural authority.

It is this relative irrelevance of pre-modern Western historiography that marks it off from the Chinese tradition.[9] Regardless of whether we see

---

7 For Ranke's historiography, see Rüsen 1993: 18–113.
8 Bambach 1995.
9 For a brilliant analysis of Chinese historiography, see Chevrier 1987.

traditional Chinese views of history as cyclical, linear or regressive,[10] the recording of history always occupied a central position in the Chinese cosmic and socio-political order.[11] The belief in this order centered around the ruling clan and the worship of its ancestors, the ritual enactment of which served as the paradigmatic expression of this order. The legitimacy of the ruling clan, rationalized as the Mandate of Heaven (*tianming* 天命), consisted of religious and political elements that were closely intertwined (*zheng jiao heyi* 政教合一).

The predecessors of the later historiographer (*shi* 史), whose early function can best be described as scribe or archivist, recorded the words and deeds of the rulers and thus served as witnesses to the ritual enactment and re-presentation of the cosmic order. This order was not conceptualized as a pre-ordained given, but rather had to be actualized through the performance of sacred rituals and via moral example. The Mandate of Heaven thus provided legitimacy for the ruling clan, but also the rationale for a change of dynasty. It stands for the conviction that men can and do deviate from the moral principles of the cosmic order (*dao* 道), and that the gap between what is and what ought to be is part of the human condition. To bridge this gap through establishing and maintaining a polity as close as possible to the normative order is the mission of the ruling clan and the political elite, and it is the task of the historiographer to keep a record of these attempts for posterity.[12] It is whence he derives his eminently powerful ethico-political position. This position was further reinforced by the fact that in Chinese culture the normative order was not perceived as accessible to humans through divine revelation, but was posited as once having been realized during the Three Dynasties in actual history. History thus acquired a quasi absolute status as the field of human activity providing privileged if not exclusive access to heavenly truth.[13] This "heavenly" sanctioned centrality of the recording of history[14] in Chinese culture translated into the institutionalization of historiography

[10] Huang 2006. Evidence for all three positions can be found depending on what we look at. For a strong argument favoring an understanding of traditional Chinese historiography as cyclical see Kwong 2001: 160–165. For discussions of Chinese concepts of time, see Chen 2006, Fraser and Haber 1986, Huang and Zürcher 1995, Liu Shu-hsien 1974, 2006, Sivin 1966, Wang Edward Q. 1995.

[11] Schwartz 1996. See also Ng and Wang 2005: esp. chapter 4.

[12] Schwartz 1985: esp. chapter 1 and 2.

[13] Sato 2007. For recent discussions of the nature of Chinese historical thinking, see Huang 2007, Rüsen 2007, Wang Edward Q. 2007.

[14] Given this constitution of the 史 it becomes clear that strictly speaking we should not speak of the writing of history or historiography in early China, which would imply a

within the imperial bureaucracy, a process that culminated during the Tang Dynasty in the establishment of the *History Office* (*Shiguan* 史官).[15]

Ultimately the *shi* 史 thus fulfilled two complementary functions, closely linked but not without internal tension: he recorded history in a historiographical mode leaving behind a record of the past as truthful as possible, yet he also manifested the cosmic order (*dao* 道) by expressing praise and blame (*baobian* 褒貶) and thus fulfilling a pivotal role in the aforementioned mission of bridging the gap between the is and the ought. In having the power to laud or to condemn, the historiographer found himself in an exposed position sometimes risking his life in fulfilling this—what Yves Chevrier calls—historiological duty.[16]

When this tradition of thinking about and writing of history was challenged in the 19th century, the Western views of history initially having the profoundest impact were not those linked to the aforesaid crisis of historicism, but rather the evolutionary, Darwinian view of history and its linear, mechanical concept of time.[17] This notion of time, which severely limited if not removed the human factor from history, was either combined with demands to modernize the discipline of history along the lines of the natural sciences,[18] or went hand in hand with the methodology of history as pronounced by Ranke in the understanding of Bernheim or Langlois and Seignobos.[19] This late historicism had discarded all notions of ideas at work in particular historical epochs, and had limited itself to

somehow subjectivist, auctorial view of the 史, but rather of the recording of history in its process of unfolding.

15 For the *History Office* see Twitchett 1992.

16 Chevrier 1987. Given this centrality of history in China it comes as no surprise that quite in contrast to their Western counterparts many modern Chinese intellectuals had a very strong interest in history or even did research on history, such as e.g. Kang Youwei, Liang Qichao, Zhang Taiyan, Liu Shipei, Hu Shi, Fu Sinian, Gu Jiegang, Chen Duxiu, Li Dazhao, Lu Xun, Mao Zedong just to mention a few. History in China was and still is a type of discourse the larger importance of which reaches far beyond the confines it is subject to in a Western context. As a consequence many important debates in China during the 1980s and 1990s were conducted by discussing history or by reevaluating previous historical judgments. See Unger 1993.

17 On the reception of linear time in China see Kwong 2001, Wang Fansen 2008, Yu Nan 2006, Zhang Ying 2003, Zhang Yue 2003.

18 Especially influential for this view was the translation of Henry T. Buckle (1857), *History of Civilization in England*. London: J. W. Parker and son. On the influence of this book in China see Li 2007: chapter 2 and 3. On the influence of Buckle on Fu Sinian, see Schneider Axel 1997 and Wang Fansen 2000.

19 Bernheim 1908, Langlois and Seignobos 1898. On the reception of German historiography in China, see Li 2007: chapter 7.

the positivistic, evidential side of Ranke's methodology as expressed in his dictum that the task of history is "to show what actually happened".[20]

Facing the desperate historical situation of imperialistic aggression during the late 19th century and under the influence of these new theories, Chinese intellectuals and historians faced formidable challenges. They had

a) to develop a new understanding of Chinese history as part of a coherent vision of world history, thus making it possible
b) to envision China as an at least equal member in the international community. At the same time they had
c) to safeguard, if necessary to create some sort of continuity with the past in order to provide a basis for a Chinese identity. They also had to
d) conceptualize history in a way that would make it possible for China to catch up if not surpass the West, and, last but not least, this
e) had to be achieved in the context of a tradition placing the historian in an elevated political position, thus putting him under considerable pressure to come up with a viable solution.

Nation, History and Ethics: The Complexities of Post-Imperial Chinese Historiography

In their research on modern Chinese views of history and historiography scholars have hitherto focused on a range of topics directly related to the aforementioned challenges: China's national identity and its position in world history, issues of historical continuity, developments in historical methodology, the institutionalization of modern history, and the political functions of historiography. However, the question of the ethical function of the writing of history—a central if not *the* core aspect of traditional Chinese historiography—has conspicuously been ignored. This lacuna is all the more astonishing if we consider that ethics were central for historiography not only from a traditional Chinese perspective. Even from a purely Western perspective the crisis of late historicism positioned the question of ethics right at the centre of twentieth century Western doubts about modernity.

---

[20] On this influential image of Ranke and the extent to which it misrepresents his historiography see Iggers 1962. On the reception of Ranke in Japan, see Tanaka 1993 and Mehl 1998. On Ranke in China, see Schneider Axel 2001b and Yi 2006.

What has happened to the centrality of ethics? How did modern Chinese historians reposition ethics as such a) vis-à-vis a concept of linear, mechanical progress that did not leave much space for human agency or b) vis-à-vis a relativist notion of history devoid of overarching, eternal principles? What sort of ethics did they, if at all, argue for? What happened to traditional Confucian ethics? What role did it play in the definition of Chinese national identity? And what were the implications of that for the sort of polity these historians imagined?

A survey of the names of those historians well-researched by Western scholars reveals an additional blind spot. Most historians who in one way or the other argued for substantial continuity with orthodox tradition and who hence were opposed to far-reaching modernization, have hardly been researched: Wang Guowei (王國維 1877–1927),[21] Chen Yinque (陳寅恪 1890–1969),[22] Qian Mu (錢穆 1895–1990),[23] Liu Yizheng,[24] just to mention a few. Why are these "conservative" historians—many of whom have made important, widely acknowledged contributions to modern Chinese scholarship—not as well-researched as their more "liberal" or "leftist" colleagues?

A very concise, purely exemplary overview of a few selected approaches to the relationship between history and ethics in modern China reveals how difficult it seems to have been to combine an orthodox Confucian notion of a universal ethical core of history with a modern linear concept of time.

Kang Youwei (康有為 1858–1927) still inherited some aspects of this traditional view of history by substantially reinterpreting it under Buddhist and Western influence. He postulated a linear and teleological view of history based on the New Text theory of the *Three Ages* (*sanshi* 三世) ultimately aiming at an ethically defined *telos* (the *Datong* 大同), i.e. with history still structured around a given ethical core. However, his position already indicated a major departure from the orthodox view insofar as he postulated a quasi necessary, linear development towards a goal that,

---

[21] Bonner 1986, Kogelschatz 1986.

[22] Schneider Axel 1997.

[23] Dennerline 1989. This monograph does not delve deeply into Qian's view and methodology of history.

[24] So far only two articles on Liu Yizheng have been published: Moloughney 2001, Hon 2004. The dissertation by Kuo (2002) is still unpublished. For an overview of the historians linked to the school of historical geography under Liu Yizheng see Peng 1995. The number of Chinese publications on these historians is substantial, however, even here it is telling that the bulk of publications on them has appeared only after 1992/93.

once mankind has reached it, also represents the end of history. Besides, the concrete goal he advocated was incompatible with the ethics of orthodox Confucians.[25]

With Liang Qichao's (梁啟超 1873–1929) early evolutionary and nationalist history, the full impact of a linear and in his case also non-teleological view of history becomes obvious. Devoid of ethical lessons that go much beyond the citizen's loyalty to the nation, time becomes the abstract and empty flow of linear progress. The historical process is ruled by cause and effect and thus is much less subject to the influence of human agency. Ethics is neither part of a larger cosmic process nor is it embedded in history through human agency. There is no resemblance left with orthodox Confucian ethics. However, while this concept allows him to integrate China into world history and envision a path to wealth and power, the price he pays for this concept of history is considerable: for one reason, because in such a universal historical progress particular national identity is defined in relation to the allegedly universal Western model, for another reason, because lacking behind in a linear and seemingly necessary historical process carries the risk of never being able to catch up.[26]

With Marxism entering China in the early 1920s a view of history developed that again reintegrated ethics into history by defining a *telos* that embodied ethical aspects and gave some room to human agency. However, the process towards this *telos* ultimately is pre-determined and scientifically verified leaving only limited space for human agency. And as with Kang Youwei, the ethical vision inspiring Marxism was not in line with Confucian ethics.

These examples suggest that a modern linear view of history in China is incompatible if not with ethics as such, then at least with traditional Confucian ethics. History is envisioned as either devoid of moral lessons and meaning, or leading toward a utopian goal defined in ethical terms at odds with Confucian orthodoxy. Human agency is either negated or substantially reduced, and Chinese identity is defined primarily as the particular manifestation of an allegedly universal model of development that *de facto* is—at least with Liang Qichao and the Marxists—to a considerable extent of Western origins.

---

25 Thompson 1958: 37–57, Hsiao 1975.

26 Schneider Axel 2008.

A first step towards answering the aforementioned questions on history and ethics is to investigate the views and theories of history of those hitherto neglected historians who were skeptical of modernization. How did they relate history and ethics? Which concrete ethical values did they advocate? What theory of historiography did they put forward and how did they deal with the challenge posed by notions of progress, causality and linear time? How did they try to establish continuity with the past and thus provide a basis for a Chinese national and/or cultural identity? What view of politics was linked to that?

The Case of Liu Yizheng[27]

Liu Yizheng, who belongs to this group of hitherto neglected historians, was affiliated with the *Critical Review Group* (*Xueheng pai* 學衡派). He wrote one of the first histories of Chinese culture in the form of a modern narrative and comprehensive history.[28] Liu was born in 1880 in Dantu county, Jiangsu province, to a family and cultural context where the examination system still was the main ladder to success. Liu thus received a thorough classical education, and, in 1900, at age 20, started working for the philologist Miao Quansun (繆荃孫 1844–1919) in the compilation and translation bureau of Jiangsu and Hubei provinces. After a short mission to Japan in 1902, his only visit abroad, Liu returned to China deeply influenced by the new, Western inspired way of writing history in Japan. He participated in the development of the new educational system and published in 1902 a history textbook under the title *A Brief Historical Account of Different Periods* (*Lidai shilüe* 歷代史略),[29] which was an adaptation of Naka Michiyo's (那珂通世 1851–1908) *A General History of China* (*Shina tsūshi* 支那通史) of 1899.[30] In the following years Liu Yizheng's participation in educational reform provided him with an opportunity to witness the debates on the introduction of local representative government during the late Qing period. Inspired by these debates he later researched the history of local self-rule in China, using this opportunity to express his views on a variety of political issues pertaining to China's modernization.

---

27 For an overview of Liu's life see Sun 1993, Wang Xintian 1999, Hon 2004.

28 See Liu Yizheng 1926d. For a summary of the views of this group, see Schneider Axel 2001a. For longer studies, see Rosen 1969, Hou 1980, Shen 1985, Kuo 2002.

29 Liu Yizheng 1902.

30 Naka 1899.

In 1922 he participated in the establishment of the *Critical Review Journal*, in which he published numerous articles. Besides, he founded the *Journal for Historical Geography* (*Shidi xuebao* 史地學報)[31] and was a leading figure in the *Journal for National Custom(s)* (*Guofeng banyuekan* 國風半月刊).[32] His *History of Chinese Culture* (*Zhongguo wenhuashi* 中國文化史) was a major success and became a textbook used at universities. After having appeared in the *Critical Review*, it was published in 1932, and had to be reprinted in 1935 and again during the 1940s. After 1949 it has been reprinted several times in Taiwan and recently also in mainland China.[33] Beginning in the 1920s Liu taught mainly at National Southeastern University (*Guoli dongnan daxue* 國立東南大學) in Nanjing. From 1927 till 1937 he was director of the Jiangsu Library for National Studies (*Guoxue tushuguan* 國學圖書館), a position he reassumed in 1945. In 1948 he was elected academician of the prestigious Academia Sinica. And from 1949 until his death in 1956, he was a member of the Shanghai Committee Overseeing Historical Artifacts (*Shanghai wenwu guanli weiyuanhui* 上海市文物管理委員會).

To date the evaluation of Liu Yizheng's historiography has been ambivalent.[34] He is often classified as a cultural conservative,[35] yet he is also described as a nationalist and even as a modernizer, at least with regard to his ideas on history and historical methodology.[36]

In the introduction to his influential *History of Chinese Culture* he states that his aim is to expound the differences between China's culture and that of Europe and India.[37] He sees the Chinese nation as defined by its particular culture and history, thus deviating from the at that time still domi-

[31] *Shidi xuebao*, Nanjing 1921–1926.

[32] *Guofeng banyuekan*, ed. Liu Yizheng, Nanjing 1932–1936.

[33] Liu Yizheng 1990, 2001. I am using the edition in three volumes published in Taiwan in 1990.

[34] In the following pages I will limit myself due to restrictions of space to an analysis of Liu Yizheng's theory of history, ethics, and the implications for his vision of Chinese identity and his political ideals. For Liu's concrete interpretation of Chinese history see in more detail Hon 2004, Kuo 2002, and Peng 1995. Collected writings used for his article are Qiao 1970, Liu and Liu 1991a, 1991b.

[35] See e.g. Schneider Lawrence A. 1976, Furth 1983, Liu Lihong 2006, Bian and Sun 1999, Wang Xinkai 2004. For a critique of this classification see Wang Jiafan 2006 and Kuo 2002, who puts him in the context of the May Fourth Movement.

[36] See Peng 1995, Tian 2004. For an analysis of Liu's historiography criticizing the classification as a nationalist without reverting to a interpretation of him as conservative see Hon 2004.

[37] Liu Yizheng 1990: I:13.

nant racial definition.[38] Chinese culture is characterized by a cohesion that, given its internal, especially racial diversity, is amazing. It is distinguished by its *assimilative* (*tonghua* 同化)[39] powers and *longevity* (*niansi zhi jiuyuan* 年禩之久遠).[40] Liu wants to understand why this is so.[41]

In addressing these issues he seems to fit our image of a modern, nationalist historian. He speaks about the particularity of China[42] as expressed by its culture and embodied in what he calls the *national spirit* or *national psyche*,[43] thus providing a basis for a Chinese national identity. He presents an interpretation of the Golden Age of the Three Dynasties that is thoroughly demythologizing and historicizing this crucial period of Chinese history, which stood at the center of historiographical debates during the 1920s.[44] Rather than ascribing the cultural advances of that period to mythological Sage Kings as had been characteristic for most traditional Chinese views of history, he interprets them as the product of collective and accumulative efforts of the *people*. These advances have then via a process of sedimentation become part of the Chinese *national spirit*.[45] Last but not least, he addresses issues akin to modern Western notions of popular sovereignty.[46]

---

38 Hon 2004: 508–509. For Liu's criticism of a racist, statist or economist view of history and his definition of China as a *civilized* (*wenming* 文明) nation emphasizing *justice* (*zhengyi* 正義) and *humanity* (*rendao* 人道), see Liu Yizheng 1990: I:48.

39 Liu Yizheng 1990: I:13.

40 Ibid. I:3–9. Liu substantiates this feature of assimilative powers across his monograph by pointing out that Chinese history during the "Medieval Ages" (*zhonggu* 中古) is a history of the close exchange and mutually enriching interaction between the China and foreign nations (see Hon 2004: 518–520). In this respect, his position is close to that of Chen Yinque.

41 It is important to keep in mind that while the book was published in serialized form between 1926 and 1931 and then as monograph in 1928 and in 1932, it is based on and goes back to lectures Liu Yizheng gave at National Southeastern University between 1919 and 1921, which were then revised for publication.

42 Liu Yizheng 1990: I:2, here he uses the expression wu Zhongguo juyou teshu zhi xingzhi 吾中國具有特殊之性質. Other terms used are character of the nation/the people (guominxing 國民性, Zhongguo minxing 中國民性) and character of the nation/the state (guoxing 國性), see Liu Yizheng 1990: I:48–50.

43 Liu Yizheng 1923: 178. He uses the terms *minzu jingshen* 民族精神 and *minzu xinli* 民族心理.

44 See Schneider Lawrence A. 1971, Wang Fansen 1987, Peng 1991 for an analysis of the Gushibian Movement (guoshibian yundong 古史辨運動) and the varying interpretations of the history of the Three Dynasties. For descriptions of the debate on the interpretation of the Great Yu between Liu Yizheng and Gu Jiegang, whom Liu criticized for his iconoclastic exaggerations, see Hon 1996, Hon 2004: 513–514.

45 Liu Yizheng 1990: I:26–27, he uses the terms *min* 民, *renmin* 人民, and *minzu* 民族.

46 Liu Yizheng 1923: 179, 194. He uses terms such as *minquan* 民權, *minzhi* 民治 in the context of discussing local rule.

However, Liu Yizheng is not putting forward a view of Chinese history as the unified, glorious progress of the Han nation. On the contrary, he is well aware of the internal contradictions, ruptures, and the constant intermingling with and assimilation of external influences.[47] Moreover, in spite of historicizing the Golden Age and its Sage Kings and in spite of providing a national history centered around China's culture and its people, he defines that culture with the help of traditional ethical ideals such as the Golden Mean, loyalty and piety. He thus seems to turn them into national characteristics.[48] On top of that, he juxtaposes this ethical vision of Chinese culture with a picture of a morally degenerate, individualistic West that can only be saved with the help of Chinese ethical values.[49]

In describing the tasks of the historian he operates with a terminology that appears to be modern too. The historian has to uncover *general laws* of human development while at the same time understanding national particularity.[50] He has to discern *causality*[51] in order to expound *evolution*[52] and *progress*.[53] History is conceptualized as being a universal, law-like process, in which cultures differ only insofar as they are positioned at distinct points within the time line of universal progress.[54]

The impression emanating is that of a historian adopting core notions of the modern academic discipline of history. He rewrites the Chinese past into a modern, national (hi)story of progress modeled in structure[55] with core assumptions along Western lines, albeit combined with a substantial definition of Chinese culture that in some aspects appears to be rather traditional and anti-Western.[56] However, when going beyond these

---

[47] See e.g. his characterization of the Three Dynasties in Liu Yizheng 1990: I:110.

[48] Ibid. I:110–116, he uses terms such as *zhong* 中, *zhongdao* 中道, *zhong* 忠, and *xiao* 孝. Liu also uses the notion of the *Golden Mean* (*zhongyong* 中庸) for criticizing modern notions of historical progress, see Hon 2004: 524.

[49] Liu Yizheng 1924a, Liu Yizheng 1924c: 228. This position is in line with common conservative stock of the early 1920s, see Alitto 1979, 1986, Furth 1976. There is by now a plethora of Chinese publications on conservatism, see Zheng Dahua 1994, Yu Dahua 2001, Liao 2001, He 2006, Liu Lihong 2006.

[50] Liu Yizheng 1990: I:Xulun 1, he uses the expressions rendao yanjin zhi tongze 人類演進之通則 and ming wu min duzao zhi zhenji 明吾民獨造之真際.

[51] Ibid. I:1 and 97 (*yinguo* 因果).

[52] Ibid. I:27, 36, 41 etc. (*jinhua* 進化).

[53] Ibid. I:423 (*jinbu* 進步), and I:421 (*yanjin* 演進).

[54] Ibid. I:41, on page 99 he uses the term *jinhua zhi lü* 進化之律.

[55] Liu Yizheng employs the typically modern, Western tripartite division of history, which he had adopted from his early Japanese models (*shanggu* 上古, *zhonggu* 中古, *jinshi* 近世). On this topic see Moloughney 2001, Hon 2004: 516.

[56] Together with his classical style of writing this is probably the main reason why he often is classified as "conservative".

historiographical texts and including more theoretical treatises and his expositions on ethics,[57] he hints at a different vision. This vision is developed over the years into a new, coherent theory of historiography, and a view of history and politics that, while much more indebted to traditional Chinese notions of the connection between history and ethics, does not simply represent a return to the past.

Already as early as 1922/23 we can find indications of a view of history that contradicts the aforementioned impression of a modern nationalist promulgating a linear, causal view of history. The view Liu expounds is that of a bifurcated history. On the one hand, China's history is characterized by progress. However, this progress seems to be limited to the material and organizational side of life[58] and is far from homogeneous. In some fields progress can be observed, in others stagnation or even regression is characteristic.[59] On the other side, Liu's view of history is one of decline.[60] Ethical ideals define China's culture, give meaning to history ever since they have been formulated, and provide practical guidance.[61] However, Chinese history since the Qin Dynasty is a history of the gradual decline of these core ideals culminating in the centralized, bureaucratic state of the late imperial period.[62] But his questioning of key assumptions of a modern, nationalist view of history goes even further. He directly criticizes the notion of progress as such and redefines history as not primarily being characterized by causality.

In a review of Lu Maode's *History of the Philosophy of the Zhou and Qin Dynasties*,[63] Liu Yizheng directly attacks the author and, among others, also Hu Shi (胡適 1891–1962) for their naïve belief in teleological progress as the constant improvement of life till a final, almost utopian stage of utmost perfection has been reached.[64] Referring to the cosmology of

---

57 See e.g. Liu Yizheng 1922b, 1924b, 1925, 1926a, 1926c, 1932, 1934, 1935, 1947, 1948a, 1948b.

58 I.e., material improvements and changes as well as socio-political structures getting more and more complicated. Liu Yizheng 1990: I:57, 59. For a later repetition of this view see Liu Yizheng 1948b: 35.

59 Liu Yizheng 1990: I:2.

60 Liu Yizheng 1923: esp. 198.

61 Liu Yizheng 1990: I:50. Liu speaks of the sedimentation of these ideals in the *national psyche minzu xinli* 民族心理, *national spirit* (*minzu jingshen* 民族精神, Liu Yizheng 1923: 178), or the *character of the people* (*minxing* 民性, Liu Yizheng 1990: I:48).

62 Liu Yizheng 1923: 198. He identifies these as the *rule by virtue* (*dezhi* 德治) realized in the tradition of *local self-government* (*xiangzhi* 鄉治) as opposed to the *rule by law* (*fazhi* 法志) typical for Western countries.

63 Lu 1923, reviewed by Liu Yizheng 1924b.

64 Liu Yizheng 1924b: 233, 236, 239–242; see also Yu Nan 2006: 69–71.

the *Book of Changes* (*Yijing* 易經), Liu Yizheng calls this attitude wishful thinking, oblivious of the fact that with all the progress in modern times, evil and suffering have progressed correspondingly:

> ...to claim that today we see civilization, and that the past was barbaric, that civilized people are happy, and that barbarians suffer, all this is utter nonsense. During barbarian times we saw the most civilized spirit, and during civilized times the most barbaric actions were committed. Human suffering and happiness come and go without [us being able] to discuss this in one comprehensive [manner].[65]

According to Liu, the *Book of Changes* captures this principle best with its notion of the constant change and interaction of *Yin* and *Yang*, a cosmology that Confucianism transformed into the notion of the *Golden Mean* (*zhongyong* 中庸).[66]

In a text published in 1925 this critique of progressivism is further underpinned by a careful redefinition of the nature of history. He begins his *History of Chinese Culture* claiming that

> The science of history most cherishes causality. Human affairs can not have a cause and yet no effect.... on the one side, we are striving [to uncover] the general laws of human progressive development, on the other side we want to understand the reality of our nation's unique creations.[67]

Yet around the same time he proclaimed in a text discussing the nature of historical knowledge that some aspects of history underlie causality, whereas others are characterized by contingency. This culminates in him concluding that humans should only try to influence those aspects of life that do not underlie the laws of causality in order to adapt to the environment and to history: "Therefore history ultimately is without causality; one has to rely on oneself to create causality."[68]

While it remains unclear what exactly he has in mind, it is obvious that he opposes any notion of history as a linear, progressive and teleological development. These doubts about causality and his ultimately ethical definition of the core of Chinese culture develop in the following years

---

[65] Liu Yizheng 1924b: 240. In another text of 1924 Liu links this reasoning directly to his evaluation of the West, its individualistic culture and the theory of evolution, which fosters competition and fight for naked survival, ultimately leading to Word War I. See Liu Yizheng 1924a.

[66] Liu Yizheng 1924b: 240–241.

[67] Liu Yizheng 1990: I:1.

[68] Liu Yizheng 1925: 84.

into a view of history and ethics that can not simply be described as either "conservative" or "nationalistic".

In the following years Liu develops a more detailed elaboration of the relationship between history and ethics, which integrates earlier positions into a larger, more coherent whole.[69] Liu's concept of an all-encompassing heavenly order[70] as the foundation for historical and cultural particularity[71] is the key to his theory of ethics and history.

This heavenly order is the basis from which ethics (*lunli* 倫理), human relations (*renlun* 人倫) and human nature (*ren zhi xing* 人之性)[72] arise. Human relations,[73] however, is still a rather abstract notion. These relations find their particular expression in the Five Relations (*wulun* 五倫), the rituals (*li* 禮) and, even more concretley, in the ceremonies (*yijie* 儀節) and folk customs (*minsu* 民俗) that are historically constituted and hence the particular expression of the heavenly order.[74] From here Liu proceeds in two closely interlinked directions. Firstly, he develops a theory and methodology of history that grants the researcher access to historical phenomena and human thought vastly different from his own world, while at the same time empowering him to reach via history beyond the particular and back to the universal *heavenly order*. Secondly, he unfolds a vision of ethics that is firmly rooted in the abstract, universal notion of *human relations*, while at the same time providing a basis for a particular Chinese identity.

---

69 Relevant texts are Liu Yizheng 1932, 1934, 1947, 1948a, 1948b. For reasons of space I will in the following present a systematic picture of Liu's final position abstracting from the development over time of this position starting in the mid/late 1920s and culminating in the publication of his *Main principles of national history* (Liu Yizheng 1948b).

70 The terms that belong to the wider field of terms denoting universal features are *heavenly nature* (*tianxing* 天性, Liu Yizheng 1925: 81, Liu Yizheng 1948b: 10), *heavenly order* (*tianxu* 天敘 or *tianran zhi zhixu* 天然之秩序 also equated with *tianli* 天理, Liu Yizheng 1948b: 10–11, Liu Yizheng 1947: 624), and, applied to the issue of human nature, *fundamental character* (*benxing* 本性, Liu Yizheng 1925: 81).

71 The terms that belong to the wider field of terms denoting *particularity* (*texing* 特性, Liu Yizheng 1925: 81) are *particular character* (*teshu zhi xingzhi* 特殊之性質, Liu Yizheng 1990: I:2), *distinct character* (*biexing* 別性,), *habitual character* (*xixing* 習性, Liu Yizheng 1925: 81), *folk custom* (*minsu* 民俗, Liu Yizheng 1948b: 10), *character of the people/national character* (*minxing* 民性, Liu Yizheng 1990: I:50), and *character of the nation* (*guominxing* 國民性, Liu Yizheng 1990: I:48).

72 Liu Yizheng 1990: Bianyan 2.

73 Interchangeably he uses the terms *Five Relationships* (*wulun* 五倫) and at times the already more concrete *Three Bonds and Five Constants* (*sangang wuchang* 三綱五常 or *gangchang* 綱常, see Liu Yizheng 1948b: 11).

74 Liu Yizheng 1934: 8, Liu Yizheng 1948b: 10–11, see also Zheng Shiqu 1998a: esp. p. 62.

The habitual character is, in contrast to the heavenly nature of mankind, the product of history,[75] i.e. it is particular in time and space. However, as it is the expression of a heavenly order, in which all humans partake, it must be accessible to the historian. Liu thus posits a methodology of research that is based on human empathy (*renlei de tongqingxin* 人類的同情心) and human rationality of self-understanding (*renlei you liaojie ziji de lixing* 人類有瞭解自己的理性). However, actual understanding still is understanding of particular habitual character. Pointing beyond theory, Liu adds that respect for each country's distinct national character (*minxing* 民性), which can only be discerned via historical research, is the basis for mutual understanding and fruitful exchange between cultures.[76] But Liu Yizheng reaches beyond the particular. The link back from history to the heavenly order, implicit in his ideas about their relationship, is made explicit by stating that the heavenly order is not directly discernable, but can be perceived only via expressions in the people and their folk customs.[77]

The close interconnection between a timeless, ethical heavenly order and particular history reveals how we have to understand Liu Yizheng's aforementioned partial rejection of causality. Liu Yizheng emphasizes that history is characterized by causality (*yinguo* 因果) and contingency (*ouran* 偶然). However, the ultimately-defining characteristic is contingency, which he identifies as the point of entry for human influence on the course of history, i.e. human actions (causes) and their effects in time.[78] Causality is thus not an impersonal, quasi-mechanical principle postulating that "whenever A then B". Liu thus conceptualizes history as the interplay between aspects of human life that are subject to causality and human actions, which in themselves should reflect the heavenly order. In other words, China's national culture is the particular manifestation of the heavenly order under the conditions of partial causality, and human actions and their effects.

Given this conception of history, it comes as no surprise that Liu Yizheng puts the historian firmly back into a central socio-political position. It is he who records the words and deeds of human beings and thus provides testimony of human efforts to actualize the heavenly order. The historian

---

75 Liu Yizheng 1923, Liu Yizheng 1948b: chapter 10. See also Wang Jiafan 2006: 175.
76 Liu Yizheng 1925: 80–81.
77 Liu Yizheng 1948b: 10.
78 Liu Yizheng 1925: 83.

focuses on the concrete rites and customs of the people,[79] yet through them he gets a glimpse of that order. No wonder then that, according to Liu, the writing of history is always highly political and geared towards application in life.[80] Liu embeds this vision of history and the historian's ethico-political responsibility in the orthodox Song Neo-Confucian interpretation of the *Great Learning* chapter (*daxue* 大學) of the *Liji*, thus ultimately anchoring it in a concept of personal self-cultivation that goes far beyond the confines of the pure recording of past events.[81] Accordingly, he opposes the pedantry of a purely fact-oriented, ivory-tower academe,[82] admonishes the historian to engage in moral self-education (*xiushen* 修身)[83] and to contribute to the moral transformation (*hua* 化)[84] of society.[85]

Turning to ethics and to Liu Yizheng's more historically concrete references to it reveals how he intends to establish national identity and what vision of politics and modernization he advocates. He claims that the Five Relationships and the Three Bonds and Five Constants are, quite in contrast to Western individualism, characterized by cooperation (*huzhu* 互助).[86] They give expression to the core ethical values of moderation (*zhong* 中),[87] yielding (*rang* 讓), loyalty (*zhong* 忠) and piety (*xiao* 孝).[88] It is through rituals (*li* 禮) as the concrete socio-political embodiment of these values that order is established.[89] These values and the rituals embodying them constitute the spirit of Chinese culture and it is their continuity and strength that helps to answer the questions concerning the cohesion, assimilative powers and longevity of Chinese culture that Liu

79 Liu Yizheng 1948b: 9, 218, see also Wang Jiafan 2006: 174.

80 Liu Yizheng 1925: 83, Liu Yizheng 1926b, Liu Yizheng 1929, Liu Yizheng 1948a: 558–559.

81 Liu Yizheng 1990: Bianyan, 1–3. The *Bianyan* was added to the 1947 edition, but from other texts of the 1920s we can see that he advocated similar positions already during the 1920s, see e.g. Liu Yizheng 1924a.

82 Liu thus argues against pure evidential scholarship (*kaozheng* 考據), see Liu Yizheng 1922a: 22.

83 Liu Yizheng 1948b: chapter 5, see also Wang Jiafan 2006: 172.

84 On Liu Yizheng's transformative notion of "culture" (*wenhua* 文化) see Hon 2004: 518–520.

85 Liu Yizheng 1923, Liu Yizheng 1990: Bianyan 2–3, Liu Yizheng 1948b: chapter 7.

86 Liu Yizheng 1924a.

87 He defines "China" *Zhongguo* 中國 as *wenming zhi guo* 文明之國, i.e. as country that is civilized because of 中, see Liu Yizheng 1990: I:50.

88 Ibid. I:109–116.

89 Liu Yizheng 1948b: 18, see Hon 2004: 524, 528–529. On the importance of rituals in the Confucian concept of society see Ebrey 1991. This once again reveals why for Liu Yizheng rituals are at the heart of Chinese historiography. See Liu Yizheng 1948b: 5.

Yizheng had asked.[90] Ethics are thus timeless and rooted in a heavenly order, and yet they are also embedded in the national psyche through a long process of sedimentation. They are manifest in Chinese history in a paradigmatic way, constitutive for a particular Chinese identity, and ultimately even justify Liu's claims to superiority over the West.[91]

Liu Yizheng's peculiar position on ethics and national identity can be further clarified by comparison with Chen Yinque, a contemporary historian, who like Liu defined Confucian ethics as the Chinese national spirit. Chen's reference to the Three Bonds and Five Constants is thoroughly particularizing them as the heart of Chinese national culture. He makes no attempt to link them back to universal ethics. While he refers to the universality of abstract ideals (*chouxiang lixiang zhi tongxing* 抽象理想之通性), these remain vague and abstract. In my interpretation this reference means nothing else but that each culture is committed to abstract ideals, which in substance differ from culture to culture. In contrast to Chen, Liu sees the substance of ethics (*lunli* 倫理) paradigmatically expressed in China in the Five Relations (*wulun* 五倫), which ultimately enables him to take them as a standard to judge other cultures.[92]

These conceptions are most clearly expressed in Liu's texts on ethics,[93] and it is especially in his research on the history of local self-government in China that the political implications for his own times are disclosed.[94] He anchors his analysis in the notion of the rule by virtue (*dezhi* 德治), a concept he juxtaposes to the rule of law (*fazhi* 法治), which is characteristic for the individualistic, utilitarian and egoistic West.[95] He clearly acknowledges that China had too little rule by law and now has to adopt some of its aspects, but the basis should remain the rule by virtue. He situates the rule by virtue in the long Chinese tradition of local self-government (*xiangzhi* 鄉治) starting during high antiquity. Admittedly, this tradition was later pushed into the background and sometimes even lost, but it remained the ethical and political standard throughout Chinese history and serves as the basis for Liu to relate to modern challenges such as democracy and individualism. Liu refutes the claim that China's ethical tradition and its emphasis on hierarchical relations is incompatible with

---

[90] Liu Yizheng 1932, Zheng Shiqu 1998b: 31–38.

[91] Liu Yizheng 1923: 228–229.

[92] For a detailed analysis of Chen Yinque's historiography, see Schneider Axel 1996, 1997.

[93] Liu Yizheng 1924a.

[94] Liu Yizheng 1923, Liu Yizheng 1990: III:229–245, Liu Yizheng 1947.

[95] Liu Yizheng 1924c: 228–229, see also Wang Jiafan 2006: 174.

the spirit of democracy. He claims that integration into hierarchies takes place in any political system and is part and parcel of human nature.[96] Besides, the principle of moderation will make sure that mutual yielding and the spirit of compromise will predominate.[97]

Addressing criticism of Confucian family ethics by protagonists of the May Fourth Movement, Liu claims that ideals such as piety have been misinterpreted throughout Chinese history as pertaining only to the family. Rather than limiting piety to the family, Liu wants it to be understood as extending to all of society and serving as the basis for solidarity and the willingness to make sacrifices for the larger collective.[98] Contrary to the protagonists of the May Fourth Movement Liu clearly does not see China's tradition as an obstacle to new political forms.[99] He emphasizes that the basis of the polity must remain in the rule by virtue, amended by some elements of the rule by law.[100] In fact, he admits that China needs to learn from the West how to strengthen the legal aspects of rule to avoid overly personalistic power relations. All of that, however, has to happen in the localities and should not come from the central modernizing nation-state.[101] It is a bottom-up process rooted in a local moral community that very much reminds us of the traditional Confucian ideal of the local covenant and its rule by virtue, an ideal that during the 20th century also inspired e.g. Liang Shuming (梁漱溟 1893–1988). Besides, next to adopting some elements of the rule of law, China's tradition offers sufficient resources akin if not superior to Western models. One example Liu mentions are the control powers of the traditional historiographer, who by recording the words and deeds of the ruler, can counter-balance the emperors powers, comparable to the modern Western notion of a division of powers.[102]

---

[96] Liu Yizheng 1924a.

[97] On Liu's notion of China as a culture of moderation as contrasted to the West as a culture of extremes, see Hon 2004: 523–525.

[98] Liu Yizheng 1990: I:114–115.

[99] He contrasts *popular rule* (*minzhi* 民治) to *bureaucratic rule* (*guanzhi* 官制), which according to Liu is at the root of China's moral decline. See Liu Yizheng 1923: 194. Liu uses at other places the term *popular rights* (*minquan* 民權) with a negative connotation as opposed to the positive *popular rule*, see ibid. 179. On Liu's experience in local self-government during the last decade of the Qing Dynasty, see Hon 2004: 527–531.

[100] Liu Yizheng 1923: 202–203.

[101] Ibid. 222–223.

[102] Liu Yizheng 1948b: chapters 2 and 7.

Liu Yizheng: Nation, History and Ethics

A reading of Liu Yizheng in light of the dual horizon as outlined at the beginning of this article will help us to understand his peculiar position. It will enable us to move towards an interpretation of modern Chinese historiography that goes beyond the hitherto well-established patterns of seeing it in terms of nationalism, professionalization and modernization under the influence of Western examples. Liu Yizheng's theory of history reminds us of the long tradition of writing history in China and the ways this tradition has shaped the modern transformation. This process of shaping has to be understood not only in terms of methodological models, such as the tradition of evidential scholarship (*kaozheng* 考證), but as reflecting fundamental assumptions about the nature of history and its position in the larger Chinese cultural framework.

It seems to me that one of the, if not *the* most fundamental difference between the development of history as a discipline in the modern West and in China is the position of ethics. In the West, historicism drew reason and ethics *into history*, thus relativizing and undermining them, ultimately leading to a fundamental epistemological and moral crisis, which is in a way at the root of current post-modern concerns. In pre-modern China, ethics were central to history and history was central to ethics, because it was through history that ethical principles could be grasped. In modern China, ethics were either driven *out of history* with the result that modern Chinese understandings of history were comparable to a positivist understanding of history in the West in that all normative judgments were to be excluded from history.[103] Alternately, ethical ideals were conceived of as the final point of a teleological and rather deterministic historical process, thus depriving them of much of their potential for practical guidance, because only in a situation of choice[104] can ethical ideals fully unfold this potential.[105]

It is precisely this elimination or transformation of ethics that Liu Yizheng refuses to accept, a decision that shapes his view and theory of history. Neither a linear history à la Liang Qichao, devoid of ethical value,

[103] As was the case with Fu Sinian's theory of history during the 1920s and 30s, see Schneider Axel 1996.

[104] Choice presupposes the acknowledgment that humans are capable of both, acting good and evil.

[105] This tension, already encapsulated in Marx's view of history, became even more obvious in the way Chinese Marxists appropriated it. See Wakeman 1973.

nor a teleological history that due to the necessity of the envisioned process ultimately seems to deprive human beings of their moral responsibility were acceptable to him. Fully aware of human propensity to evil, he insists that historians have to rely on historically "revealed" standards of the Good to counter what he identifies as the evil trends of (Western) modernity: individualism, utilitarianism, hedonism.[106]

Does this justify classifying Liu as conservative? Using a concept of classicist conservatism[107] I argue that on the level of basic assumptions about human life Liu Yizheng was a conservative. He opposed the ethical relativism typical for the modern liberal society of the West and its tendency to privatize the Good, as a yardstick for gauging conservatism in the realm of thought and especially ethics.[108] Ethics for Liu are rooted in the heavenly order and are fundamental to history and historiography. The ethical crisis of late historicism does not play a role for him. He is concerned with timeless ethics. In this, he is similar to modern Western conservatives such as Leo Strauss, yet his position is also fundamentally different in that he does not rely on divine revelation to become aware of the right ethical principles.[109] He links back to a tradition of reasoning that assumes that eternal ethical principles are manifested in history and thus accessible to human understanding. Therefore, while Liu does not believe that ethics are *drawn into history* and hence subverted, they nonetheless continue to *reside in history* and are thus confirmed.

---

[106] See Alitto 1986 for an analysis of the binary logic of many cultural conservatives.

[107] As stated above conservatism is a concept notoriously difficult to define. As an historian of ideas I operate with a distinction of two types of conservatism: classicist and historicist conservatism. Classicist conservatives argue that there exists a set of universal moral standards that although can be apprehended, can not be altered and adjusted to human needs. On the contrary, the human condition is defined by these standards. They point out that the essence of these ethical standards is to exert self-control, to control one's desires, to value duty, and to act in such a way as to further the interests of the collective and not of the individual. However, these aims can not be enforced or ensured via some grand socio-political design. The utmost human beings can do, is to set up socio-political structures that are conducive to the realization of these standards. However, due to the inherent fallibility of human beings this is facing serious limitations and will time and again face failure. Historicist conservatives, however, start out from the assumption that all human institutions are historically relative. They argue that universal standards may perhaps exist, however, they can not be apprehended. Human beings should thus rather rely on particular, historically grown traditions. They should inherit, continue and further develop tradition, i.e. the wisdom of the forefathers. For a more detailed analysis see Schneider Axel 2005. On conservatism in general see Mannheim 1980, 1984, 1986, Allen 1981, Muller 1997, Kekes 1997, Greiffenhagen 1974, Berlin 1998.

[108] To use an expression coined by Alistair MacIntyre. See MacIntyre, Kommers and Solomon, 1990.

[109] McAllister 2006.

Yet, this conservative resistance does not lead to an outright rejection of all things new and modern, nor does it mean that he returns to a pre-nationalist approach to writing history. Liu Yizheng departs in many ways from a traditional view of history. He historicizes the Golden Age and turns his attention to the people and the nation as the main protagonists of Chinese history. He relies on the concrete embodiment of ethics in Chinese culture to ground a particular national identity, which at the same time is firmly anchored in universal ideals that are *not* of Western origin.[110] Liu writes a new type of history of the culture of the nation using modern concepts such as historical particularity, causality and progress, albeit at times in a rather truncated way. He also argues in favor of history as a professional, modern academic discipline going beyond traditional distinctions (such as that between Han- and Song-learning).[111] And, last but not least, he is influenced by Western notions of the rule by law and popular sovereignty, albeit projecting these back into the Chinese past and, again, subordinating them to the higher ethical ideal of the rule by virtue.[112]

In Liu Yizheng's thought history and ethics are not the same, they have to be carefully distinguished, and yet they are inseparable. Ethics constitute the human condition, but they can only be apprehended via history. History is thus both a gateway to the human condition *and* the foundation for a particular national identity. It is true and mine, *verum* and *meum*. For this reason, I argue that Levenson was wrong in postulating that history is nothing but the locus of an emotional commitment to what happens to be one's particular life story, while commitment to values reflects a rational search for truth. Here he might have fallen victim to his Neo-Kantian inspirations, arguing that history is meaningless and value is transcendental. While this Neo-Kantian position might make sense against the background of the profound crisis of historicism in late 19th century Europe, which it is supposed to solve, in China it is out of

[110] Naoki Sakai has convincingly argued that for late-developing nations emphasizing national particularity more often than not means subjecting oneself to Western universality, see Sakai 1997.

[111] Luo 2003: 376.

[112] Whether these modern elements, some of which are clearly influenced by nationalism, others reflect the changed thinking about politics, would have been adequate for solving China's problems in case Liu would have had a chance to put them into practice, is a question that is beyond the scope of this article.

place, because here history had, and for Liu Yizheng, continues to have, a different status.[113]

And what about the nation? Liu writes a national history, but not one of a unified national subject progressing through time. His delineation of the borders of the nation relies on historically manifested culture and yet also points beyond it to a vision of the human condition. He rejects the modern West on the basis of this vision, and yet he also demands respect for the particularity of other cultures. He clearly is trying to strike a balance between his universalistic ethical convictions and his awareness of historically grown particularity. It is thus obvious that Liu's concept of history does not have to be rescued from the nation.[114]

References Cited

Alitto, Guy (1979), *The Last Confucian: Liang Shu-ming and the Chinese Dilemma of Modernity*. Berkeley: University of California Press.

—— (1986), *Wenhua shoucheng zhuyi lun* 文化守成主義論 (On Cultural Conservatism). Taibei: Shibao wenhua chuban shiye youxian gongsi.

Allen, David Y. (1981), "Modern Conservatism: The Problem of Definition," in *The Review of Politics* 43, 4 (1981), 582–603.

Bambach, Charles R. (1995), *Heidegger, Dilthey, and the Crisis of Historicism*. Ithaca: Cornell University Press.

Berlin, Isaiah (1998), "The Counter-Enlightenment," in Berlin Isaiah ed., *The Proper Study of Mankind: An Anthology of Essays*. New York: Farrar, Straus and Giraux, 243–268.

Bernheim, Ernst (1908), *Lehrbuch der historischen Methode und der Geschichtsphilosophie*. Leipzig: Duncker & Humblot.

—— (1937); Chen, Tao 陳韜 (Trs.), *Shixue fangfalun* 史學方法論 (Methodology of Historical Research). Shanghai: Shangwu yinshuguan.

Bian, Xiaoxuan 卞孝萱 and Sun Yongru 孫永如 (1999), "Shixuejia Liu Yizheng de xueshu gongxian yu daode fengfan" 史學家柳詒徵的學術貢獻與道德風范 (The Moral Paragon and Academic Contribution of the Historian Liu Yizheng), in *Ningbo daxue xuebao (Renwen kexueban)* 3 (1999), 69–74.

Bonner, Joey (1986), *Wang Kuo-wei an Intellectual Biography*. Cambridge/Mass.: Harvard University Press.

Buckle, Henry T. (1857), *History of Civilization in England*. London: J. W. Parker and son.

Chen, Chi-yun (2006), "Immanental Human Beings in Transcendent Time: Epistemological Basis of Pristine Chinese Historical Consciousness," in Huang, Chun-chieh and Henderson, John eds., *Notions of Time in Chinese Historical Thinking*. Hong Kong: The Chinese University Press, 45–73.

---

113 For Levenson's position on history and value see Levenson 1953a, 1953b. For an analysis of the Neo-Kantian position on history in the context of the crisis of historicism see Bambach 1995. For the potential Neo-Kantian background of Levenson's thought see Schneider Axel 2008.

114 Compare Duara (1995), who presents a rather one-sided picture of modern Chinese historical thinking.

Chevrier, Yves (1987), "La servante-maîtresse: condition de la référence à l'histoire dans l'espace intellectuel chinois," in *Extrême-Orient, extrême-occident, Cahiers de recherches comparatives IX no. La référence à l'histoire* (1987), 117–144.
Dennerline, Jerry (1989), *Qian Mu and the World of Seven Mansions*. New Haven: Yale University Press.
Duara, Prasenjit (1995), *Rescuing History from the Nation: Questioning Narratives of Modern China*. Chicago: University of Chicago Press.
Ebrey, Patricia Buckley (1991), *Confucianism and family rituals in imperial China: a social history of writing about rites*. Princeton, N.J.: Princeton University Press.
Fraser, Julius T. and Haber, Francis C. eds. (1986), *Time, Science, and Society in China and the West*. Amherst, MA: University of Massachusetts Press.
Furth, Charlotte ed. (1976), *The Limits of Change: Essays on Conservative Alternatives in Republican China*. Cambridge/Mass.: Harvard University Press.
Furth, Charlotte (1983), "Intellectual change: from the Reform Movement to the May Fourth Movement, 1895–1920," in Fairbank, John K. ed. (1983), *The Cambridge History of China, vol. 12, Part I, Republican China, 1912–1949*. Cambridge: Cambridge University Press, 322–405.
Greiffenhagen, Martin (1974), "Das Dilemma des Konservatismus," in Schumann, H.G. ed. (1974) *Konservatismus*. Köln, 156–198.
Guofeng banyuekan 國風半月刊, ed. Liu Yizheng, Nanjing 1932–1936.
He, Xiaoming 何曉明 (2006), *Fanben yu kaixin: jindai Zhongguo wenhua bashou zhuyi xinlun* 反本與開新: 近代中國文化保守主義新論 (Back to the Roots and Reform: On Modern Chinese Cultural Conservatism). Beijing: Shangwu yinshuguan.
Hon, Tze-ki (1996), "Ethnic and cultural pluralism: Gu Jiegang's vision of a new China in his studies of ancient history," in *Modern China* 22, 3 (1996), 315–339.
—— (2004), "Cultural identity and local self-government, a study of Liu Yizheng's *History of Chinese Culture*," in *Modern China* 30, 4 (2004), 506–542.
Hou, Jian (1980), *Irving Babbitt in China*. New York: SUNY, unpublished PhD dissertation.
Hsiao, Kung-chuan (1975), *A Modern China and a New World: K'ang Yu-wei, Reformer and Utopian, 1858–1927*. Seattle: University of Washington Press.
Huang, Chun-chieh (2006), " 'Time' and 'supertime' in Chinese historical thinking," in Huang, Chun-chieh and Henderson, John eds. (2006), *Notions of time in Chinese historical thinking*. Hong Kong: The Chinese University Press, 19–41.
—— (2007), "The defining character of Chinese historical thinking," in *History & Theory* 46 (2007), 180–188.
Huang, Chun-chieh and Zürcher, Erik (1995), "Cultural notions of time and space in China," in Huang, Chun-chieh and Zürcher, Erik eds. (1995), *Time and space in Chinese culture*. Leiden: Brill, 3–16.
Iggers, Georg G. (1962), "The Image of Ranke in American and German Historical Thought," in *History and Theory* 2 (1962), 17–40.
—— (1968), *The German Conception of History; the National Tradition of Historical Thought from Herder to the Present*. Middletown, Conn.: Wesleyan University Press.
Kekes, John (1997), "What is Conservatism?," in *Philosophy* 72 (1997), 351–374.
Kelley, Donald R. (2003), *Fortunes of History: Historical Inquiry from Herder to Huizinga*. New Haven: Yale University Press.
Kogelschatz, Hermann (1986), *Wang Kuo-wei und Schopenhauer: eine philosophische Begegnung. Wandlung des Selbstverständnisses der chinesischen Literatur unter dem Einfluß der klassischen deutschen Ästhetik*. Stuttgart: Steiner Verlag.
Koselleck, Reinhart (1975), "Geschichte, Historie", in Otto Brunner, Werner Conze and Reinhart Koselleck eds. (1975), *Geschichtliche Grundbegriffe, Historisches Lexikon zur politisch-sozialen Sprache in Deutschland*. Stuttgart: Klett-Cotta, 593–717.
Kramer, Lloyd (1997), "Historical Narratives and the Meaning of Nationalism," in *Journal of the History of Ideas* 58, 3 (1997), 525–545.

Kuo, Ya-pei (2002), *The Crisis of Culture in Modern Chinese Conservatism, The Case of the Critical Review*. University of Wisconsin-Madison, unpublished PhD dissertation.

Kwong, Luke S.K. (2001), "The Rise of the Linear Perspective on History and Time in Late Qing China c.1860–1911," in *Past and Present* 173 (2001), 157–190.

Langlois, Charles and Seignobos, Charles (1898), *Introduction aux études historiques*. Paris, 1898.

—— (1926); Li Sichun 李思純 (Trs.), *Shixue yuanlun* 史學原論 (Principles of Historical Research). Shanghai: Shangwu.

Levenson, Joseph R. (1953a), " 'History' and 'Value': The Tensions of Intellectual Choice in Modern China," in Arthur F. Wright eds. (1953), *Studies in Chinese thought*. Chicago: University of Chicago Press, 1953), 146–194.

—— (1953b), *Liang Ch'i-ch'ao and the mind of modern China*. Cambridge/Mass.: Harvard University Press.

Li, Xiaoqian 李孝遷 (2007), *Xifang shixue zai Zhongguo de chuanbo (1882–1949)* 西方史學在中國的傳播 (1882–1949) (The Spread of Western Historical Science in China, 1882–1949). Shanghai: Huadong shifan daxue chubanshe.

Liao, Mei 廖梅 (2001), *Wang Kangnian: cong minquanlun dao wenhua baoshou zhuyi* 汪康年: 從民權論到文化保守主義 (Wang Kangnian: From People's Rights to Cultural Conservatism). Shanghai: Shanghai guji.

Liu, Lihong 劉黎紅 (2006a), " 'Wusi' shiqi liang zhong zhengli guogu huodong de bijiao" '五四' 時期兩種整理國故活動的比較 (Comparison of two Movements to Sort Out the National Past during the Period of the May Fourth Movement), in *Dongfang luntan* 東方論壇 3 (2006), 102–107.

—— (2006b), *Wusi wenhua baoshou zhuyi sichao yanjiu* 五四文化保守主義研究 (Research on Cultural Conservatism during the May Fourth Period). Beijing: Zhongguo shehui kexue chubanshe.

Liu, Shu-hsien (1974), "Time and Temporality: The Chinese Perspective," in *Philosophy East and West* 24, 2 (1974), 154–153.

—— (2006), "On the Formation of a Philosophy of Time and History through the Yijing," in Huang, Chun-chieh and Henderson, John eds. (2006), *Notions Of Time In Chinese Historical Thinking*. Hong Kong: The Chinese University Press, 75–94.

Liu Yizheng 柳詒徵 (1902), *Lidai shilüe* 歷代史略 (*Brief Account of the Past*). Nanjing: Zhongxin shuju.

—— (1922a), "Gu shi xueshu, fu Chen Di 'Maoshi guyin kaoxu' " 顧氏學術, 附陳第《毛詩古音考序》, in *Xueheng zazhi* 5 (1922), reprinted in Liu Zengfu and Liu Dingsheng 1991b, 20–34.

—— (1922b), "Xiushi siyi" 修史私議 (Personal Deliberations on Compiling History), in *Shidi xuebao* 1, 4 (1922), reprinted in Qiao, Yanguan 1970, 191–196.

—— (1923), "Zhongguo xiangzhi zhi shangde zhuyi" 中國鄉治之尚德主義 (The Moralism of Chinese Local Rule), in *Xueheng zazhi* 17, 21, 36 (1923), reprinted in Liu Zengfu and Liu Dingsheng 1991b, 178–223.

—— (1924a), "Ming lun" 明倫 (Understanding Ethics), in *Xueheng zazhi* 26 (1924), 1–5.

—— (1924b), "Ping Lu Moude Zhou-Qin zhexueshi" 評陸懋德周秦哲學史 (A Critical Appraisal of Lu Moude's *History of Zhou and Qin Philosophy*), in *Xueheng zazhi* 29 (1924), reprinted in Liu Zengfu and Liu Dingsheng 1991b, 231–242.

—— (1924c), "Zhongguo wenhua xibei zhi shangque" 中國文化西被之商榷 (Discussion of the Extension of Chinese Culture into the West), in *Xueheng zazhi* 27 (1924), reprinted in Liu Zengfu and Liu Dingsheng 1991b, 224–230.

—— (1925), "Lishi zhi zhishi" 歷史之知識 (Historical Knowledge), in *Shidi xuebao* 3, 7 (1925), pp. 19–21, reprinted in Liu Zengfu and Liu Dingsheng 1991a, 80–84.

—— (1926a), *Shixue gailun* 史學概論 (*Outline of Historical Research*), reprinted in Liu Zengfu and Liu Dingsheng 1991a, 97–117.

—— (1926b), "Shu she" 述社 (On Society), in *Xueheng zazhi* 54 (1926), reprinted in Liu Zengfu and Liu Dingsheng 1991b, 273–289.

—— (1926c), "Zhongguo shixue zhi shuanggui" 中國史學之雙軌 (The Double Track of Chinese Historiography), in *Shixue yu dixue* 1 (1926), 3–6, reprinted in Liu Zengfu and Liu Dingsheng 1991a, 92–96.

—— (1926d), "Zhongguo wenhuashi" 中國文化史 (*History of Chinese Culture*), in *Xueheng zazhi* 49, 50, 51, 52, 53, 54, 56, 58, 61, 63, 64, 67, 70, 72, 75 (1926–1931).

—— (1929), "Yu mou jun lun yanjiu Zhongguo jingjishi zhi fangfa" 與某君論研究中國經濟史之方法 (Discussing the Methods for Research on Chinese Economic History), in *Shixue zazhi* 1, 4 (1929), 1–6, reprinted in Liu Zengfu and Liu Dingsheng 1991b, 335–343.

—— (1932), "Kongxue guanjian" 孔學管見 (My Humble opinion on Confucianism), in *Guofeng banyuekan* 1, 3 (1932), 11–19.

—— (1934), "Duiyu Zhongguo wenhua zhi guanjian" 對于中國文化之管見 (My Humble Opinion on Chinese Culture), in *Guofeng banyuekan* 4, 7 (1934), 1–9.

—— (1935), "Jiang guoxue yi xian jiang shixue" 講國學宜先講史學 (Who Wants to Talk about National Studies Ought to Talk first about Historiography), in *Guangbo zhoukan* 25 (1935), reprinted in Liu Zengfu and Liu Dingsheng 1991a, 492–503.

—— (1947), "Zhongguo lisushi fafan" 中國禮俗史發凡 (Introduction to the Rites and Custom of China), in *Xueyuan* 學原 1, 1 (1947), reprinted in Liu Zengfu and Liu Dingsheng 1991b, 610–651.

—— (1948a), "Yu qingnian lun dushi" 與青年論讀史 (Discussion with the Youth on Reading the Histories), reprinted in Liu Zengfu and Liu Dingsheng 1991a, 549–559.

—— (1948b), *Guoshi yaoyi* 國史要義 (*Main Principles of National History*). Shanghai: Shangwu.

—— (1990), *Zhongguo wenhuashi* 中國文化史 (*History of Chinese Culture*) 17 ed. 3 vols. Taibei: Zhengzhong.

—— (2001), *Zhongguo wenhuashi* 中國文化史 (*History of Chinese Culture*) 2 vols. Shanghai: Shanghai guji.

Liu, Zengfu 柳曾符 and Liu Dingsheng 柳定生 eds. (1991a), *Liu Yizheng shixue lunwen ji* 柳詒徵史學論文集 (*Collection of Historiographical Writings of Liu Yizheng*). Shanghai: Shanghai guji.

—— (1991b), *Liu Yizheng shixue lunwen xuji* 柳詒徵史學論文續集 (*Collection of Historiographical Writings of Liu Yizheng, Volume B*). Shanghai: Shanghai guji.

Lu, Moude 陸懋德 (1923), *Zhou Qin zhexueshi* 周秦哲學史 (*History of Zhou and Qin Philosophy*). n.p.: Jinghua yinshuju, 1923.

Luo, Zhitian 羅志田 (2003), *Guojia yu xueshu: Qing ji Min chu guanyu 'guoxue' de sixiang lunzheng* 國家與學術: 清季民初關于 '國學' 的思想論證 (State and Scholarship: Discussion of Ideas on "National Learning" during the Qing and the Early Republic). Beijing: Sanlian.

MacIntyre, Alasdair; Kommers, Donald P. and Solomon, W. David (1990), "The Privatization of Good: An Inaugural Lecture," in *Review of Politics* 52, 3 (1990), 344–377.

Mannheim, Karl (ed. and introduced by David Kettler) (1986), *Conservatism, a contribution to the sociology of knowledge by Karl Mannheim*. London: Routledge & Kegan Paul.

—— (ed. David Kettler and Volker Meja) (1980), *Strukturen des Denkens* (Structures of Thought). Frankfurt am Main: Suhrkamp.

—— (ed. David Kettler) (1984), *Konservatismus: ein Beitrag zur Soziologie des Wissens* (Conservatism: A Contribution to the Sociology of Knowledge). Frankfurt am Main: Suhrkamp.

McAllister, Ted V. (2006), *Revolt against Modernity: Leo Strauss, Eric Voegelin, and the Search for a Postliberal Order, American political thought*. Lawrence, Kansas: University Press of Kansas.

Mehl, Margaret (1998), *History and the State in Nineteenth-Century Japan*. New York: St. Martin's Press.

Moloughney, Brian (2001), "Nation, narrative and China's new history," in Roy Starrs ed. (2001), *Asian Nationalism in an Age of Globalization*. Richmond, UK: Japan Library, 205–222.

Moloughney, Brian und Zarrow, Peter eds. (2012), *Transforming History: The Making of a Modern Academic Discipline in Twentieth-Century China*. Hong Kong: Chinese University Press, 271–302.

Muller, Jerry Z. (1997), "What is Conservative Social and Political Thought?," in Muller, Jerry Z. ed. (1997), *Conservatism, An Anthology of Social and Political Thought from David Hume to the Present*. Princeton: Princeton University Press, 3–31.

Naka, Michiyo 那珂通世 (1899), *Shina tsūshi* 支那通史 (General History of China). Shanghai: Zhongxi shiyin shuju.

Ng, On Cho and Wang, Edward Q. (2005), *Mirroring the Past: The Writing and Use of History in Imperial China*. Honolulu: University of Hawaii Press.

Peng, Minghui 彭明輝 (1991), *Yigu sixiang yu xiandai Zhongguo shixue de fazhan* 疑古思想與現代中國史學的發展 (Doubting Antiquity and the Development of Modern Chinese Historiography). Taibei: Shangwu.

—— (1995), *Lishi dilixue yu xiandai Zhongguo shixue* 歷史地理學與現代中國史學 (Historical Geography and Modern Chinese Historiography). Taibei: Dongda.

Pöggeler, Otto (1973), *Hegels Idee einer Phänomenologie des Geistes* (Hegel's Idea of a Phenomenology of Spirit). Freiburg, München: Alber.

—— (1995), *Eine Ende der Geschichte? Von Hegel zu Fukuyama* (The End of History? From Hegel to Fukuyama). Opladen: Westdeutscher Verlag.

Qiao, Yanguan 喬衍琯 ed. (1970), *Liu Yimou xiansheng wenlu* 柳翼謀先生文錄 (Writings of Mr. Liu Yimou). Taibei: Guangwen shuju.

Rosen, Richard B. (1969), *The National Heritage Opposition to the New Literature and New Culture Movements of China in the 1920s*. Berkeley: University of California, unpublished PhD dissertation.

Rüsen, Jörn (1993), *Konfigurationen des Historismus, Studien zur deutschen Wissenschaftskultur* (*Configurations of Historicism, Studies on the German Culture of Science*). Frankfurt a.M: Suhrkamp Taschenbuch.

—— (2007), "Crossing Cultural Borders: How to Understand Historical Thinking in China and the West," in *History & Theory* 46 (2007), 189–193.

Sakai, Naoki (1997), "Modernity and its critique: the problem of universalism and particularism," in Sakai, Naoki ed. (1997), *Translation & Subjectivity: On "Japan" and Cultural Nationalism*. Minneapolis: University of Minnesota Press, 153–176.

Sato, Masayuki (2007), "The Archetype of History in the Confucian Ecumene," in *History & Theory* 46 (2007), 218–232.

Schnädelbach, Herbert (1974), *Geschichtsphilosophie nach Hegel, Die Probleme des Historismus* (*The Philosophy of History after Hegel, The Problems of Historicism*). Freiburg/München: Verlag Alber.

Schneider, Axel (1996), "Between Dao and History: Two Chinese Historians in Search of a Modern Identity for China," in *History and Theory Theme Issue 35 Chinese Historiography in Comparative Perspective* (1996), 54–73.

—— (1997), *Wahrheit und Geschichte: Zwei chinesische Historiker auf der Suche nach einer modernen Identität für China* (*Truth and History: Two Chinese Historians in Search of a Modern Identity for China*). Wiesbaden: Harrassowitz Verlag.

—— (2001a), "Bridging the Gap: Attempts at Constructing a 'New' Historical-Cultural Identity in the PRC," in *East Asian History* 22 (2001), 129–144.

—— (2001b), "Reconciling history with the nation?—Historicity, national particularity, and the question of universals," in Starrs, Roy ed. (2001), *Asian Nationalism in an Age of Globalization*. Curzon, 223–233.

—— (2005), "The One and the Many: A Classicist Reading of China's Tradition and its Role in the Modern World," in *Zhongguo wenxue lishi yu sixiang de guannian bianqian guoji xueshu yantaohui lunwenji* 中國文化歷史與思想的觀念變遷國際學術研討會論文集. Taibei: Taiwan daxue wenxueyuan, 311–373.

—— (2008), "Shijie lishi yu lishi xiangdui zhuyi de wenti—1919 nian yihou Liang Qichao de shixue" 世界歷史與歷史相對主義的問題—1919 年以后梁啟超的史學 (World History and the Problem of Historical Relativism—The Historiography of Liang Qichao after 1919), in Schneider, Axel (2008), *Zhenli yu lishi: Liang wei Zhongguo shijia dui Zhongguo xiandai rentong de qiusuo* 真理與歷史: 兩位中國史家對中國現代認同的求索. Beijing: Shehuikexue wenxian, 238–259.

Schneider, Lawrence A. (1971), *Ku Chieh-kang and China's New History, Nationalism and the Quest for Alternative Traditions.* Berkeley: University of California Press.

—— (1976), "National Essence and the New Intelligentsia," in Furth, Charlotte ed. (1976), *The Limits of Change: Essays on Conservative Alternatives in Republican China.* Cambridge/Mass.: Harvard University Press, 73–89.

Schwartz, Benjamin I. (1985), *The World of Thought in Ancient China.* Cambridge/Mass.: Belknap Press of Harvard University Press.

—— (1996), "History in Chinese Culture—Some Comparative Reflections," in *History and Theory, Theme Issue 35 Chinese Historiography in Comparative Perspective* (1996), 23–33.

Shen, Songqiao 沈松僑 (1985), *Xuehengpai yu wusi xin wenhua yundong* 學衡派與五四新文化運動 (*The Critical Review Group and the New Culture Movement*). Taibei: Taida wenshi congkan.

Shidi xuebao 史地學報, Nanjing 1921–1926.

Sivin, Nathan (1966), "Chinese Conceptions of Time," in *Earlham Review* 1 (1966), 82–92.

Sun, Yongru 孫永如 (1993), *Liu Yizheng pingzhuan* 柳詒徵評傳 (Critical Biography of Liu Yizheng). Nanchang: Baihua zhouwen chubanshe.

Tanaka, Stefan (1993), *Japan's Orient: Rendering Pasts into Histor.* Berkeley: University of California Press.

Thompson, Laurence G. (1958), *Ta T'ung Shu. The One-World Philosophy of K'ang Yu-wei.* London: George Allen and Unwin.

Tian, Liang 田亮 (2004), "Liu Yizheng de minzu zhuyi shixue sixiang" 柳詒徵的民族主義史學思想 (The Nationalist Historiography of Liu Yizheng), in *Shixueshi yanjiu* 2 (2004), 16–21.

Twitchett, Denis C. (1992), *The Writing of Official History under the T'ang.* Cambridge: Cambridge University Press.

Unger, Jonathan (1993), *Using the Past to serve the Present: Historiography and Politics in Contemporary China.* Armonk, N.Y.: Sharpe.

Wakeman, Frederic E. (1973), *History and Will: Philosophical Perspectives of Mao Tse-tung's Thought.* Berkeley: University of California Press.

Wang, Edward Q. (1995), "Time Perception in Ancient Chinese Historiography," in *Storia della Storiografia* 28 (1995), 69–86.

—— (2001), *Inventing China through History: the May Fourth Approach to Historiography.* Albany, NY: State University of New York Press.

—— (2007), "Is there a Chinese Mode of Historical Thinking? A Cross-cultural Analysis," in *History & Theory* 46 (2007), 201–209.

Wang, Fansen 王汎森 (1987), *Gushibian yundong de xingqi, yi ge sixsiangshi de fenxi* 古史辨運動的興起, 一個思想史的分析 (The Rise of The Gushibian Movement: An Intellectual History). Taibei: Yunchen.

—— (2000), *Fu Ssu-nien: a Life in Chinese History and Politics.* Cambridge, UK, New York: Cambridge University Press.

—— (2008), "Jindai Zhonggo de xianxing lishiguan—yi shehui jinhualun wei zhongxin de tantao" 代中國的線性歷史觀－以社會進化論為中心的討論 (Linear Views of History in Modern China—A Discussion Focusing on Social Darwinism), in *Xin shixue* 19:2 (June 2008), 1–46.

Wang, Jiafan 王家范 (2006), "Liu Yizheng 'Guoshi Yaoyi' tanlü" 柳詒徵《國史要義》探津 (Explorations of Liu Yizheng's *Main Principles of National History*), in *Shilin* 6 (2006), 169–176, 191.

Wang, Xinkai 王信凱 (2004), " 'Xueheng' zhong de 'Liu Yizheng' "《學衡》中的『柳詒徵』(The Liu Yizheng of the Critical Review [Period]), in *Zhongguo lishi xuehui shixue jikan* 35 (2004), 251–294.

Wang, Xintian 王新田 (1999), "Liu Yizheng xiansheng nianpu jianbian" 柳詒徵先生年譜簡編 (Short Chronological Biography of Mr. Liu Yizheng), in *Zhongguo wenzhe yanjiu tongxun* 9, 4 (1999), 147–155.

Yi, Lan 易蘭 (2006), *Lanke shixue yanjiu* 蘭克史學研究 (*Ranke's Historical Research*). Shanghai: Fudan daxue chubanshe.

Yu, Dahua 喻大華 (2001), *Wan Qing wenhua baoshou sichao yanjiu* 晚清文化保守思潮研究 (*Research on Cultural Conservatism During the Late Qing*). Beijing: Renmin.

Yu, Nan 于楠 (2006), "Shilun 'Wusi' qianhou xueshujie dui 'jinhua shiguan' de fanxing—Yi Zhang Taiyan, Liang Qichao, Liu Yizheng de shixue sixiang wei li" 試論 "五四" 前后學術界對 "進化史觀"的反省——以章太炎、梁啟超、柳詒徵的史學思想為例 (On Reflections Concerning "Evolutionary History" around the Time of the May Fourth Movement—The Example of the Historiographical Ideas of Zhang Taiyan, Liang Qichao and Liu Yizheng), in *Shoudu shifan daxue xuebao* (*shehui kexueban*) 2 (2006), 65–71.

Zhang, Ying 張英 (2003), "Zhongguo jindai jinhua shiguan tanjiu" 中國近代進化史觀探究 (Research into Evolutionary Concepts of History in Modern China), in *Lilun tantao* 2 (2003), 38–40.

Zhang, Yue 張越 (2003), "Jinhua shiguan dui Zhongguo shixue zhuanxing de cujin he yingxiang" 進化史觀對中國史學轉型的促進和影響 (The Influence of Evolutionary History on and its Promotion of the Transformation of Chinese Historiography), in *Qiushi xuekan* 1 (2003), 111–115.

Zheng, Dahua 鄭大華 (1994), *Liang Shuming yu Hu Shi: wenhua baoshou zhuyi yu xihua sichao de bijiao* 梁漱溟与胡適: 文化保守主義與西化思潮的比較 (*Liang Shuming and Hu Shi: A Comparison of Cultural Conservatism and the Intellectual Trend of Westernization*). Beijing: Zhonghua shuju.

Zheng, Shiqu 鄭師渠 (1998a), " 'Gujin shi wu shu, dongxi ji qi liang'—lun Xuehengpai de wenhuaguan" '古今事無殊, 東西迹豈兩'—論學衡派的文化觀 (No Difference between the Past and the Present—On the View of Culture of the Critical Review Group), in *Jindaishi yanjiu* 4 (1998), 55–88.

—— (1998b), "Xuehengpai shixue sixiang chutan" 學衡派史學思想初探 (Initial Explorations of the Historiographical Thought of the Critical Review Group), *Beijing shifan daxue xuebao* (*shehui kexueban*) 4 (1998), 31–38.

CHAPTER FOUR

READING TAKEUCHI YOSHIMI AND READING HISTORY

Sun Ge

In 2005, an edition of the selected works of the Japanese thinker Takeuchi Yoshimi was published in China. This collection—the first book of translations of Takeuchi that has followed copyright procedures—was created by selecting pieces from the seventeen volumes of Takeuchi's collected works.[1]

Takeuchi is the founder of the study of Chinese literature and the Organization for the Study of Chinese Literature. Therefore, he has always been seen as a scholar of Chinese literature. However, throughout his intellectual life, the study of Chinese literature and thought was not actually his real goal. He used this study as a medium in order to establish Japanese thought. Therefore, the above-mentioned book of translations is not a collection of essays on the study of Chinese literature. Instead, it describes the intellectual trajectory of a strong thinker who constantly negates and re-creates himself. It also presents an outline of a Japanese thinker who shoulders social responsibility while searching amidst reality.

In mainland China, there exists a basic epistemological fallacy, which makes people see thought as an absolutely "correct" concept that does not change. Especially because the development of the media has led mass society to seek the common, the intellectual milieu cannot avoid the problems of simplifying history and relying on already existing conceptual frameworks. This has made it extremely difficult for intellectuals to understand that reality is complex and today it has the most diverse meanings. Therefore, the present is a period that silently entails many different possibilities. When intellectuals try to put complex reality into simple frameworks that are easy to understand, reality leaks outside of discourse.

---

[1] Takeuchi 2005. In the early 1980s, there was a translation (by Li Xinfeng) published in China of Takeuchi Yoshimi's "Lu Xun"; this work was greatly influential. *Overcoming Modernity* has retranslated this long text and has added ten smaller works.

Takeuchi has been translated precisely at this critical juncture. The Chinese translation of Takeuchi contains his most politically incorrect piece, "The Greater East Asian War and Our Determination," a manifesto published in January 1942 supporting the Pacific War. At the same time, it contains a text that is most difficult to understand, namely "Overcoming Modernity," which was written in 1959 in order to find elements that could initiate transformation in the midst of Japanese intellectuals' attempt to reflect on Japan's own historical trajectory. Against the postwar trend to simply negate all intellectual activity during the war, this essay attempts to analyze why various projects during the war failed. At the same time, it attempts to find elements that could be transformed into positive intellectual resources in what was called fascist ideology during the postwar period. Because of the limitations of space, the volume of Chinese translations did not include two other aspects of his work, namely his analysis of Chinese history and the Chinese revolution and his analysis of Japanese pan-Asianism, which began to emerge in the Meiji period. These two aspects are also important parts of Takeuchi Yoshimi's thought and if one looks at them directly, they seem to be contradictory: on the one hand, Takeuchi highly evaluates the subjectivity of modern China, but on the other hand, he does not consequently simply negate the pan-Asianism of the early Meiji period, although, after 1890, the latter was inextricably linked to invasion in a complex formation, and finally fell into fascist ideology.

Precisely because Takeuchi has such a complex face, the editors of the Chinese volume hesitated in choosing which essays to include. To put it simply, when the Chinese translated Takeuchi Yoshimi, they had extremely intellectualized considerations. With the start of the intellectual liberation movement in the 1980s, the scholarly community in China began to break away from the simplified thinking of the Cultural Revolution. Scholars attempted to establish a method of thinking the subject and history and use this method to avoid simplified ideological thinking. However, because it is extremely difficult to break free from a simplified dualistic method of thinking and because of the way in which Western theory was introduced, the scholarly community in China did not have the foundation to break free from simplified thinking. Moreover, in the midst of twenty years of transformations of contemporary history, a number of intellectuals became anxious and worried. They took thought as reality and concepts as the process of history. This type of epistemological mistake made it such that the instincts of the scholarly and intellectual communities to seek political correctness (and that too an extremely simplified understanding of this term) exponentially increased. In the midst of big theories and great politics, without any modifications, reality was made to fit into its assigned seat. Therefore, scholarship lost an opportunity to develop. In this context, Takeuchi Yoshimi was brought in as a type of medium. Through his non-conceptual historical thinking and intellectual practice that does not try to be a wise man after the fact, the editors hoped to pierce some of the blind spots in a theoretical framework among Chinese intellectuals that lacks historical understanding. From here, they hoped to find themselves and develop new possibilities. Because Takeuchi Yoshimi does not take a self-righteous attitude in the face of great historical events, the question he constantly pondered was how to establish a Japanese subjectivity and not once again trod down the old path of fascism. This made him suspect that some Japanese leftists would simply negate the history of the Japanese invasion. From his perspective, this standpoint of political correctness is too external and too removed from the common people. Therefore, there is no way that it could effectively enter history. Takeuchi does not have the ability to propose an intellectual program. He only tried to establish the subjectivity of Japanese society by approaching the way of thinking of most of the Japanese people. He always held the belief that a society that does not have subjectivity can be easily and dangerously manipulated by politicians. Therefore, a lack of social subjectivity is threatening to international society. Thus during the war, he tried to use some possibilities among the intellectuals who formed the conditions of the war. He hoped to create certain elements of subjectivity and persevered in the project of "taking chestnuts from the fire." Because of this, Takeuchi is far from political correctness and far from the possibilities of pre-given knowledge and a priori sensibilities.

The publication of the volume of Chinese translations of Takeuchi's *Overcoming Modernity* seems to have provoked quite a reaction. The readers of the book were primarily scholars and graduate students in Chinese and history departments and many people began to cite it. But these reactions were only superficial and in the citations it was extremely difficult to find instances of creative and productive imagination. However, a bit later, some "critics of Takeuchi" grasped the risky part of his intellectual practice and plainly stated that he was a fascist. Among the essays criticized were not only "The Great East Asian War and Our Resolution," in which he endorsed Japan's war against the United States, but also the round-table discussion on "Overcoming Modernity," which played a similar fascist tune. They even criticized his famous work, "Lu Xun."

In comparison to the content of these criticisms, it is much more important to pay attention to their form. One should say that these criticisms basically reflect the insistence on political correctness and then judge a concept to be incorrect. The standard is thus whether or not a concept accords with the end of history.

Once an event has already taken place, we can then look back and examine the whole process and especially look closely at those elements that are not in accord with the outcome of this event. From this perspective, we can easily identify the "mistakes" at every level. The historical interpretation that we see most often, "namely an analysis of causes and effects," relies on precisely such a logic. From a superficial point of view, an analysis of causes and effects seems to proceed from cause to effect, but actually, if we pay attention to the process of thought, there is no way to proceed with such an analysis unless one moves from an effect back to a cause. If one inhabits an event that is in the process of developing, the situation is completely different. An analysis of causes and effects is completely useless in such a situation. In the complex unfolding of an event, it is extremely difficult to make a correct judgment. Because of incomplete and unreliable information, any judgment must have the character of a "bet" or "gamble." Because it is a risky resolution, perhaps correct and mistaken each are fifty percent valid and there is no way to move from the effect to judge what the "cause" is; not to mention that between correct and mistaken, there are even more "gray areas" and each point has many possibilities.

With respect to history and the student of history, there are so many camps and so many definitions that it is now impossible to determine in a simple manner "what is history and what is not." However, that said, we can still find a basic axis from which to discuss a method for reading and discussing history. First, history has its own limitations. Any historical event cannot leave the situation of its period. Using contemporary standards to judge history is anti-historical. In other words, a historical attitude aims to deal with the difference between the present and the past and does not force present standards on the past. However, this does not mean that one unconditionally recognizes history's own limits; it does not mean that "the actual is the rational." To analyze historical events and personages is one of the tasks of the study of history. The only thing is, such an analysis must respect historical principles and cannot make them up at will. This is precisely the place where historical studies require extremely skillful use of imaginative powers. Secondly, there are a number of different methods for studying history. For example, one can study a person or not study individual people, one can study a long or short period, and one can take an event as an object of study or not study events. However, regardless of what the case may be, the study of history must broach the question of quality. One cannot simply make specious theories. Thirdly, the study of history is an action that "transcends experience" and it cannot rely on "common knowledge." Just as no one can grasp the weather in order to control a flight, the process of history stems from a balance of a number of different forces, and cannot be grasped by the direct experience of an individual. The development of a society's interpretation expresses whether this society's mind has imaginative power and creativity that goes beyond direct experience. One can say that the profundity of reading history depends on the sedimentations of the intelligence of the whole society. Put differently, the quality of historical studies has a direct impact on the intelligence of society. Thus, should not reading history profoundly imply a standard? Given the fragmented nature of the discussions about the study of history today, it is difficult to have a unified standard. Nonetheless, at least we can say that reading history profoundly and forcing already existing concepts and ideological thinking onto history are two fundamentally opposed methods. Reading history profoundly requires that one make a discovery outside ready-made conceptual frameworks.

What is a "discovery" and how can one make a "discovery?"

When an event is unfolding, the effects of each choice cannot be evaluated at the same time. One must wait until the event has come to its end before one judges whether a specific choice was correct or misguided. This means that when one makes a risky choice, there are at least two possible qualities—from the perspective of the end of the event, it is either correct or mistaken. If one only grasped the end of the event, then one could only induce from there each juncture of the process and judge whether it was correct or mistaken. But in each moment when one chooses from the process, the process will be ossified, as if everything was determined by the end and there was only one choice. In this manner, the extremely tense action of the moving subject making a choice becomes simplified into a simple step which is either correct or mistaken. This simplification takes place because the end fixes the moving subject which chooses.

The process of history is just like a person who plans to take an airplane and encounters a thunderstorm. This person must wait until the thunderstorm abates before taking off. If tumultuous and ever-changing weather can symbolize the process of history itself, in a process that no individual can control, the action of the subject is extremely risky and his "choices" are limited. At best, the subject can only decide whether to abstain from voting and be responsible for the effects of abstaining or not abstaining from voting. Opposed to this limited choice is the position of playing it safe and not entering the field of action. This option avoids risk, but makes it impossible for the subject to enter history. At the end of each period, there will always be some wise men after the fact, who look at the results of an event and make judgments. The most salient characteristic of such judgments is that they overlook the indeterminacy of the process of the event. They falsely turn history into a fait accompli. If history is already determined, then the risky choices of the subject have no meaning. The only choices that have value are those which are in accord with the end of history.

This is of course an old problem in history. However, this old problem is forever new, since it has not been resolved. It remains a problem. In the history of philosophy, almost no one still adheres to the view that history is determined and develops in a linear fashion. But this does not stop people from making judgments about history as wise men after the fact. Think about this. If you make a judgment about a period of history, does this not imply that you take yourself to be more intelligent or more "advanced" than the historical characters of that period? More importantly, the manner of reading history as wise men after the fact overlooks the indeterminacy of history and takes history as a fixed object. At this point, starting from the outcome of an event and judging whether it was correct or mistaken becomes the main aim of reading history. This "politically correct" way of reading history has a serious shortcoming, namely that it is difficult to make discoveries why reading history. There is a theory that if one finds mistakes in the choices of historical characters, it will lead us to avoid making similar mistakes today. This is an extremely superficial understanding because even if history could repeat itself, this would be in the manner of a tragedy becoming a comedy. How could you avoid making "the same mistake" when the nature of the play has changed? Regardless of whether one is judging historical characters or imitating them, these are both ahistorical ways of reading history. This is because both these attitudes confirm the absolute validity of the outcome of history and negate the fluidity of the historical process.

How then does one find "discoveries" in the midst of reading history? I believe that reading Takeuchi Yoshimi provides good training.

In his works, Takeuchi never took history to be something fixed and unchanging. At the same time, he did not simplistically believe that the discussions of intellectuals could directly influence social reality. With such presuppositions, Takeuchi refuses the attitude of being a "wise man after the fact" and he also refuses the strategy of avoiding risk for the sake of safety. In his whole life of discussion and debates, many of his ideas still seem to have involved questions that continue to exist today, including those politically incorrect questions that progressive intellectuals do not want to touch. In Japan, Takeuchi encouraged at least two generations: those who grew up during the postwar period and those who were born after the war. The latter were the important part of the student movement during 1968. Even today, if one talks to these people from these two generations, they will still say that Takeuchi formed the starting point for their understanding of China and the Chinese revolution.

However, if one reads the works of Takeuchi, you will find that his real perspicacity does not consist in telling his contemporaries what they should do; it does not consist in prescribing medicine. Rather, his perspicacity lies in his acute sensitivity to the fact that history moves. Takeuchi tried to sketch the contours of a modern Japanese history full of indeterminacy and full of transformative junctures. Moreover, compared to people who tried to affirm the junctures of rational transformation for their Japanese contemporaries or who had been trained in Western theories of modernity and were accustomed to using Western theory as a standard to judge Japanese experiences, Takeuchi's work was riskier. This makes it such that we have some documents to examine the complexity of that period in history. Moreover, Takeuchi's contemporaries witnessed the real relationship between Takeuchi's writings and this period that is already gone.

Takeuchi's vision of China does have the problem of "idealization" and although the approach is different, the effects are similar to those produced by some of the Japanese progressive intellectuals who "idealized" the West. They all used a certain strategy in a particular historical period in order to reevaluate their own culture. If Takeuchi's vision of China lacks a grasp of the fluidity of Chinese history, then it will be difficult for one to separate his position from the common people preaching "friendship between China and Japan." Actually, in the 1950s there were some malicious rumors that he was "being subsidized by the Chinese Communist Party." Later some capricious scholars who did not read carefully, focused only on his idealization of China and totally overlooked his "vision of history." However, Takeuchi's vision of history is not like any of his "preconfigured" correct or incorrect conclusions. That is, he did not present a fixed position on his vision of history. We can carefully construct his vision of history only by carefully reading his texts and establishing their historical context.

This requires a certain skill in reading history and Takeuchi is the best object to help us train to read history.

In Japanese contemporary intellectual history, there are few other thinkers who have been evaluated in so many different ways as Takeuchi. In Japan as well as in the international scene, from Takeuchi's period to the present, it is extremely difficult to find a complete picture of Takeuchi. Some people claim he was a pro-Chinese intellectual, some say he was a modern proponent of Japanese pan-Asianism, some say he was an "antimodern" nativist, some claim he was another type of modernist or postmodernist, some say he was a nationalist, some say he was a profound cosmopolitan, some say he was a leader of the democracy movement and the movement against the imperial regime, and some say he was an advocate of the imperial system. The list goes on.

The reason why so many conflicting interpretations of Takeuchi abound is that he was different from his contemporary rightist and leftist intellectuals. It is impossible to categorize him in terms of the political left or right, but at the same time, he never was outside real political life. He dared to deal with those questions that only conservative and rightist intellectual touched, such as the idea of "Asianism" (in Takeuchi's view, his theme in intellectual history is not "the Greater East Asia Co-Prosperity Sphere" but rather something that once existed as an ideal and then was qualitatively transformed into the military ideology of "Asianism"). He also wished to seriously question the boundaries of thought of the "group of Japanese romantics which had direct connections with the ideology of Japanese nativism and the emperor system. Through this questioning, he aimed to develop a strategy of attack that would "go deep into the place where the object was produced and develop an internal critique." Moreover, he stressed that this is "the only way to defeat the enemy."[2] (According to Hashikawa Bunzō, this essay that Takeuchi wrote in the early 1950s, was the real juncture of a critique of the Japanese school of romanticism. Moreover, Hashikawa's *A Prefatory Attempt to Critique the Japanese School of Romanticism*, remains today a classic study of the subject.) In a context where East Asian intellectuals almost instinctively accepted the Western ideal of "democracy," Takeuchi bravely compared "democracy" to the Meiji emperor's

2 Takeuchi 1980–1982.

"Five Point Oath." Such statements provide materials for quoting Takeuchi out of context and have caused some scholars to attack one point without looking at the rest. On the other hand, those who want to use Takeuchi as an intellectual resource often avoid facts that are not beneficial to Takeuchi's image and use the model of "his achievements are seventy percent"[3] and his misdeeds make up only thirty percent to focus on his main contributions. One could say that in both China and Japan, our intellectual training has not provided us an effective means of dealing with Takeuchi. In the end, if a person has been split up into a number of different and contradictory elements, the problem is probably not in this fragmented object, but in our own epistemology.

It is precisely in this basic context that Takeuchi has yet to be dealt with in a complete manner. With respect to those who quote him out of context and those who superficially support him on the "seventy-thirty" model, such practices cannot end up becoming "Takeuchi studies." The problem does not lie in whether one establishes a perspective of "Takeuchi studies"; rather, it lies in whether we have just missed an opportunity for epistemological adjustment. In the scholarly milieu of China and Japan, people have recently begun to pay attention to Takeuchi Yoshimi. But they have not begun to adjust their epistemological frameworks. One should rather say that, to a large extent, Takeuchi has been absorbed into their existing epistemological frameworks. This framework is actually an evolutionary framework and it gives scholars the false impression that they are at the highest point of history, a point from where they can get a bird's-eye view and judge everything about previous characters in history. With this attitude, it is as if one does not need to think about the choices and resolutions of past people in a historical process. The problem with this framework is not only that it leads to making one's epistemology of history idealist and ideological. More importantly, it leads to an idealized concept of reality. Contemporary history is still history. The idealized conception of reality precludes understanding Takeuchi's mode of reading history and also blinds us to contemporary history. The mode of thought that tries to find contributions or mistakes in Takeuchi will also only find fixed "conclusions" in the analysis of contemporary society. Moreover, in some cases, because one cannot grasp the fluidity of the situation, this

[3] Translator's note: Here the author invokes Mao Zedong's statement about Stalin based on the seventy-thirty ratio, namely that Stalin was seventy percent correct and thirty percent wrong.

method of only looking at conclusions makes one oblivious to the real problem. As a result, one displaces responsibility onto unrelated people or events.

In modern Japanese intellectual history, an evolutionary vision of history has been dominant and a group of Marxist scholars also have this tendency. Among these, especially those who work on Japanese history self-consciously used Japanese-style Marxist political analysis to examine history, and also used it study themes linked to periodization. After Japan lost the war, in the specific intellectual context of Japan, the influence of Japanese Marxism was pervasive in the academic realm. This created a situation in which the Japanese Communist Party, which did not play a leading role in actual society, began to play a leading role in academia. From this perspective, looking at the role that the Japanese Communist Party as a marginal party played in the formation of scholarly thought is a topic that requires extreme sensitivity to proportion. I cannot fully discuss this issue here, but will limit myself to pointing out a basic fact: in the realm of postwar scholarship, in the intellectual context of postwar Japan, the scholarly position of Marxism in the background of the Japanese Communist Party (of course, from a political perspective, there were a number of rifts in this party and it was not completely unified) attracted a number of extremely talented intellectuals. These intellectuals also produced a number of important works. Moreover, because these intellectuals needed to deal with a specific ideology related to party politics, they met with more problems when they dealt with scholarly issues. Such problems were caused by the contradiction between the creativity and freedom required when dealing with scholarly problems and the type of judgment related to the real politics associated with the struggle between political parties. Hence these talented intellectuals attempted to use "science" in order to unify and resolve this contradiction. In addition, some of them were able to free themselves from political ideology when producing scholarship. Two examples are Ishimoda Shō and Tōyama Shigeki. Such types of intellectuals reflected on the problem of idealization in Japanese Marxist reflections on society. But, in the end, most scholars could not develop the type of marginal consciousness that such an extremely difficult form of reflection demanded. So most Japanese Marxist historians were content to "criticize from the higher plane of principle" in their intellectual production. Moreover, even with respect to the most sophisticated historians, when they dealt with extremely sensitive political issues, they would at times choose the path of "political correctness." In this way, certain problems that should have been discussed at length were prematurely cut off.

Takeuchi got into debates with many different types of intellectuals, including with Japanese Marxists. In this context, he was criticized and made some counter-criticisms. Of course, because in the Marxist camp it was common to develop an ideological mode of scholarly discourse, Takeuchi had to face some non-scholarly "great criticisms." These criticisms opposed scholarly common knowledge and were caused by a specific political context and by "misreadings" that were popular in society for a certain time. (More precisely, these "mis-readings" are really "failures to read." The critics do not have the patience to read the words in their object of criticism. They just quote some passages out of context and label someone as guilty and then attack vehemently.) These attacks clearly infuriated Takeuchi and this anger influenced the scholarly depth of his response. This is extremely regrettable. Among the various texts containing Takeuchi's debates, the most representative was published in 1966—"On the Responsibility of Scholars."[4] This essay was originally written in response to some of the queries of Tōyama Shigeki. Unfortunately, because a number of "great criticisms" by other people of a lower level gripped his attention, he did not make his dialogue with Tōyama's the main part of the essay. This influenced the completeness of this essay. However, even in this case, if one reads this essay together with Tōyama's, one can still establish an important ground for discussion.

In the early 1960s, Takeuchi once commemorated the Meiji Restoration and intended to revisit this event. He believed that Meiji nationalism was a "state without a nation" and thus a failed project. This is intimately related to the later turn to militarism. Through this new discussion of the Meiji Restoration, he hoped to uncover resources in Japanese intellectual history that could be transformed to guide the future. In other words, he wanted to find elements in what was simplistically called "extreme militarism" in order to attack militarism. However, his hope was not understood. On the contrary, in literary circles, the trend to completely affirm the Meiji Restoration dominated, to the extent that the government and the public collaborated to "Commemorate the Centennial of the Meiji Restoration." Takeuchi could only announce that he wanted to cancel his plans for commemorating the Meiji Restoration and withdrew from

[4] Takeuchi 1980–1982b.

the process. In 1965, the Marxist historian Tōyama Shigeki published an essay entitled "Studying the Meiji Restoration and Social Responsibility." This essay takes the study of the Meiji Restoration as its guiding thread in order to pose issues about principles of scholarship, the social responsibility of intellectuals, and intellectual collaboration. He extremely perceptively positioned Takeuchi's proposal with respect to the commemoration of the Meiji Restoration in the following manner: "Takeuchi Yoshimi anticipated that when 1968 arrives, the government and the leftists would both enthusiastically participate in the centennial commemoration of the Meiji Restoration. Hence he proposed that he take over the power to control the discourse and then reevaluate the Meiji Restoration."[5] As an exemplary historian in the Marxist camp, Tōyama had to fight two battles. On the one hand, he was aware of the paucity of results among Japanese Marxist history and the speedy demise of Marxism in literary circles. Thus he actively sought nourishment from non-Marxist intellectuals (including Takeuchi). Moreover, he subtly and indirectly criticized the strong tendency of parochialism and the neurotic application of labels within the Marxist camp. On the other hand, he also wanted to return the political potency of scholarship and thought to "belonging to a party" and advocated such belonging. (He stressed that belonging to a party is a choice within politics in reality and does not stem from the perspective of the small group that attempts to use the party for its own sake. Hence he especially separated belonging to a party from the struggles between various factions of a party.) Precisely because of this, Tōyama became one of the few Marxist historians whom Takeuchi respected. When Takeuchi wrote his response, he did not criticize Tōyama in the same way that he lambasted other Marxists, and this shows his sincerely "sympathetic understanding" of Tōyama's position.

Takeuchi could "sympathetically understand" Tōyama because they both felt a sense of crisis in relation to contemporary reality. This means that one "realizes that in the study of history, the whole discipline has a tendency towards becoming waste, and because of this I think about the responsibility of scholars (I do not think one needs to say that it is a social responsibility)."[6] Takeuchi not only separates "the responsibility of scholars" and "social responsibility" into two different categories, he also mentions the problem of the "scholarly realm becoming a wasteland."

5 Tōyama 1965: 14.
6 Takeuchi 1980–1982b: 246.

These two points subtly separate his position from Tōyama's. Although Tōyama was reflexive and critical of the way Marxist historians attach labels to people, he still avoided a thorny question, namely whether one could directly use the model of real political parties for scholarship. Put differently, if one stresses the effect of political culture, does it imply that this culture has a direct and reciprocal relationship with real politics? Within the Marxist camp, there was always the label-attaching model of intellectual production. Later this model travelled beyond the Marxist camp and became a problem of leftists around the world. This has today already roughly had the side effect of producing the problem of "political correctness." Like the way Marxists faced a problem during the 1960s, "political correctness" also stems from a faulty epistemology. It takes scholarly activity simplistically as real political activity. It simplistically reduces the complexity of intellectual production to the "results" of real struggles and dualistic solutions (If you are not this then you must be that). Although Tōyama was an excellent scholar who looked at the indeterminacy and complexity of historical processes, he directly linked the scholar's responsibility to social responsibility. This actually intensified an instinctually dualistic thought pattern: one either makes social problems into problems of scholarship in order fulfill one's social responsibility or one leaves reality and hides in an ivory tower. This makes it such that when one sees those problems that go against "political correctness," one will instinctively suddenly go against one's training as a historian and make a simplistic judgment.

From Takeuchi's point of view, this model of thinking cannot establish an effective tradition of thought and struggle. On the contrary, such a mode of thinking will only lead to the "scholarly world turning into a wasteland." When he encountered Tōyama, who also felt that the scholarly world was turning into a wasteland and wanted to turn this situation around, his feeling was perhaps more of futility than anger. This is because in the above-mentioned essay, Tōyama brought up some important epistemological questions and made some constructive suggestions. But at the same time, he used a paradigmatic attitude of political correctness to dissolve these problems. This is pardonable. In this essay, Tōyama attempted to deal with the problem of how Takeuchi had to annul his earlier plan because of a change in the situation: the commemoration of the centennial of the Meiji Restoration thus became a question of how one evaluates Asianism in modern Japanese history. In their contemporary intellectual context, such questions were basically discussed by Japanese conservative intellectuals, and Takeuchi's idea of "critically continuing" such ideas basically did not have a market. At the same time, because he proposed a parliamentary motion, he did not have time to develop his thoughts. (In 1965, in his essay "Impressions of 'Meiji Fever,'" he mentioned that at the same time that he withdrew from the project to commemorate the centennial of the Meiji Restoration, he traced the outlines of ideals through writing essays such as "Overcoming Modernity," "Japan and Asia," and "The Hope of Asianism.") More importantly, in the intellectual context of the time, people did not have the foundation to understand Takeuchi's outline and his method. Therefore, when Tōyama dealt with these questions, he was at a loss as to how to deal with Takeuchi. In his essay "The Study of the Meiji Restoration and Social Responsibility," on the one hand, he acknowledged Takeuchi's work and did not follow other critics in claiming that Takeuchi was "speaking for Japanese imperialism." He even correctly pointed out that "when one embarks on extremely creative thinking and scholarship, it is difficult to avoid misunderstandings and the malicious misuse of one's work."[7] But even as a Japanese Marxist who criticized Marxist ideology, Tōyama refused Takeuchi's plan to "take chestnuts from the fire." He wrote:

> Takeuchi Yoshimi has consistently adhered to the following idea: In Asia, Japan is the only imperialist country and after the war, it also restored the imperialist line. This situation has caused some problems for Japanese people. As citizens, the Japanese people have a very weak sense of subjective responsibility. Because of this, they lost the opportunity to excavate a possibility from the situation. In order to get clear about the reason behind this problem, he traces history. He uses his skill to examine when and under what conditions the Japanese chose imperialism and lost the ability to recognize Asia. I really understand Takeuchi's consciousness of the problem. However, when one begins from such a consciousness of the problem and defining Asianism as "pointing to the solidarity of various Asian nations (without asking whether this is through invasion)," this makes one question the meaning of such an understanding of the problem.[8]

This passage represents Tōyama's ideas in the same essay and he appears to contradict himself. With Tōyama's profound skill as a historian, he should not so blatantly misread texts. In Takeuchi's "Japanese Pan-Asianism,"[9] his definition of Asianism is not so simple, not to mention that he was discussing Asianism and solidarity in a historical context and was not making

---

[7] Tōyama 1965: 15.
[8] Tōyama 1965: 27–28.
[9] Takeuchi 1980–1982c.

a value judgment. Moreover, Takeuchi forcefully stressed that one makes the distinction between early and late Asianism only after late Asianism became linked to imperialist expansion. By making this point, he appears to uncover the "solidarity" concealed in Japanese Asianism. The difference between Takeuchi and Tōyama lies in that when tracing history, one focuses on what elements should be criticized, while the other focuses on what elements have possibilities. Tōyama makes the above judgment because of his epistemology. In the same essay, when he criticizes the Marxist camp's methods of ideological struggle through attaching labels, he also explains his epistemology:

> In the process of a debate about relationships and gaps between one's subjective intent, the realization of this intent and the social reaction to theory, one must approach the subjective intent, enter the content of the given theory, and develop an internal analysis and critique. If one does not exert sufficient efforts in this direction, and only attaches labels outside of theory and thought, then one will make ideological enemies of people who should have been allies. This opens a path for the enemy's wicked intentions and makes them fixed. This is the objective effect of such actions.[10]

But at the same time, he stressed something else. That is, when one analyses, compared to the subjective intent of the object, the objective effects that the object brings about are more important. Because of this, Tōyama could not avoid a problem that he could not adequately treat. When he uses the principle "objective effects are more important than subjective intent" to make judgments, he cannot make sense of the various historical objects where the objective effects are out of joint with respect to their subjective intent. Hence in the end, he names his heroes based on defeat or victory. This is especially the case with people who have been negated posthumously or after their historical period. Tōyama actually does not want to apply his principles of approaching subjective intent and "developing an internal analysis and critique." Among Marxist historians, there is a basic epistemological standpoint that becomes effective in practice (not just giving lip-service), namely they are never able to "learn from one's enemies." To learn from one's enemies implies that before one makes moral judgments one must sympathetically interpret, regardless of whether the object is a friend or foe. This sympathetic interpretation completely faces the indeterminacy of the historical process, completely considers the internal logic of the object, and then makes the appropriate

[10] Tōyama 1965: 16.

evaluation. In this process, "naming heroes based on victory or defeat" greatly simplifies the depth of one's analysis. This is because with respect to indeterminate elements in a historical process and in situations which are not transparent, such judgments become insignificant. People who make such judgments only discuss phenomena that conform to their value system based on results; they do not consider those phenomena that appear negative with respect to results or effects. Consequently, they conceal many elements with positive potential. This is the bottom line of Japanese Marxist scholarship in history. Tōyama is no exception. Although he did not so callously ignore texts and quote out of context, just like other Japanese Marxists he did not want to develop a charitable interpretation when faced with a historical object with a different political standpoint.

When Takeuchi responded to Tōyama, he obviously felt a sense of both anger and futility when faced with such an epistemological framework. He wrote: "What can I do to make Mr. Tōyama understand?" He felt that he must establish some common problems in order to ensure that there was a meeting point for the discussion. In his essay "On the Responsibility of Scholars," he made the following basic point, in order to develop a common problem:

> In other to truly examine something, regardless of whether the object is a devil or a snake, one must follow the necessary procedures, read the necessary materials and hear what must be heard.... The Marxist historians' concepts of "social responsibility" or "belonging to a party" cannot be denied. However, before stressing these things, I propose that one must first regain respect for scholarship and the responsibility of scholars.[11]

Here Takeuchi speaks of his real impressions. When countering scholars who quote his work out of context in order to give him a bad name, he sarcastically wrote: "If one breaks a text into pieces and then puts it back together in different ways, one can do all kinds of things." Faced with all different types of misinterpretations, Takeuchi could only stress the principles associated with quotation. "This principle involves finding out one's object's main point. One has just to glance at the quotations and one will know the level of the person doing the quoting. If you meet someone who misquotes in order to make his point and distorts your original meaning, this is a disaster."[12]

---

[11] Takeuchi 1980–1982b: 266, 268.

[12] Ibid., 249.

If such disasters are abundant, then there is no more respect for scholarship. This is true in China and abroad, from ancient times to the present. Throughout Takeuchi's life, he met many such "disasters" and quite of few of these where caused by leftist misinterpretations.

However, Takeuchi knew that Tōyama was not like these other people. Therefore, Takeuchi took Tōyama as a serious partner in a conversation. Thus he did not scold and make fun of him as he did when dealing with others. One could say that Takeuchi's response to Tōyama could lead to a more profound understanding of the problem because Tōyama was a scholar who also respected the rules of scholarship. Moreover, precisely because of this, this dialogue is more stimulating for people from later generations like us. For example, in response to Tōyama's claim that with respect to Japanese pan-Asianism one should consider both solidarity and invasion, Takeuchi responded by asserting: "I say the opposite. I am suspicious of a starting point that separates solidarity and invasion."[13] This suspicion has its origins in Takeuchi's historical epistemology. Using a binary opposition to separate qualitatively different or even opposed elements of a historical process can only be effective when one stands at the endpoint of this process. On the other hand, if one does not take as one's point of departure the end of a particular process or stage in history, then it is difficult to find meaning in such binary oppositions. This is because in a historical process, qualitatively different elements must necessarily combine together. One can only separate them using a historical method. Takeuchi described such a method in the following manner: "My problem consists in examining different ways of combining solidarity and invasion."[14]

Takeuchi criticized Tōyama and continued: "I felt that historians read historical documents by taking shortcuts. One cannot ask them to penetrate behind the paper, but at least their eyes should be able to go into the paper." Even if one only grasps the superficial dimension of words and is oblivious of the constraints of the "paper," one can still engage in scholarship, but one will need to rely on elements outside of the text, which are already given. For example, one could rely on ideologies prevalent in society, or a certain consciousness of crisis at a given period, or various specific hopes in society, and so on. This way of doing scholarship encourages a superficial reading of the text and thus there is no obstacle to quoting out of context. Compared to malicious misquoting, this flippant manner

[13] Ibid., 272.
[14] Ibid.

of reading might not lead to "disaster"; however, it at least constructs a field of intellectual production in which disasters can be "legitimized."

Nonetheless, this type of criticism is inappropriate when applied to Tōyama. Takeuchi's difference from Tōyama is a difference in epistemology and does not concern a difference in reading skills. Takeuchi's further remarks really get to the heart of the matter.

> According to Tōyama, people's motivation and means can be clearly distinguished. Moreover, he believes that they can be grasped as a transparent substantive totality. But in my opinion, human's motivations and means are fluid, situational, and form an object in which it is difficult to clearly separate the self from the other. In Tōyama's view, history is a heavy given object, but for me history is a constructed object that one can trace and dissect. I believe that this is the difference between Tōyama and me.[15]

With respect to reading history, the question of whether one enters history is not mere wordplay. It refers to the idea that the perspective and standards of reading history respect different models. With regard to participating in one's contemporary history, this is not merely a subjective intent. This project also is constrained by epistemology and the object. Takeuchi spoke of an object that is "fluid, situational and in which it is difficult to separate self and other." One can understand the above point only when one is clear about the opacity of historical processes and therefore stresses that every decision of the subject is ambiguous. Otherwise, Takeuchi's phrase will just be another "theory." When history is seen as a heavy "given object," it is hard to avoid a situation in which progressive intellectuals can only take the attitude of "external criticism." "Political correctness" is entwined with the question of what type of attitude one must take towards this "given object." Political correctness is not concerned with how one analyses the object and changes it from the inside. In the end, history is not only about the past, it connects to the present and points to the future in a nonlinear manner.

In the mid-1960s, when his health was not as good as he would have liked, Takeuchi did not develop his debate with the excellent Marxist historian Tōyama to the level he should have. This is a great shame. This of course shows the limitations of Takeuchi's mode of scholarship. However, more importantly, the level at which the debate remained is also connected to the limited epistemological self-awareness of Japanese historians (and not only Marxist historians). Actually, there is another historian

[15] Ibid., 273.

who influenced the realm of Japanese historians but could not change the state of Japanese scholarship, namely the excellent historian Ueno Senroku. The ideological method of reading history was not a practice limited to Marxists. The method of thinking that "looks at the process from the result" has always been the dominant method in reading history, especially in periods filled with crisis. In such periods, history is greatly simplified and made transparent. Situations are greatly fixed and externalized. This type of method actually easily finds a social base. The old problem that Takeuchi posed appears yet unresolved. One cannot deny the importance of "social responsibility" or "belonging to a party" but before stressing these things, one must first regain respect for scholarship and responsibility for scholarship. To put it at a fundamental level, when we today revisit the problem of epistemology, it is not to understand Takeuchi, and it is not even just to profoundly read history; rather, the problem of epistemology is directly linked to our scholarly production both in the present and in the future.

Translated by Viren Murthy

References Cited

Takeuchi Yoshimi (1980–1982a), "Kindaishugi to minzoku no mondai" (Modernism and the problem of the nation), in *Takeuchi Yoshimi zenshū* (The complete works of Takeuchi Yoshimi), vol. 6. Tokyo: Chikuma shobō.

—— (1980–1982b), "Gakusha no sekinin ni tsuite" (On the responsibility of scholars), in *Takeuchi Yoshimi zenshū* (The complete works of Takeuchi Yoshimi), vol. 8. Tokyo: Chikuma shobō.

—— (1980–1982c), "Nihon no ajiashugi" (Japanese Pan-Asianism), in *Takeuchi Yoshimi zenshū* (The complete works of Takeuchi Yoshimi), vol. 8. Tokyo: Chikuma shobō.

—— (2005), *Overcoming Modernity* (*Jindai de chaoke*), Sun Ge ed., Li Dongmu et al. (trs.). Beijing: Sanlian shudian.

Tōyama Shigeki (1965), "Meiji isshin kenkyū no shakai sekinin" (The study of the Meiji Restoration and social responsibility), in *Tenbō* (Dec. 1965). Tokyo, 14.

THE BURDEN OF THE PAST AND THE HOPE FOR A BETTER FUTURE

CHAPTER FIVE

AN ESCHATOLOGICAL VIEW OF HISTORY: YOSHIMI TAKEUCHI IN THE 1960s

Takahiro Nakajima

> The Japanese people's experience of war cannot overcome *The Tale of Heike* (*Heike monogatari*) or *The Ten-Foot-Square Hut* (*Hōjōki*): this was Hideo Kobayashi's pioneering argument on the experience of war. We should not yield this glory to Kobayashi. In order to get out of the experience of war made into season words, our task is to discover the origin of the criticism of the state.
>
> —Takeuchi Yoshimi, " 'Sensō-taiken' zakkan" (Impressions of the "Experience of War")

Generalization of the Experience of War

When Yoshimi Takeuchi (1910–1977) reflected back after one and a half years on the movement protesting the revision of the Japan–United States Security Treaty, which he regarded as the "mass movement of 1960," he considered it to be the "fruition of the experience of war":

> With regard to the generalization of the experience of war, I think the method of *overlapping* the experience of war upon that of the postwar period has been pretty effective, but now it is time for us to take a further step and think of the *problem of method* in a more definite way. We already have the common experience of 1960. I think it is possible to use a method of considering this experience to be the *fruition* of the *experience of war*, and tracing back from there to the experience of war in a reverse way. I think it is not only possible, but also necessary. We should begin to explore this method, for otherwise the experience of war might be decisively aborted, and we cannot be emancipated from being a *captive of naturalism* despite our every effort. We could think that the experience of 1960 was an outbreak after fifteen years of what should have happened during the war. It was a type of resistance that should emerge in a period of a war against fascism. From this perspective, we could think reversely that the war had not ended and its experience continued until this period in Japan. Such an *intellectual operation* is not only possible but also necessary unless we accept the

> fatalism of naturalism. In this way, I think we can fill the gap between generations, and also reversely seize the moment to overcome naturalism.[1] (Every emphasis added by T. Nakajima, unless otherwise stated).

What is important here is to communicate the "experience of war" to the younger generation which did not experience the war, by making it a "true experience." Yet it is not easy, because actually there has always been a Japanese naturalism that refuses to let experience exist as experience, but considers it by means of "season words," as it is represented by Hideo Kobayashi; there has never been a "method" that would make such communication possible. "Naturalism" has continued without being destroyed either by the war or by postwar intellectuals. In this tendency, the "experience of war" has been "buried in the real-life experience," and only discussed as a "product of narcissistic and feudalistic consciousness of beauty."[2] "It was a matter of course that the younger generation was reluctant to see" the "experience of war."[3] This is the very reason why Takeuchi thought it necessary to invent a new method (and a new consciousness of beauty) different from the "naturalistic way of recognition and consciousness of beauty."[4] Without it, it will not be possible to deal with the war intellectually.

In this deadlock, it was the protest movement against the revision of the Japan-United States Security Treaty as the "common experience of 1960" that became a moment to refuse the "experience of war" as "buried in real-life experience" and "naturalism." Takeuchi, in the above passage, considered the "experience of war" to be an "outbreak after fifteen years of what should have happened during the war" or a "type of resistance that should emerge in a period of a war against fascism," and declared that the "war had not ended until that period in Japan." Then he continued that by passing through this "common experience," the "experience of war was established as an experience for the first time after the war ended fifteen years ago, or at least it found a clue to be established."[5]

What Takeuchi pursues here is not to find out a new method somewhere else and apply it to the experience. Rather, he seems to deepen the method of "overlapping" the present upon the past by an "intellec-

[1] Takeuchi, "On the Generalization of the Experience of War" (Sensō-taiken no ippanka ni tsuite) (November 1961), in Takeuchi 1980–1982: 31–32 (emphasis added).

[2] Ibid., 30.

[3] Ibid.

[4] Ibid.

[5] Ibid.

tual operation," which requires a historical perspective of "tracing back" from the present to the past. We may define this type of historical method as a hetero-genealogical historiography. However, what are the conditions needed for this method? Focusing on this point, we will trace back Takeuchi's discourses on the method of "overlapping."

The Method of "Overlapping"

For example, let us take a look at Takeuchi's paper titled "Overcoming Modernity" (November 1959), which he wrote a year before the outbreak of the "common experience of 1960." Postwar Japan has been haunted by the roundtable discussion "Overcoming Modernity" in 1942, which Takeuchi described as something "elusive like a specter, yet problematic for living people."[6] It has never been exorcized and has appeared repeatedly in different forms without being subjected to repentance. If "Overcoming Modernity" were a simple "relic of the past," "it would be enough to let the past bury it."[7] However, if it is haunting us as a specter even now, we must inherit it *as a legacy*, and be responsible for it *as we would for a tradition.* In order to do so, we should "examine what problem 'Overcoming Modernity' had, how it resolved the problem concretely, or how it did not."[8]

In the first place, "Overcoming Modernity" was a "concentration of aporia in modern Japanese history."[9] Its intellectual task was to resolve "such essential dichotomies in the tradition as reactionaryism and restoration, royalism and exclusionism, isolation and the opening of the country, nationalism and civilization, East and West."[10] But, in fact, it did not reach a resolution, and it ended with merely a "dissolution of aporia"[11] and "could not become even an ideology of war and fascism."[12]

In that case, how can we inherit this polemic "Overcoming Modernity," which ended up in such "failure"? Takeuchi argued that in order to do so, it is necessary to trace back through history and resolve the aporia.

---

6 Takeuchi, "Overcoming Modernity" ("Kindai no chōkoku), in Takeuchi 1980–1982: vol. 8, 6.

7 Ibid., 12.

8 Ibid.

9 Ibid., 64.

10 Ibid.

11 Ibid., 65.

12 Ibid., 7.

However, it is not enough to trace back just chronologically. If we do so, then it will be a simple "view of history based on a flowing model," which situates causes according to effects, and makes a "history of literature" or a "history of philosophy or thought" by reversing this cause-and-effect relationship.[13] Rather, it is necessary to trace back through history *in a different way*. Such a way constitutes the method by which we overlap plural dates and moments upon each other and connect them from an allegorical perspective, and save the past in the present.

Let us examine this concretely. Takeuchi began the paper "Overcoming Modernity" by introducing criticism of "Overcoming Modernity" by Makoto Jinna and Hideo Odagiri, and he concluded it with the criticism by Jun Etō. This was because Takeuchi thought that the real figure of "Overcoming Modernity" could not be captured unless it passed through the "contemporary situation of problems" raised by such postwar critics:

> Now, following the order of discussion, we must examine "Japanese Romanticism," which remains to be considered lastly, but I would like to think of it not by deducting it from [Yojūrō] Yasuda's thought but by concentrating on the role it played in the discussion of "Overcoming Modernity." In other words, *I would like to consider Yasuda by means of what Yasuda left behind*. This examination has as a matter of course something to do with exploring the intellectual origin of "Overcoming Modernity."
>
> In order to prepare for it, let us return again to the contemporary situation of problems. *By tracing back through history from here, we can trace it more easily*.[14]

Yojūrō Yasuda did not participate in the roundtable discussion "Overcoming Modernity." But it is far more important to "consider Yasuda by means of what Yasuda left behind" than to consider "Overcoming Modernity" by "deducting it from Yasuda's thought." In the same way, Takeuchi is trying to examine "Overcoming Modernity" from a perspective of what it left behind in the postwar period.

We must here pay attention to an issue: Takeuchi at the same time uses another "overlap." To put it in another way, Takeuchi traced the origin of "Overcoming Modernity" in the other moments: the situation of the "Sino-Japanese War" as of 1941; the "double structure of the state of Meiji"; the "unequal treaties in the Ansei period." In this way, he understands "Overcoming Modernity" as a place where the contradiction in these moments blasted out. What was questioned in each moment

---

[13] Ibid., 10.
[14] Ibid., 52.

was the "double structure" of modern Japan, by which resistance to the West becomes at the same time suppression of Asia. The challenge of "Overcoming Modernity" was to resolve the "unresolvable" "eternal war" since the Sino-Japanese War, by "driving a wedge into the double structure of the war and transforming its character."[15] However, it was not resolved; it just ended up with the "dissolution of aporia," as mentioned above. It is "today" that "we must discover when logical mistakes were made, by tracing back history in a reverse way."[16] What was reflected in Takeuchi's eyes was an overlapping of the moments of aporia, which had not been resolved from the "unequal treaties in the Ansei period" until "today." An allegorical perspective which penetrates this overlapping: that was Takeuchi's method.

Eschatological View of History: Yoshimi Takeuchi and Walter Benjamin

Walter Benjamin declared: "in allegory the observer is confronted with the *facies hippocratica* of history as a petrified, primordial landscape."[17] For Benjamin, the method of allegory was nothing but a new attitude toward history and a particular perspective on the past.

> Everything about history that, from the very beginning, has been untimely, sorrowful, unsuccessful, is expressed in a face—or rather in a death's head . . . This is the heart of the allegorical way of seeing, of the baroque, secular explanation of history as the Passion of the world; its importance resides solely in the stations of its decline.[18]

The method of allegory separates out the "stations of decline" as "*facies hippocratica*" and requires "history as the Passion of the world," which ties them together. In order to do so, it is necessary to see objects in the "allegorical way of seeing" and make them "dead."

Thus, how is a history to be written that ties together the moments taking on *facies hippocratica*, which was separated out by this way of seeing? Benjamin, using the term "historical materialism," when facing his own death said:

---

15 Ibid., 49.
16 Ibid., 51.
17 Benjamin 1998: 166.
18 Ibid.

> Materialist historiography, on the other hand, is based on a *constructive principle*. Thinking involves not only the movement of thoughts, but their arrest as well. Where thinking suddenly *comes to a stop* in a constellation saturated with tensions, it gives that constellation a shock, by which thinking is crystallized as a monad. The historical materialist approaches a historical object only where it confronts him as a monad. In this structure he recognizes the *sign of a messianic arrest* of happening, or (to put it differently) a *revolutionary chance* in the fight for the oppressed past. He takes cognizance of it in order to blast a specific era out of the homogenous course of history; thus, he blasts a specific life out of the era, a specific work out of the lifework. As a result of this method, the lifework is both preserved and sublated *in* the work, the era *in* the lifework, and the entire course of history *in* the era. The nourishing fruit of what is historically understood contains time in its *interior* as a precious but tasteless seed.[19]

What is necessary is a "constructive" approach by which the past should be emancipated from historicism and separated as a monad, and the "entire course of history [should be preserved and sublated] *in* the era."

Benjamin's "historical materialism" is not different from Takeuchi's hetero-genealogical method of "overlapping." For example, consider Takeuchi's comment on Taijun Takeda's *Sima Qian: The World of Shiji* (*Shiba Sen—Shiki no sekai*):

> Even though history may sometimes seem to be stagnant, or the trajectory of its circulation to be immovable, it might reserve inside an unexpected triggering power. If more attention had been paid to this point, it [Takeda's work] could have emphasized more not only continuation but also revolution. It must not be unfair to imagine that if *Sima Qian* had been written by supposing *an apocalyptic world behind, which opposes against a continuous world*, it would have been completed with more profoundness as a work. Of course, this is an excessive expectation. Considering the period it was written, he did the best in order to oppose *the view of history based on a flowing model or the model of an unbroken line of emperors.*[20]

Taijun, opposing historicism, picked out moments of "stagnancy" in history and illustrated its structure. Takeuchi demands here an "eschatological view of history" penetrating the moments, which further pursues the possibilities of Taijun's discussion.[21] Takeuchi's comment here shows that he was ultimately trying to save every event whether major

---

19 Benjamin 2003: 396, emphasis added.

20 Takeuchi, "Note on Volume 9 of *Complete Works of Taijun Takeda*" ("Taijun Takeda zenshū" Dai 9-kan kaisetsu) (February 1961), in Takeuchi 1980–1982: vol. 12, 161.

21 Ibid.

or minor, just as Benjamin did by finding messianic time in "historical materialism."[22]

However, the messianic "Eschaton/End," when human beings are emancipated, will only arrive through people in the present who emancipate the past. In order to achieve this, Takeuchi had to save the "tradition" by drawing out as much power as possible from and even exhausting the "present," and tracing back through history.[23]

Becoming Other and Morality

Let us trace back ten more years from the paper on "Overcoming Modernity." Takeuchi had written up an article titled "The Energy of Literary Revolution" (February 1949) in Japan, a country without revolution, in the year when the Chinese Communist Party gained power. In this article, Takeuchi said with regard to the renovative nature of Chinese literature that "from the point of view of tradition and renovation, the Literary Revolution was a movement which revived tradition by denying it completely."[24] To put it in another way, true revolution is not to counterpose something new (literature or thought) borrowed from somewhere outside against the old tradition, but to hold tradition as something thoroughly identical to oneself, even though it is a bad tradition, or precisely because it is a bad tradition, and then deny it completely. It is not until one goes through this process that the self transforms itself into the other,

---

[22] In the preface of *Legacy of Disobedience* (*Fufukujū no isan*) edited as a "document of the 1960s," Takeuchi aimed to leave the document to be used in the future (or in the Eschaton/End) by juxtaposing minor and major facts. Takeuchi "wanted to be a witness of history" and he "dug" "things" out of a "pile of junk" and tried to collect them (*Legacy of Disobedience* [July 1961], in Takeuchi 1980–1982: vol. 9: 5). This project is resonant with Benjamin's words from far away: "The chronicler who narrates events without distinguishing between major and minor ones acts in accord with the following truth: nothing that has ever happened should be regarded as lost to history. Of course only a redeemed mankind is granted the fullness of its past—which is to say, only for a redeemed mankind has its past become citable in all its moments. Each moment it has lived becomes a *citation à l'ordre du jour*. And that day is Judgment Day" (Benjamin 2003: 390).

[23] It is from Chinese literature, especially Lu Xun, that Takeuchi received such a messianic and "eschatological" view of history. Toramaru Itō indicated this point early on: "It might be a misunderstanding, so I might even be laughed at to say the following, but I think that Mr. Takeuchi's postwar criticism of 'modernism,' which we have examined here, was an effort to make postwar 'scientism' (or Marxism) a 'thought from within on its own will,' by introducing a 'moment of eschatology' he learned while writing 'Lu Xun' " (Itō 1957: 293).

[24] Takeuchi, "Energy of Literary Revolution" (Bungaku kakumei no enerugī), in Takeuchi 1980–1982: vol. 7: 40.

the new.

> It is said that "even a mouse driven into a corner will bite a cat" ("even a worm will turn") or "counteract a poison with a poison" ("fight one evil with another"). An ultimately cornered situation would be as such. Conversion of value will occur only in such an ultimate place. Unless every way of transforming itself into the other was blocked, *the self cannot be transformed into the other by being itself.*[25]

Chinese literature (and China) could "be transformed into the other by being itself." It is because China had a method of forming a subject as resistance by making tradition experienced *as tradition*, and of transforming itself self-consciously by detaching itself from tradition. But Japanese literature (and Japan) could not do so in the same way. At the end of this article, Takeuchi wrote: "Japanese literature has not experienced a literary revolution, that is, an *interruption of history*."[26]

However, Japanese literature should also have been ultimately cornered, and made tradition be experienced in a certain way. Even so, if there is a decisive reason that separates Japanese literature from Chinese literature, what is it? Summing up Takeuchi's discussion, it is the lack of its own judgment independent from the other, and above all, the lack of "morality."

In "Spirit of New China" (Shin Chūgoku no seishin) (December 1949), Takeuchi argued:

> The only way to completely remove authority is to eliminate its exteriority by making oneself an authority. Only by subjectivizing what was imposed heteronomously and by internalizing external authority will destruction be completed and the true freedom of people be realized. The morality of the Chinese Communist Party came into existence by such an overturn. Today, Confucian norms do not binding the for people, but an internal law of freedom. From this point of view, it is misleading to criticize the Chinese Communist Party as a destroyer of the nation's tradition; rather the Chinese Communist Party embodies the highest morality of the nation. This morality is the source of the power of the new China.[27]

Here, China (or the Chinese Communist Party) and its "morality" are overly idealized in order to make a contrast with Japan. We should think of the consequences of Takeuchi's discussion which considered "morality" to be the "source of power" and took it positively.

---

25 Ibid., 139.
26 Ibid., 141.
27 Takeuchi 1980–1982: vol. 4, 101.

To begin with, "power" is a power of "resistance" against violence. To this extent, "power" is supported by "morality" and it should be acceptable. It is the "original motive power" or the "original violence," by which a base is constructed.[28] Thus, if such "resistance" were to succeed, and if "destruction [were] completed and the true freedom of people [were] realized" as he said, what would happen?

Even though the realization of such a situation is impossible in Japan, for Takeuchi, destruction should be completed and the "End" should arrive in China, or in a *possible* China. Have "morality" and "power" been finally synthesized there? Takeuchi predicted the Eschaton/End, arguing that "if a base is maximized, the enemy's military potential will be minimized; so armed struggle will disappear. This is the extreme type of peaceful revolution."[29] However, he did not answer whether China realized this Eschaton/End. In actual China, "morality" was continuously produced and "power" was also generated in the same course; "original violence" was continuously repeated. We cannot measure from Takeuchi's argument how great a disaster will result from continuing to "resist" even if there is no more to "resist," or even by inventing something to "resist." It is because the tighter "morality" and "power" are tied together, the more difficult it becomes to criticize the violence which sneaks into "morality" while transforming itself.

In case we are obliged to resort to "morality," how can we avoid being deceived by its ruse? Even if "violence as triggering power" is inevitable, how can we ceaselessly imagine its "End"? Opposing Takeuchi, but along with Takeuchi, we would like to pose these questions.

Takeuchi has often been criticized as "nationalistic," because he always mentioned the "independence of the country" and "patriotism" if allowed the opportunity. In what sense was Takeuchi a "nationalist?" To consider this question, let us look at his essay "Modernism and the Problem of Nation" (September 1951).

Takeuchi said that "it is necessary to wash out a stain with one's own hands"[30] in order to interrupt history. In other words, he argued that the Japanese as a "bloodstained nation" should face the fact that their

---

[28] "Mao Zedong: A Critical Biography" (Hyōden Mō Takutō) (April 1951), in Takeuchi 1980–1982: vol. 5, 310–11.

[29] Takeuchi 1980–1982: vol. 5, 310.

[30] Takeuchi, "Modernism and the Problem of Nation" (Kindai shugi to minzoku no mondai), in Takeuchi 1980–1982: vol. 7: 37.

nationalism culminated in ultra-nationalism and confront it for themselves not depending on others, and pay for it:

> From one point of view, the mentality of avoiding a confrontation with nationalism illustrates *the lack of consciousness of responsibility for the war*; in other words, the lack of conscience. The lack of conscience is based on the lack of courage. Because people do not want to be hurt, they try to forget a bloodstained nation, and hesitate to cry out that I am Japanese. However, forgetting can never purify the blood.[31]

Takeuchi's standpoint as a "nationalist" is clear. It is nothing but the "lack of consciousness of responsibility for the war," the "lack of conscience," and the "lack of courage" that make the Japanese "[avoid] the confrontation with nationalism." "Nationalism" is necessary in order to take responsibility for themselves, and also for a minimum of their "morality." Such "nationalism" and "morality" are indispensable for a "literary revolution" and revolution in general.

Such "morality" should not be a kind of norm that would end with self-justification or self-satisfaction, because it is a responsibility which, for others, and facing others, radically transforms the self while being itself. Here, independence from the other is an attitude of taking one' own responsibility unilaterally at the beginning while accepting the asymmetry of the relationships with the other.

This is why Takeuchi, making the "common experience of 1960" a "true experience" by an allegorical method, attempted to extend this experience toward the two others. Takeuchi's attempt here may be called two solidarities.

Bridging the Gap between Generations

One of these solidarities extended toward the younger generation, which proclaimed their separation from the "experience of war."

> We must accept as a fact that there is a group of people who persist in the experience of war and another group of people who proclaim their separation from it, apart from asking if each of them is based on the difference among generations or individuals. I think both of them have their own reason for existence. What is the common term shared by both of them? This question should be without doubt an immediate task for a discussion on the experience of war. I do not have a ready-made solution, but I can imagine

[31] Ibid., 36.

> that the key is to discover *a method of experiencing other people's experience*. Such a method should not come out from the tradition of the "I" novel (*shi shōsetsu*) or that of scholarship just inheriting what predecessors said; a new framework is necessary. I hope that the younger generation, which is going to depart from the experience of wartime evacuation, discovers a method of seizing the experience of schoolchildren's evacuation from the points of view of teachers, parents, and schoolchildren at the same time.[32]

As mentioned above, Takeuchi struggled to bridge the gap between the younger generation and the older generation, which persisted in the experience of war, because he was sure that the "very mentality of separating oneself from the war has something to do with one's [experience of the] trauma of the war."[33] In order to do so, it is necessary to "discover a method of experiencing other people's experience." Such a method was that of overlapping, just as Takeuchi mentioned a "method of seizing the experience of schoolchildren's evacuation from the points of view of teachers, parents, and schoolchildren at the same time."

In order to confirm this point, let us go back to May 1960 again. On May 27 Takeuchi participated in a roundtable discussion with Masao Maruyama and Ken Kaikō, who belonged to the younger generation, while "hearing singing voices in a demonstration outside the window at Nagatachō." This roundtable discussion was published in the journal *Chūō kōron* in July 1960.[34]

In this discussion, both Takeuchi and Maruyama asked Kaikō about the "motive for action" of the younger generation who were participating in the resistance movement in 1960. However, according to Takeuchi's memory, Kaikō responded: "I would like to think more about it later on,"[35] and he "did not answer directly."[36] Takeuchi noticed that the participants in this discussion "were not on the same wavelength, and it was inconsistent to some extent," and perceived a gap between the younger generation and himself.[37] However, the cause of this inconsistency resides in "our [Takeuchi's] generation," which failed to generalize the "experience

---

32 Takeuchi, "Impressions of Discourse on the Experience of War" (Sensō-taiken-ron zakkan) (January 1960), in Takeuchi 1980–1982: vol. 8, 221–22.

33 Ibid., 218.

34 Here let us remark that the article we cited at the first paragraph, "On the Generalization of the Experience of War" (July 1961), began with recalling this round-table discussion.

35 Takeuchi, "On the Generalization of the Experience of War," in Takeuchi 1980–1982: vol. 8, 223.

36 Ibid., 224.

37 Ibid.

of war." Takeuchi felt remorse for his failure in liquidating a legacy to be inherited and leaving it to the younger generation, who were fighting with "refusing legacies."

> Given that the experience of war is exclusively closed, it is reasonable that a part or a large number of the younger generation would be indifferent to or refuse the former generation's experience of war. But if they think that they can separate themselves from the generation of the experience of war by refusing it subjectively, it illustrates that they are not emancipated from *the trauma of the war*, and they are also victims of the particularization of the experience of war. The attitude of refusing the legacy itself makes them captives of the legacy. I do not object to cutting history artificially, but it should be cut by using a *method*. Such a *method* could not be found without recognition of the war.[38]

Takeuchi mentions here again the "trauma of the war." What is necessary is a "method" of "cutting history artificially" by inheriting the legacy as a legacy. He raised the literary criticism by Takaaki Yoshimoto, Shūgo Honda, and Hotsuki Ozaki as examples that succeeded to a certain extent. Takeuchi thought they succeeded because they "adopted consciously or unconsciously" a "method of dealing with the experience of war by overlapping it upon the postwar experience."[39]

Yet, it was not easy close the gap because Takeuchi could not keep an alliance even with Takaaki Yoshimoto, who was mentioned as a successful example. Takeuchi described a discussion with Yoshimoto at the end of May 1960 as "overheated": "we concluded that each of us should go our own way, and I left the place during the discussion."[40] Such was the difficulty of communicating the "experience of war."

Takeuchi continued to stick to the "trauma of the war," because the war had not ended. Here, we should extend our discussion toward another other: China, which was the greatest other for Takeuchi.

---

38 Ibid., 227–28.

39 Ibid., 230–31.

40 Takeuchi, "Notes by Author" (Chosha kaidai), in Takeuchi 1980–1982: vol. 9, 502–3. Shortly thereafter, Takeuchi published "Four Proposals" (Yottsu no teian) and "Democracy or Despotism" (Minshu ka dokusai ka), and claimed that it was an urgent "program" to rebuild democracy opposing despotism, before resolving the problem of the Security Treaty. In contrast, Yoshimoto responded that "something almost beyond comprehension is discussed here" (Yoshimoto Takaaki, "The End of Fiction" [Gisei no shūen] [October 1960], in Yoshimoto 1986: 94) and argued that the "movement of civil democracy as is discussed by Yoshimi Takeuchi or Masao Maruyama" was "another form of fictional progressivism," and "has nothing to do with the real movement of avant-gardes, intelligentsias, workers, and citizens" (Ibid. 98–100).

The China Issue and the Law

To begin with, what was the most important reason Takeuchi protested against the new Japan–United States Security Treaty? The old Security Treaty was concluded when Japan was under the rule of the United States and not independent. But now, after Japan's "independence" granted by the San Francisco Peace Treaty, revising the old Security Treaty would make Japan "voluntarily" hostile toward China and keep it in a state of war with China. Takeuchi spoke in favor of the overall peace treaty during the period of the conclusion of the San Francisco Peace Treaty, so for him putting a *legal* end to the war against China constituted the heart of the "moral responsibility of the Japanese."[41] Even as of May 16, 1960, he contended that:

> Such a situation [where Japan and China are still legally in a state of war] is not consistent with the hopes of Japanese people. Most Japanese people feel the largest war responsibility for China and hope to pay for this responsibility by the restoration of diplomatic relations with China, and think that the restoration should be the core of Japan's foreign policy.[42]

The pacification and the restoration of diplomatic relations with China were not a question of a simple political technique or tool. Rather, they were, as Takeuchi later argued, the "minimum ethical condition for living as an individual and as a nation."[43] Therefore, approving and ratifying the new peace treaty meant nothing but nullifying the law concerning this "minimum ethical condition for living." Even though the decision was made in the name of another law, it was a radical violence against the law as long as it destroyed the condition of the law.

As is well known, Takeuchi immediately resigned from Tokyo Metropolitan University after the forcible passage of the new Peace Treaty in the middle of the night of May 19, 1960. He then wrote a statement of the reason for his resignation, and published it through the media. The reason was that he could not remain in office in a situation where the heads of the legislative and administrative powers ignored the constitution, because he, as a public servant, had pledged to respect it.

---

41 Takeuchi, "The Future of Japan-China Relations" (Nicchū kankei no yukue) (March 1960) in *Legacy of Disobedience*, Takeuchi 1980–1982: vol. 9, 47.

42 Takeuchi, "Japan-China Relations Facing a Breakup" (Hakyoku ni chokumen suru Nicchū kankei) (May 1960), in Takeuchi 1980–1982: vol. 9, 96.

43 Takeuchi, "After Reading Ikeda's Talk" (Ikeda kōen wo yonde) (November 1968), in Takeuchi 1980–1982: vol. 11, 342.

Takeuchi appealed to the constitution, and resigned in the name of the constitution.[44]

Takeuchi pretended to think that a "flash of heavenly inspiration" made him appeal to the constitution then. He said that because the constitution "was cultivated in everyday life," and "latent in the subconscious," it spouted out in a state of emergency and made him decide.[45] But it was not the first time he referred to the constitution; he already mentioned it before at a critical moment.

Pick up the Discarded Constitution!—The Constitution Resisting Violence

When in 1960 Takeuchi considered the core issue of the new Security Treaty to be China, he overlapped it upon the injustice of the conclusion of the San Francisco Peace Treaty. Let us go back to 1952, when the San Francisco Peace Treaty (which had been concluded in 1951) was ratified.

In that year, Takeuchi published a book titled *Japanese Ideology* as if he had inherited the legacy of Jun Tosaka's *A Discussion of Japanese Ideology* (1935). The fourth chapter of "A Letter to a Young Friend" was given the title "The Constitution and Morality."

> Why has it [the new constitution] penetrated into our body? It is in part because of the lapse of time. But it is not everything. While I observed the new constitution was cruelly violated within view of others, I began to wonder whether this situation was acceptable. Can I observe such a scene as if it were someone else's affair? Doesn't it belong to me? After a little while, I began to feel a pain on my own body.... In my case, the governor's ignorance of the constitution conversely brought me the intention of defending it. Seeing this process from the perspective of the law of history I have just proposed, I could not deny the given constitution, but fortunately or unfortunately, because people in power violently destroyed it, I gradually came to accept it as mine, by using such destruction as leverage. This is a kind of wisdom of the weak.[46]

The new constitution did not appear after the Japanese people eliminated the old constitution by themselves. That is why Takeuchi declared:

---

44 See "Statement of the Reason for My Resignation" (Jishoku riyūsho) (May 21, 1960), in *Legacy of Disobedience*, Takeuchi 1980–1982: vol. 9, 99–100.

45 Takeuchi, "Our Sensitivity toward the Constitution" (Watashitachi no kempō kankaku) (June 1960), in *Legacy of Disobedience*, Takeuchi 1980–1982: vol. 9, 132–33.

46 Takeuchi, "A Letter to a Young Friend" (Wakai tomo e no tegami) (May 1952), in Takeuchi 1980–1982: vol. 6, 43–44.

"I thought it was a good constitution, but I felt it to be like someone else's affair; it was kind of too bright."[47] But he changed his mind while the new constitution was "cruelly violated" by people in power, and he began to try to pick it up in his own arms and make it his own. Such is the "way of life of the weak," whose archetype was represented in Lu Xun. In other words, it was "to become to be hated by someone I hate."[48] The reason for picking up the constitution is precisely because it was discarded: this kind of the defense of the constitution was the "wisdom of the weak."

Takeuchi continued to assert this kind of the defense of the constitution. Those who gave rise to the constitution, as the law at the base of every other law, violently destroyed it. Takeuchi expressed his anger at this violence against the law, saying: "I felt an emotion like 'damn it!;' " "I got angry."[49] He tried to firmly defend the law in the name of law, opposing this violence.

Here might be a faith in primordial relationships in general concerning the law, and "universality" that supports the law. That is, the law is always and already given by force and by compulsion. But the force that gave the law tramples it without understanding it. In contrast, the *resisting* "subject" forms itself as a "subject" by picking up the initial law and its original violence in its arms. That is, it defends the initial law in the name of "universality" of law, opposing the secondary violence, and makes the law its own. This is the "principle of Asia" and the "creation of universality by the reversal of value."[50]

But what is it for? Takeuchi's answer to this question was that it is for something prior even to the life: "I think the constitution takes priority over life."[51] The constitution here is not just the law of law, which supports positive law. It is, if we use an already cited expression, the "minimum ethical condition for living as an individual and as a nation."

---

47 Ibid., 42.

48 Ibid., 45.

49 Takeuchi, "Our Sensitivity toward the Constitution," in Takeuchi 1980–1982: vol. 9, 132; Takeuchi, "Defense of the Constitution Precedes Everything" (Kempō yōgo ga issai ni senkōsuru) (May 1954), in Takeuchi 1980–1982: vol. 6, 372.

50 Takeuchi, "Asia as a Method" (Hōhō toshite no Ajia) (January 1960), in Takeuchi 1980–1982: vol. 5, 115.

51 Takeuchi, "Defense of the Constitution Precedes Everything," in Takeuchi 1980–1982: vol. 6, 373.

Patriotism: The State Aimed at Dissolution

Let us keep in mind that Takeuchi discussed "morality" and "ethics" not only with regard to individuals but also with regard to "nations." In the same manner, when he discussed the "subjectivization" and "internalization" of the constitution, he always mentioned its "nationalization": "If we want to be free people, we should internalize the constitution and make it penetrate into our body. In other words, there must be the 'nationalization' of the constitution. We should make it our own; not something given, but something generated from within ourselves."[52] It *should* be safe to say that for Takeuchi the "nation" or "nationalization" must not be understood in a "naturalistic" way. However, his terminology was always accompanied by such "naturalism." For example, let us take a look at the following passage discussing "nation" and "patriotism":

> The word "patriotism" was once treated cautiously, but I think patriotism is still important. Facing this state of emergency such as has never been experienced in the glorious history of the Japanese nation, *let us rewrite history* by mobilizing all the energies of Japanese people. I would like to go along with you hand-in-hand in this struggle, in which we should take action honorably as Japanese before our descendants.[53]

Takeuchi here was just like the Takeuchi who was criticized as a "nationalist" to the extent that he discussed "patriotism" through the "glorious past of the Japanese nation" and the "glorious history of the Japanese nation."[54]

---

[52] Takeuchi, "Our Sensitivity toward the Constitution" (June 1960), in Takeuchi 1980–1982: vol. 9, 137.

[53] Ibid., 138.

[54] Toramaru Itō, citing Taijun Takeda, pointed out that Takeuchi's way of illustration of Lu Xun, especially his way of abstraction of Lu Xun's individual thought which refuses the system, was influenced by Hideo Kobayashi (Itō 1957: 255–58). We should not overlook an irony that Takeuchi was entangled in Kobayashi-style "naturalism" even while criticizing him.

For example, let us take a look at Takeuchi's criticism of *Country of the Dawn* (*Yoake no kuni*) (Iwanami Films 1967), a film of the Cultural Revolution. Takeuchi wrote as follows about the scene at the Fushun coal mine in the later part of this film: "Along an inner slope of an open-pit mine at Fushun, there were laid almost innumerable rails, on which long trains of freight cars travel this way and that as if they were superimposed on each other. Such a *natural* landscape (there is no other way to say it) itself *impresses* me. I admit that I am sentimental, but then my impressions are out of control, and it is difficult to explain their content" ("Country of the Dawn" [Yoake no kuni] [November 1967], in Takeuchi 1980–1982: vol. 4, 424).

Of course, Takeuchi must have known this coal mine had been a large income source for the South Manchu Railway Company and was a place where many Chinese workers were subjected to slave labor. Moreover, there is a scene in this film where the "myriad graves"

But it is still worth underlining that this "nation" or "patriotism" is not fully understood as only "naturalistically." This is because Takeuchi tried to cut off completely the "eternal continuity" (*tenjō-mukyū*) of the "national entity" (*kokutai*). We can find a clue in the postscript of *Legacy of Disobedience*: "I would like to propose that Japan as a state should also make provision for dissolution of the state. Otherwise, people would not have patriotism."[55] This is a passage in the last section titled "On Eschatology." We can understand here that "patriotism" in Takeuchi is, so to speak, an *intellectual emotion* that emerges by supposing an "Eschaton/End" of an association, a state, or a civilization as a group of people. It could not be a natural emotion or an emotion comprehended "naturalistically."[56]

What is a state aimed at its dissolution?

> One of my fancies is that we divide Japan into a few parts, and have each of them build a state based on its plan. This is an experimental idea that seeks an autonomous division, not a heteronomous one as in Korea or Germany. I think this is better than conducting a uniform and irreparable experiment using power on a nationwide scale.[57]

Takeuchi proposed that Japan divide itself into a few parts on its own, and that each part "build a state." Thus, for him, a state is not an "eternal continuation" or something "natural." It is something to be constructed and it has its own beginning and end. Moreover, its construction is not a naturalistic one either; rather, it follows a "law" abstracted from experiences by a reformed scholarship. To put it in a reverse way, a state that does not have such a scholarship or does not follow such a "law" is not a state. People who gather there cannot be called a "nation."

> It might have been too much to say that unless national literature is established, Japan will be ruined; but I rather think it is not an exaggeration.

("bannin no haka"), in which these workers were buried, continue on and on. However, Takeuchi surprisingly wrote: "Current movie fans, who love extremes, might feel hypothermal to this scene. They would say that it is nothing compared to Auschwitz. I admit this too" (ibid. 425) and assimilated the scene of this film into "everydayness."

55 Takeuchi, "Postface" (Atogaki), in *Legacy of Disobedience*, Takeuchi 1980–1982: vol. 9, 294.

56 On Takeuchi's "emotion" from the point of view of deep relationships between romanticism and nationalism, see Richard Calichman, "Literature and Its Dangers," in Calichman 2004. In addition, Sun Ge, even in a different context, pointed out that Takeuchi raised the problem of emotion penetrated into the memory of war and the difficulty of sharing it, and she attempted to explore it using a concept of "emotional memory" (*kanjō kioku*) (Sun 2005: 204–5, Sun 2002: 68–77).

57 Takeuchi, "Is 1970 the Goal?" (1970 nen wa mokuhyō ka?) (October 1964), in Takeuchi 1980–1982: vol. 9, 386.

> I think the Japanese nation is now ruined. In the future, it has a possibility of being reborn, but I think the status quo is the ruination of the country.[58]

For Takeuchi, a scholarship that makes an experience an experience and abstracts a "law" from experience is first and foremost "national literature." But in postwar Japan, such "national literature" had not emerged. This means that a moral subjectivity had not been established by radically transforming itself, and that it could not give itself its own beginning and end. Therefore, it is inevitable to say that the "Japanese nation is ruined" and the "status quo is the ruination of the country."

A Song for a Ruined Country

Throughout the 1960s, Takeuchi wrote an essay at the beginning of each year to report where we had been and to express his hope for where we were going. The last one of them was "the Seventh and Last Report on the 1960s" (January 1967). At its very end he said:

> People of a ruined country do not have anything to do but sing a song for a ruined country. As an old haiku runs: Come and play with me, a sparrow without parents!
>
> There will be a subterranean fire; a subterranean fire will never be extinct. But I think I have to give up seeing for myself its eruption. I would rather play with a sparrow.[59]

With what emotion did Takeuchi cite this haiku? Takeuchi, like Issa Kobayashi, also got sick and damaged his health just before turning fifty-seven years old. And one year before, Takeuchi, in his essay "The Sixth Interim Report on the 1960s" (January 11, 1966) cited the *Zhuangzi* and also said, "One of the central ideas of the *Zhuangzi* is to leave life and death to nature."[60] But this stage should be attained by a "method of intelligence." He said, "Oh, I shall study, I shall study!"[61] "A song for a ruined country" should be sung afterwards.[62]

---

[58] Takeuchi, "Prevision and Mistake" (Yoken to sakugo) (June 1966), in Takeuchi 1980–1982: vol. 9, 418.

[59] Takeuchi, "The Seventh and Last Report on the 1960s" (60 nendai, 7 nenme saishū hōkoku), in Takeuchi 1980–1982: vol. 9, 428. It is not an accident that Takeuchi here borrows a "song for a ruined country" from the haiku poet Issa Kobayashi, because in that year, Takeuchi was fifty-seven years old, and Issa was also in his fifty-seventh year when he wrote this haiku.

[60] Takeuchi, "The Sixth Interim Report on the 1960s," in Takeuchi 1980–1982: vol. 9, 410.

[61] Ibid.

[62] Takeuchi declared in June 1951 in "Song for a Ruined Country," where he talked about "literature and morality": "When a country is ruined, a literary man has only to

In addition, the term "subterranean fire" mentioned here is cited from the "Foreword" to Lu Xun's *Wild Grass*.[63]

> I love my wild grass, but I detest the ground which decks itself with wild grass.
>
> A subterranean fire is spreading, raging, underground. Once the molten lava breaks through the earth's crust, it will consume all the wild grass and lofty trees, leaving nothing to decay.
>
> But I am not worried; I am glad. I shall laugh aloud and sing.[64]

Lu Xun took it positively that wild grass as literature, which "imbibe[d] ... the blood and flesh of the dead," will "decay" and be "consume[d]" by a "subterranean fire," saying "I am glad." Then, a song is sung here. However, the "subterranean fire" will not break through. He continued:

> Heaven and earth are so serene that I cannot laugh aloud or sing. Even if they were not so serene, I probably could not either. Between light and darkness, life and death, past and future, I dedicate this tussock of wild grass as my pledge to friend and foe, man and beast, those whom I love and those whom I do not love.[65]

The "seren[ity]" where the "wild grass" is in full bloom to no purpose, and the "ground" "decked" by it—in such a place, "I cannot ... sing."

> For my own sake and for the sake of friend and foe, man and beast, those whom I love and those whom I do not love, I hope for the swift death and decay of this wild grass. Otherwise, it means I have not lived, and this would be truly more lamentable than death and decay.
>
> Go, then, wild grass, together with my foreword![66]

Takeuchi sang a "song for a ruined country" in serenity where a song cannot be sung. Didn't this mean throwing away his wild grass? In January 1960, he proposed holding an event commemorating the centennial of Meiji in 1967. He tried to find a "method to objectify the state of Meiji,"

---

sing a song for a ruined country. He just has only to sing it with wholehearted passion. However, I think it is *a little too early* to begin from now" (in Takeuchi 1980–1982: vol. 7, 28). Takeuchi had embraced the ideal of Chinese literature, believing that a literary man expresses the era and his performance of expression itself has morality. For this literary man, would it be too much to say that postwar Japanese literature could not even become a "Song for a Ruined Country"?

63 Takeuchi described the work *Wild Grass* as "an essence of Lu Xun." See Takeuchi, "Notes on *Wild Grass*" (*Yasō* kaisetsu) (July 1955), in Takeuchi 1980–1982: vol. 1, 324.

64 Lu 1974: 3.

65 Ibid., 4.

66 Ibid.

but he withdrew his proposal in 1965.[67] The year 1965 came during the period when he declared that he was "not qualified as a critic," and it was the year when he judged that the " 'postwar' revolution" had definitely failed.

In May 1960, Takeuchi could still say: "When darkness deepens, it is the time brightness is approaching. Hope will be born at the bottom of despair."[68] However, just after eight years, he could only say in disgust: "In postwar Japan, I once had hope, but now there is no hope. It was just an illusion."[69] The "Eschaton/End" he wanted so much had arrived in a completely regrettable way.

In this regrettable "Eschaton/End," how can we pick up the "hope" Takeuchi discarded? This is a question for us, who inherit the legacy of disobedience, in today's Japan where the "trauma of the war" still afflicts us.

References

Benjamin, Walter (1998), *The Origin of German Tragic Drama.* Trans. John Osborne. New York: Verso.

—— (2003), "On the Concept of History," in Howard Eiland and Michael W. Jennings eds., Edmund Jephcott and others trs., *Walter Benjamin Selected Writings Volume 4 (1938—1940)*. Cambridge, MA: Belknap Press of Harvard University Press.

Calichman, Richard (2004), *Takeuchi Yoshimi: Displacing the West.* Ithaca, NY: Cornell University East Asia Program.

Itō, Toramaru (1957), *Rojin to shūmatsu-ron—kindai riarizumu no tanjō* (Lu Xun and eschatology: The birth of modern realism). Ryukoku shosha.

Lu, Xun (1974), *Wild Grass.* Beijing: Foreign Languages Press.

Oguma, Eiji (2002), *"Democracy" and "Patriotism": Nationalism and the Public Sphere in Postwar Japan.* Shin'yosha.

Sun, Ge (2002), *Asia wo kataru koto no jiremma* (The dilemma of discussing Asia). Iwanami.

—— (2005), *Takeuchi Yoshimi to iu toi* (A question called Yoshimi Takeuchi). Iwanami.

Takeuchi, Yoshimi (1980–1982), *Takeuchi Yoshimi zenshū* (Complete works of Yoshimi Takeuchi). 17 vols. Chikuma shobō.

Yoshimoto, Takaaki (1986), *Yoshimoto Takaaki zenshūsen, 3, Seiji shisō* (Selections from the complete works of Takaaki Yoshimoto, vol. 3, Political thought). Daiwa-shobō.

---

[67] Takeuchi, "The Fifth Interim Report on the 1960s" (January 1965), in Takeuchi 1980–1982: vol. 9, 92.

[68] Takeuchi, "The Time to Decide" (Ketsudan no toki), in *Legacy of Disobedience*, Takeuchi 1980–1982: vol. 9, 103.

[69] Takeuchi, "Thought in the Era of Crisis" (Kiki no jidai no shisō) (January 1968), in Takeuchi 1980–1982: vol. 9: 437–38.

CHAPTER SIX

THE CAMPAIGN TO CRITICIZE LIN BIAO AND CONFUCIUS (批林批孔) AND THE PROBLEM OF "RESTORATION" IN CHINESE MARXIST HISTORIOGRAPHY

Susanne Weigelin-Schwiedrzik*

> Our ancestor dragons have already died, but their souls are still with us[1]
>
> —Mao Zedong

Introduction

In his pathbreaking article on modern Chinese historiography, Wang Gungwu refers to the Campaign to Criticize Lin Biao and Confucius (Pi Lin Pi Kong 批林批孔) as a reflection of a new historical consciousness generated by the attempt of Marxist historians in China to bring Chinese history and world history into sync: "The implication seems to be that world history and Chinese history are as one in the same way that the world today is one with China today."[2] Wang Gungwu comes to this conclusion by describing post-1949 Marxist historiography as a process of coming to terms with a number of basic issues, which culminate in the "persistent desire to change the cultural values and the historical perceptions of the Chinese people."[3] In his mind, Marxism opened an opportunity for Chinese intellectuals to look at Chinese history as part of world history. This view, however, created a dissonance with the long-held conception of the Chinese past being "autonomous" in the sense of

* I would like to thank Arif Dirlik and Axel Schneider for their stimulating comments on an earlier version of this paper.

[1] 祖龙虽死魂犹在. This line is taken from a poem Mao wrote as a letter to Guo Morou criticizing Guo's positive attitude on Confucius and criticism of Qin Shihuang in his book 十批判书 (Ten Critical Essays), which was first published in the 1940s. The poem has been published in several different versions. The version here is from Gao and Yan 1986: 484.

[2] Wang Gungwu 1975: 23.

[3] Ibid., 24.

fundamentally different from the history of other parts of the world. As up until the 1970s the conceptions of Chinese history as part of world history and Chinese history as autonomous could not be reconciled, the writing of history vacillated between dogmatically writing Chinese history into Marxist terminology on the one hand, and the use of "traditional historicism" on the other. Historians trying to adjust to the dominance of Marxism after 1949 were reluctant to accept the new worldview as they saw the danger of belittling the Chinese past by overcoming its autonomy, while Marxist historians in the 1950s and 1960s had some difficulty in dealing with the cultural heritage of China in a convincing way. They soon realized that they had to acknowledge the importance of the Chinese heritage, but at the same time "they had to make sure that the revolution was not weighed down by the Chinese heritage and did not merely become another form of the old dynastic change."[4]

"Restoration" and the "Pull of the Chinese Past"

This is where the topic of "restoration" enters the discussion. Although this topic seemed to be merely of political importance in the context of the Pi Lin Pi Kong Campaign, it is of much more far-reaching implications for Wang Gungwu. To a certain degree, he believes the leadership of the Chinese Communist Party (CCP) under Mao was worried about the fact that "Chinese history had a tremendous pull and the power of analogy across the full length of Chinese history was immensely attractive to all knowledgeable Chinese."[5] As long as Marxism could not provide a convincing interpretation of Chinese history it remained contested by the persistence of the traditional view on Chinese history. This implies, according to Wang Gungwu, that "there was a distinct danger that the Chinese past would become overwhelming again."[6] In this sense, the Campaign to Criticize Lin Biao and Confucius aimed "to transform the accepted views of the Chinese past by instilling a more integrated Sino-Marxist interpretation of history."[7] Looking at Chinese history by way of historical analogies was finally overcome and replaced by a combination of integrating Chinese history into world history on a structural level

---

[4] Ibid., 12.
[5] Ibid., 19.
[6] Ibid., 11.
[7] Ibid., 22.

while preserving its particularity as a reservoir of "historical analogies for current political ends."[8]

The theme of "restoration" was one of the main topics discussed during the campaign. It was introduced into the debate as a world history problem with several articles about the French and the English revolutions. In a next step, articles were published discussing the theme of restoration in the context of the unification under Qin Shihuang in 221 BC. The combination of articles about world history and about Chinese history implied that the danger of restoration is imminent to any process of revolutionary change, no matter whether society moves on to feudalism or to socialism, and it implied that this is true both for world history and for Chinese history. However, by using the restoration process of the past to attract the reader's attention to the on-going restoration process in the present, the traditional method of analogy was integrated into Marxist historiography. Thus Marxist historiography was seemingly reconciled with traditional Chinese historiography, and the particularity of Chinese history and the Chinese way to discuss history was preserved without deviating from a basically Marxist approach.

In contrast to the mainstream of publications on the Campaign to Criticize Lin Biao and Confucius, which concentrate on deciphering its political aims,[9] Wang convincingly shows in his article that the discussions on Chinese history and historiography were much more than at the center of the political discourse during the Cultural Revolution. It was not by chance that the Cultural Revolution got started with an intense struggle against the "academic authorities" in the field of history; and it was also not by chance that the attempt to launch a "Second Cultural Revolution" started with the Pi Lin Pi Kong Campaign and again focused on issues of history and historiography. Discussions about the past were discussions on how to deal with the past and how to define an identity for the Chinese nation in light of the past. They were discussions on the relationship between continuity and change, universality and particularity, on tradition and modernity—and above all, discussions on the past were discussions about the future. Thus they were much more far-reaching than what a merely political understanding of the Pi Lin Pi Kong Campaign would suggest. However, while Wang Gungwu's article shows us that we have to reach deeper with our analysis of the Pi Lin Pi Kong Campaign, it leaves

---

8 Ibid., 1.

9 For an early example, see the still very interesting article by Goldman 1975.

a number of questions unanswered. The most important is: What are the origins of the "enormous pull of Chinese history" and what does it have to do with the danger that "the Chinese past would become overwhelming again?"[10] In other words: How should we understand the logic of an argument that as long as Chinese history and world history in the Marxist sense of the word cannot be reconciled the newly established social order in China is confronted with the danger of "restoration"?

Linearity and Cyclicality in History

For Wang Gungwu the idea of "traditional historicism" seems so natural that he does not bother to explain what it is. Among many things, we surely have to keep two things in mind. On the one hand, traditional Chinese historiography is characterized by what Balazs (1964) calls the "anecdotal" mode of history writing which is historiography "of bureaucrats for bureaucrats"; on the other hand, dynastic Chinese historiography sticks to a cyclical understanding of history. A dynasty goes through a cycle of upsurge, stagnation, and decline and is eventually replaced by a new dynasty going through the same cycle.

Looking at history as a reservoir of anecdotes from the past which instruct adequate moral behavior in the present is based on the idea that history and ethics are closely linked and that ethical standards are not subject to change. The moral basis of human existence is unchangeable. This, of course, does not imply that things do not change at all. It also does not imply that change cannot lead to something which we call progress. It only implies that the moral basis of human existence is unchangeable. Under this precondition, anecdotes from the past which reflect these unchanging moral principles are valid as guidance for morally appropriate behavior in any present.[11]

However, the danger of "restoration" discussed during the Pi Lin Pi Kong Campaign and the fear that "the Chinese past would become overwhelming again," as described in Wang Gungwu's paper, also have an implication which is related to the above-mentioned cyclical view of history.[12] The danger of restoration is generated by the idea that change is

---

[10] Wang Gungwu 1975: 11.

[11] I owe this understanding of traditional Chinese historiography to Axel Schneider and his contribution to this volume.

[12] Wang Gungwu 1975: 11.

reversible. In this sense, the discussion on restoration during the Pi Lin Pi Kong Campaign collides with the linear view of history that has prevailed in modern historical thought in general and Marxist thought in particular. That is where the debate becomes really astonishing: although it makes use of categories derived from historical materialism, it deviates from the basic Marxist belief in the irreversibility of change and in change as the basis of progress. In its attempt to reconcile Marxism and traditional Chinese historiography, the Pi Lin Pi Kong Campaign was, as I will argue in this chapter, overwhelmed by the Chinese past. It did adhere to principles of Marxist historiography, but its use of the anecdotal form of history writing as well as of the cyclical view of history were more dominant than its use of historical materialism.

The invention of a linear mode of historical thought is closely linked to the idea of the nation-state and a form of historiography which explains the necessity of the nation-state.[13] The establishment of the nation-state was accompanied by a fundamental rupture. With this rupture the idea of development as a process of unending change generating new and unprecedented elements had an uncontestable basis which changed the whole outlook on the relationship between past, present, and future. Simultaneously, this rupture underlined the idea of human agency implying that change is not a matter of fate, but the result of human action. On the one hand, this rupture generated the idea that the present and the future had to be different from the past; on the other hand, it generated the idea that there was no way to understand the present and foresee the future without knowing about the past. Because of this inherent contradiction, the linear mode of historical thought allows for intensive research on the past although it is based on the assumption that nothing which happened in the past can ever happen again.[14] The experience of everyday innovation makes it impossible for the past to become overwhelming. However, as the only rational explanation for the present and the future lies in the past, the past constantly overwhelms the present.

In this sense there is an implicit possibility of repetition in linear thought which is only made explicit if discussed on more abstract levels of historical development. The search of historical materialism for laws of historical development allows for repetition on the abstract level when it comes to explaining revolutionary change. However, historical materialism

[13] See Duara 2001: 360–61.
[14] Ibid., 362–63.

also sees history as progress in which yesterday is never the same as today. On the level of historical laws, the present is a repetition of the past; on the level of everyday experience, the present is always new in comparison to the past. Even if the laws of revolutionary change apply once and again they generate progress as society moves from the primitive to more advanced and finally perfect stages. The reconciliation between repetition and change is achieved by referring to the persistence of historical laws and to the concrete forms which change as part of historical development. If knowledgeable about the laws of history, mankind can claim historical agency and therefore act in freedom. If not, the laws of history become reality with necessity beyond human agency. In this sense, historical materialism is only one among many attempts to domesticate the problem of contingency in face of a present which is disassociated from the past.

However, if looked at in the specific Chinese context, the problem becomes more acute. This is because writing history into a linear evolution is a mode of European historiography imported into China via Japan, whereas the nonlinear mode of writing history is regarded as suggested by ancient Chinese dynastic thought. The orientation towards progress and innovation is not complemented by a search for the past. Instead, the ideas of linear progress and of repetition are mutually exclusive. Linear progress is associated with something from outside China which is made use of in order to get rid of Chinese particularities; and the idea of relating to the past to look for advice in the present is associated with an indigenous Chinese way of looking at history which people refer to when they want to preserve Chinese particularity. As part of this dissonance, the longing for continuity in terms of Chinese particularity is more powerful than the idea of change and innovation. That is why, after more than a hundred years of revolutions in China, Chinese historiography—as we know better today than did Wang Gungwu back in 1975—has still not found a way of explaining Chinese history that would convince the public of the possibility of irreversible change. Instead, even what is regarded as the most profound change in China by Western observers—the development in the People's Republic of China (PRC) since the late 1970s—is increasingly being described in terms of restoration! This is, of course, only a tendency among those new-left historians who criticize the ongoing process in terms of China regressing into a state of capitalism.[15] It

[15] Gao Mobo 2008.

is not a mainstream authoritative explanation as it was during the second half of the 1970s. Nevertheless, doubts about the irreversibility of change are widespread among all parts of the Chinese population. They might even be regarded as one motor of the dynamism of reform which outside observers regard as miraculous. Take your chances today, as tomorrow might be yesterday!

The doubt about the irreversibility of change does not with necessity create a pessimistic outlook. As Ban Wang shows in his analysis of Chinese film, the film audience adheres to a sort of "roundism" which makes it believe in a happy ending when the protagonists are going through the worst of trouble. This "roundism" is shaped by traditional Chinese historiography which at least in its official version canonized in the twenty-five dynastic histories writes the history of every dynasty into circles. When things are going well, this "roundism" makes us aware of the possibility of stagnation and decline, and vice versa; when going through a period of decline, people expect things to improve sooner or later. The fact that the Chinese government refers to China's "peaceful rise" as its future orientation can be interpreted as yet another form of belief in restoration. In contrast to Mao, who saw the possibility of restoration in a negative light, this form of restoration would be the first successful attempt in world history for a world civilization restoring its pole position after going through a period of decline!

This kind of repetition shows some similarity with the Marxist notion of historical laws[16] as it is a structural form of repetition that does not apply to everyday experiences. Instead, it is placed on a higher level of abstraction and allows for different forms of concreteness. It can even be accompanied by the idea of progress, albeit—in contrast to historical materialism—not in the sense that society can move from lower to higher forms of organization.

According to traditional Chinese historiography, the order of the world is based on moral principles that are eternal.[17] That is why past and present are always comparable even if things happen in the present that have never happened before. Based on this idea, the remembrance of the past has been preserved in China over the centuries in the form of a reservoir of anecdotes to which—as Wang Gungwu puts it—every knowledgeable Chinese can refer. For those knowledgeable Chinese, this reservoir of

[16] Wang Ban 2004.

[17] See Axel Schneider's contribution to this volume.

anecdotes is what they regard as the particularity of Chinese history. The Pi Lin Pi Kong Campaign referred to the logic of historical laws and integrated, as Wang Gungwu remarks, the long-held knowledge of anecdotes from the past into this logic. However, it added implicitly the idea of eternally unchanging values by reactivating the reservoir of anecdotes from Chinese history and it added the idea of restoration and thereby deviated from the idea of progress. Finally, it posited the idea of the incessant struggle between Confucianists and Legalists and thus established a structural framework for a historiography in analogies. This is why I argue that rather than suggesting how to overcome the enormous pull of Chinese history and how to avoid the past overwhelming the present—as Wang Gungwu suggests—the Campaign to Criticize Lin Biao and Confucius is yet another example of the persistence of traditional Chinese historical thinking amidst a revolution that declared itself to be cultural.

The Campaign to Criticize Lin Biao and Confucius

Many authors writing about the Cultural Revolution deplore that the outcome of Mao's Great Strategic Plan was so much the contrary of its original intentions. Arif Dirlik identifies this contradiction as the "trauma" of the Cultural Revolution,[18] and many authors both in China and outside try to explain this phenomenon by referring to Mao's betrayal of his own principles and his sudden turn around from leader of the masses organizing the fight against the bureaucratization of socialism to the leader of the party, reestablishing bureaucracy in China.[19] When reading recent biographies of Mao Zedong and publications on the Cultural Revolution from the PRC,[20] the turning point in Mao's attitude towards the Cultural Revolution is not explained but dated: it is the Lin Biao Affair that made Mao rethink his Cultural Revolution strategy.[21] From this time on he was torn between realizing that his strategy had come to a dead end and hoping for this "last revolution" to lead China into a modern world of no return. Mao had launched the Cultural Revolution to find the right successors for himself and the generation of revolutionaries that had founded the party and led the revolution to victory. That Lin Biao and Lin Liguo

[18] Dirlik 2003: 177.
[19] Weigelin-Schwiedrzik 2006.
[20] Feng and Jin 2003, Zhang and Su 2000.
[21] Liu Xiuming 2000: 533.

had made plans to assassinate him and fled the country had made it clear to him that the 9th Party Congress had chosen the wrong successor. He had also launched the Cultural Revolution to make sure that the country would not change color after his death. The closer he approached the end of his life, the less sure he was that he had reached this aim.[22]

The Campaign to Criticize Lin Biao and Confucius took place in the middle of this process of rethinking the Cultural Revolution. Forces around the group of people who were later labeled "the Gang of Four" used the movement to attack the group of "old revolutionaries" around Zhou Enlai, who again used the opportunity of Lin Biao's defection to reinstall many of those leading cadres who had been expelled from office during the earlier phase of the Cultural Revolution. The core question that they struggled over was how to evaluate the Cultural Revolution. However, the issue itself rarely popped up in public discourse. Instead the newspapers were full of articles about Qin Shihuang and Confucius, about the establishment of the Qin dynasty and the unification of the country, about Legalists and Confucianists. One of the key words that occurred in nearly every of those innumerable articles was "fubi" (复辟), which stands for "restoration." Dictionaries explained the meaning as restoring the rule of a former leader or a system of rule, positioning the word outside the Marxist discourse in China.[23] Mao himself did not use the word very frequently in what we know as his post-1949 texts. He preferred to speak of the country "changing color" (变颜色), of revisionism (修正主义), or of a reversal of verdicts (翻案). But he had prompted the discussion by making several remarks about the relationship between Confucianists and Legalists and about the necessity of studying history. Already in a Politburo meeting in May 1973 he discussed the necessity of criticizing Confucius, and after another discussion with Wang Hongwen and Zhang Chunqiao, he had a poem published in which he criticized the widespread disgust for Qin Shihuang and his burning of Confucian classics.[24]

The Campaign to Criticize Lin Biao and Confucius was launched in early 1974 as a reaction to Zhou Enlai's attempts to use the Lin Biao incident to criticize what he called "extreme leftism" and rehabilitate many of those cadres from the elder generation who had been criticized and expelled from office during the early years of the Cultural Revolution.

[22] Ibid., 547.
[23] *Xiandai hanyu cidian* 2002: 395.
[24] Gao and Yan 1986: 484–85, Feng and Jin 2003: 1657.

Preparations for this movement had already been made in late 1973 after the 10th Party Congress with Chi Qun,[25] then the director of the Group for Science and Education of the State Council, asking universities to criticize Confucianism on the basis of Yang Rongguo's report "The Fight between Confucianists and Legalists and the Reactionary Influence of Confucius' Ideas."[26] In this report as well as in his well-known article "Confucius Is a Thinker Stubbornly Safeguarding the System of Slave Society"[27] Yang Rongguo had argued that Confucius had been partial to the aristocratic regime of the Zhou dynasty and therefore opposed the establishment of a united feudal society under Qin leadership. Quite in tune with this interpretation, Mao introduced Qin Shihuang and Confucius to a foreign visitor saying: "Qin Shihuang was the first feudal emperor of China. I am also a kind of Qin Shihuang. Lin Biao criticized me for being like Qin Shihuang. In China we have always had two factions; one faction says Qin Shihuang is good, the other one says Qin Shihuang is bad. I like Qin Shihuang and dislike Confucius."[28] According to Mao, all reactionaries, including Lin Biao, tended towards a positive assessment of Confucius, while progressive forces in history supported the Legalist point of view. Legalists were in favor of uniting the country under a central leadership, of pushing development forward focusing on the present rather than the past. Confucianists, to the contrary, stressed the past rather than the present and therefore tended to head backward rather than forward. Mao stressed his admiration for Qin Shihuang[29] several times and even though he never went into any detail explaining why it was necessary to criticize Lin Biao by criticizing Confucius, he had certainly not inhibited the campaign to be launched under these auspices.

Since 1972, Mao had immersed himself in reading historical documents and texts. And as his eyesight was rapidly deteriorating, he had 86 documents printed in big letters and annotated by scholars from Beijing and Shanghai. Among these documents, he selected a number of readings from Legalist authors (which were printed and annotated between August 1973 and July 1974, i.e., during the initial months of the Pi Lin Pi Kong Campaign), including Han Feizi, Xunzi, Liu Zongyuan, Wang Anshi, Wang Fuzhi, and Zhang Binglin. He used these documents to communicate with

---

25 Wang Haiguang 2000: 1101.
26 San 2007.
27 Yang 1973.
28 Wang Haiguang 2000: 1100.
29 Wang Haiguang 2003: 1099, Feng and Jin 2003: 1657.

leading members of the Chinese political and intellectual elites on the question of a possible restoration of capitalism as well as of a lack of unity and loyalty among the provinces. Other works Mao selected for reprinting stressed the necessity of replacing the rule of rites by the rule of law and making the country strong by uniting it from within. Whatever Mao was interested in during this period was related to the idea of change and development and a criticism of a conservative approach focused on learning from the past. According to one of the annotators, Mao's main aim was to use Legalism as a resource through which to stress the necessity of change through revolution, especially the Cultural Revolution.[30] Much in contrast to what the authors of the official Mao biography from the PRC suggest,[31] these pieces of information seem to stress that Mao was the instigator of the Campaign to Criticize Lin Biao and Confucius. Later on, he added fuel to the discussion by carefully selecting readings to be distributed among the leading members of the CCP from among those authors that argued in favor of change, strong leadership, authoritarianism, and anti-Confucianism.

During the campaign, three groups of authors were most active in writing articles. The Liang Xiao Group of professors from Peking University and Qinghua University published a number of articles in the then most prominent journals and newspapers, such as *Red Flag* (红旗) and *People's Daily* (人民日报), but also in the *Beijing Daxue University Journal* (北京大学学报) and in *Historical Studies* (历史研究); the group using the pseudonym Luo Siding was located in Shanghai and used the journal *Study and Criticize* (学习与批判) as its platform. The third group was affiliated with the Central Party School and published under the penname of Tang Xiaowen.[32] The articles of these three groups of authors soon made clear that the struggle between Confucianism and Legalism was not of historical importance alone. They explained that the on-going struggle inside the party had to be understood in terms of the unending conflict between two different worldviews as historically represented by Confucianists and Legalists. In January 1974, the CCP sent out the No. 1 Document stressing that "Lin Biao was a full-fledged follower of Confucius. Like all declining reactionaries in history, he believed in Confucius and detested the Legalists. He attacked Qin Shihuang and used the Confucian way to usurp

30 Liu Xiuming 2000: 533–34, Feng and Jin 2003: 1680.
31 Feng and Jin 2003: 1692.
32 Wang Haiguang 2000: 1101.

power inside the party. The reactionary thought of Confucius was his weapon to restore capitalism."[33] The Document listed a number of proofs related to Lin Biao's belief in Confucian rites as his program of restoring capitalism in China. To reestablish the rule of rites, the argument went on, meant to negate the Cultural Revolution and restore the system of the first seventeen years after the founding of the People's Republic of China.

Mao Zedong and Qin Shihuang, Zhou Enlai and the Villains at the Qin Court

In contrast to Lin Biao and Confucius as representatives of "restoration," Qin Shihuang is described in an astonishingly positive way obviously inspired by Mao's above-mentioned poem. Already in late 1973, the Shanghai journal *Xuexi yu pipan* (Study and Criticize) published a lengthy biography of Qin Shihuang in which astonished readers are confronted with the story of a wise and humane politician who had the exceptional ability to detect all insurgents among his entourage. The reason he turned out to be victorious is that he acted in accordance with the objective law of history by uniting the country, establishing the Chinese empire and thus executing the transition from slave to feudal society. However, he met with severe criticism and resistance from Confucianists, who stubbornly insisted on the legitimacy of the Zhou rites. As they were part of the political system, they organized their resistance from inside as confidants, consultants, and ministers. This is why they were extremely dangerous for the newly established dynasty.[34]

Everything we learn about Qin Shihuang in those days is carefully selected to comply with what the Chinese public was supposed to know about Mao. Qin Shihuang is omnipresent and omniscient, but nobody is ever allowed to know where he physically resides (although Mao is on the news on a daily basis!); the emperor is fully aware of all the hardships his subjects have to deal with and is concerned with their problems, travelling through the country, talking to them and giving instructions to local administrators. Qin Shihuang is always together with his soldiers, gives them advice in difficult situations, and praises them for their victories.[35] Uniting China is a military process just as the Chinese revolution under

33 Quotation from Wang Haiguang 2000: 1104.
34 Fang 1973: 82–88.
35 Li 1974: 15–21.

Mao's leadership was a military battle over China. Qin Shihuang is described as the engineer of the unification campaign; however in order to draw an analogy to present times, Li Shi stresses: "One should not look at the unification of the six states one by one and put too much stress on individual generals. It was Qin Shihuang, who as the representative of the ascending landlord class played a decisive role because his policies were in accord with historical development."[36] This means that the army is an instrument of politics and cannot be victorious without adequate political leadership.[37] Also the army should not get involved into politics that run counter to the objective laws of history!

Among the confidants, consultants, and ministers in Qin Shihuang's entourage Lü Buwei is selected as the main target of attack. Lü is traditionally described as a typical victim of Qin Shihuang's distrust. However, in articles published during the Campaign to Criticize Lin Biao and Confucius he is shown to be a power-hungry schemer whom Qin Shihuang already had to disempower before taking over the throne. Because of this defeat, Lü organized a revolt against Qin Shihuang as the result of which he was banned from the capital. Later on he committed suicide.[38]

The next to be targeted is Li Si, a disciple of the Legalist scholar Han Feizi and loyal minister to Qin Shihuang. He was active in preparing the ground for unification and instrumental in suppressing Confucian influence during the early years of Qin Shihuang's rule. He had proposed to Qin Shihuang to burn the Confucian classics and bury 460 Confucianists alive. However, the group criticizes him in an article published by *Xuexi yu pipan* for his alleged eclecticism.[39]

Zhao Gao, advisor to Qin Shihuang, was a close ally of Li Si. His relationship to Qin Shihuang's eldest son, who was supposed to ascend to the throne after his father's death, brought him into trouble. When Qin Shihuang got into a fight with his eldest son, the latter committed suicide, where upon the youngest son was selected as Qin Shihuang's successor. In order to avoid punishment for his involvement in this affair, Zhao Gao is shown to have organized a revolt from inside the emperor's palace aimed at restoring the system of slavery that had been overthrown by Qin Shihuang. Li Si, who should have known about these plans, did not prevent him from doing so and therefore is assigned a major role in making

36 Ibid., 17.
37 Liu and Zhu 1977.
38 Fang 1973: 82.
39 Luo 1974b.

the restoration of a pre-Qin system possible. Eventually, the feudal system could not be consolidated during the rule of Qin Shihuang.[40]

It is quite obvious that the details given in the above-mentioned articles are selected according to whether or not they provoke associations with on-going politics in China. Li Si, the eclecticist, stands for Zhou Enlai and his attempt to rehabilitate cadres who were expelled from office during the earlier part of the Cultural Revolution. These are the people in Mao's entourage who could organize revolts from within the inner circle of power and thereby "restore" a system that had been overthrown in the initial stage of the Cultural Revolution. Also, even though taken from the conventional reservoir of anecdotes about the Chinese past, the stories are interpreted in a totally unconventional manner. People who were highly praised in former times are now seen as villains and vice versa. This is what later on was denounced as a "distortion" of history. As a matter of fact, it was the logical consequence of taking the "Legalist" side in history and rejecting the conventional "Confucian" interpretation of history. In the context of the Pi Lin Pi Kong Campaign the unchanging moral standards were no longer derived from Confucianism, but as exemplified by the idea of the incessant struggle between Confucianists and Legalists there were two contesting ideas about how to rule the country in existence since antiquity. They both developed a set of moral standards which are valid up until the present even if society has moved on to other stages.

This is how the articles come to the conclusion that the main question that needs to be discussed in the context of Qin Shihuang and the Qin dynasty is the "struggle between restoration and anti-restoration."[41] The forces of the old system try time and again to question the establishment of the new order by sneaking into the inner circles of the power structure. This is from where they attempt to overthrow the new regime. During the Qin dynasty three such attempts took place. The first was led by Zhao Gao; the second was a peasant uprising that was corrupted from within and therefore unable to overthrow the newly established system; the third was led by Liu Bang. Liu eventually overthrew the Qin dynasty and established the Western Han dynasty, but decided to cling to the Legalist "line" of Qin Shihuang.[42]

40 Luo 1974: 19–20.
41 Luo 1973.
42 Luo 1974.

The articles on the history of the Qin can be read as an early sign of the authors acting as mouthpieces of the Shanghai Left, anticipating that an immediate takeover after Mao's death was impossible. Instead, people should be prepared for several rounds of struggles in which the forces of restoration and anti-restoration compete with each other for dominance over China.[43] Luo Siding states: "Even in cases when one system of exploitation is replaced by another one, the efforts of one generation do not suffice. Efforts of several generations are needed and will eventually install the new system if looked at from the perspective of the overall historical process."[44] This sentence not only alerts the reader to the possibility of defeat in the near future, it also explains the long durée of what is called feudalism in the context of Maoist historiography. The whole period between the first unification under Qin Shihuang and the fall of the last dynasty in 1912 is one in which Legalism did not dominate the political thought of the ruling elite, which implies that the imminent danger of restoration was never overcome. In the case of the PRC, the situation is even more complicated, the reader is supposed to conclude. In this case the system of exploitation is replaced by one in which exploitation and oppression are supposed to have been abolished. This means that the struggle between restoration and anti-restoration is even more protracted and fierce. In the same way as in history all Confucianists joined in condemning the burning of books during the Qin dynasty, in the present they unanimously condemn the Cultural Revolution as violent, chaotic, and inhuman. Progressive forces, however, could be detected by their support for Qin Shihuang and his harsh criticism of Confucianism as well as by their applause for the Cultural Revolution.[45]

Analogy and Allusion: Using the Past to Serve the Present

At first sight, the Pi Lin Pi Kong Campaign does not seem to fit into a typically "left" context in China. It turns to the past for guidance in the present, it invokes an esoteric, if not elitist form of discourse which does not relate to "the masses," and it anticipates defeat instead of mobilizing the masses for the victory of the "left." Were the intellectuals writing for Liang Xiao, Luo Siding, and Tang Xiaowen not doing exactly the same

43 See Gregor and Chang 1979: 1084–85.
44 Luo 1974: 26.
45 Chen, Li, and Mei 1973: 58.

thing as those writers from the early 1960s such as Wu Han, Deng Tuo, and Liao Mosha, who had communicated their criticisms of the Great Leap Forward by way of historical allusions?[46] Had these writers not been reproached for reverting to traditional forms of historiography and helping to prepare the ground for a restoration of capitalism? Why did they have to die if their adversaries would avail themselves of the same methods only a few years later?

Wang Gungwu believes that the way the past was used to serve the present during the Campaign to Criticize Lin Biao and Confucius was different from the mode employed in the "Evening Chats at Yanshan."[47] He stresses that the discussion had reached a new level of understanding by the time of the Pi Lin Pi Kong Campaign. This new level of understanding is reflected by the fact that Yang Rongguo had made a combination of integrating China into the scheme of five social stages while preserving China's particularity. According to his periodization, China's last big revolution took place when the six states were united under Qin Shihuang and the country simultaneously moved from slave to feudal society.[48] Since that time until the victorious takeover of the CCP, no revolution that moved Chinese society on from one stage to the next had taken place, which is why it was legitimate to compare the two revolutions. Yang Rongguo established similarity based on structural factors that could change in content without changing in form.[49] That is why comparing Mao with Qin Shihuang, Lin Biao with Confucius, and Zhou Enlai with Li Si was no longer a political trick which required manipulating the interpretation of the past in order to fulfill one's political aims. The comparison was based on "scientific" reasoning and thus beyond any doubt.

The "scientific" aspect of this logic is the fuzzy relationship between unchanging moral standards and progress in history. For Yang, Legalism represents the thought of the feudal ruling class and is therefore more progressive than Confucian thought representing the ideology of the ruling class of the slave-holder society. That is why the struggle between Confucianism and Legalism is a fight between the reactionary forces of the past and the progressive forces heading towards the future. The analogy as form of reasoning that only collides with the idea of progress intrudes into this scheme when Yang's argument is continued into the

46 Cheek 1997.
47 Chuang Hsin Cheng 1970.
48 Gregor and Chang 1979: 1084.
49 Teoh 1985.

present. Even if Yang argues that the two revolutions can be compared on a structural level, the problem arises why politicians criticizing the Cultural Revolution should be regarded as Confucianists and those promoting the idea of continuous revolution, like Mao Zedong, would associate themselves with Qin Shihuang.

In fact, Yang Rongguo had put down his basic ideas about the history of thought in ancient China as early as in the 1940s. His book on ancient Chinese history of thought was printed in the early 1950s[50] and then again during the Campaign to Criticize Lin Biao and Confucius. Thus Yang could not be reproached of having changed his ideas so as to fit Mao's political intentions.[51] In this respect he came in much more handy than Guo Moruo, who had gradually changed his ideas about Confucius and was reproached by Mao again in the 1970s for his publications of the late 1940s in which he had criticized Qin Shihuang for his authoritarianism.[52] Yang could actually make the point that he had been in favor of Qin Shihuang since the late 1940s and that he had seen the history of thought in China dominated by the conflict between Legalists and Confucianists ever since. What was later called "historiography of allusions" (影射史学) and harshly criticized as a manipulation and distortion of history[53] was, in fact, at the time regarded as a new level of understanding Chinese history beyond Eurocentrism and dogmatic Marxism. In this sense, Yang's way of alluding to the past was indeed fundamentally different from the kind of "using the past to serve the present" which Deng Tuo, Wu Han, and Liao Mosha had displayed in the "Evening Chats from Yanshan" and the "Talks from the Three Families" back in the early 1960s.

However, it did not take long for Yang Rongguo's version of the Chinese history of thought to be deleted from memory. As soon as the esoteric form of communication of the mid-1970s had been reduced to its political contents after the fall of the "Gang of Four," it had lost its appeal.[54] Its claim to scientific reasoning could no longer be upheld. Simultaneously, the form of historical allusion which continued Yang Rongguo's argument into the present was rejected because of its political implications. Paradoxically the form of "using the past to serve the present" typical for

50 Yang 1954.

51 Teoh 1985: 361.

52 Ye 2005.

53 See Fan and Wang 1977.

54 See Fan and Wang 1977, for recent debates about "history in allusions" related to the Bingdian incident (Bingdian 2006).

the debate during the late fifties and early sixties was rehabilitated. In fact, what the public debate during the late 1970s wanted to criticize was not Yang Rongguo's approach, but the way history was used to criticize politicians like Zhou Enlai in the present; and what was rehabilitated during this debate was not so much the historiography of allusions of the early 1960s as historians like Wu Han, Deng Tuo, and Liao Mosha, who were victimized during the first months of the Cultural Revolution.

On the Use of "Traditional Historicism"

What Wang Gungwu calls traditional historicism is based on the common knowledge of all participants in the discourse about a shared reservoir of historical anecdotes the interpretation of which is related to perpetuating moral standards beyond historical change. Convention demands that Confucius be regarded as positive whereas Qin Shihuang be used as a negative example. The "historicism" of this mode of thought is linked to the fact that every argument is related to a certain historical event and constellation documented in historical sources. The "traditionalism" about this historicism is that the canon of anecdotes has more or less stayed unchanged and thus preserved an understanding of Chinese history that has persisted over time. Whenever a certain debate arises in the present, participants can draw upon this[55] set of anecdotes and freely select from them whatever serves best for the present. Under normal circumstances they will refrain from changing the canon of anecdotes and the meaning of a single anecdote within the canon. Rather, they select some and reject others in order to make their argument in the present both legitimate and convincing. During the Pi Lin Pi Kong Campaign Mao not only selected some and rejected other anecdotes, more importantly, he changed the interpretation of the anecdotes, and after the Cultural Revolution ended the conventional interpretation was resurrected.

In April 1974 Mao read a text by Han Feizi in which Han deplores the fact that it was very difficult for people like him supporting Legalist ideas to speak up in public. He therefore had to invent a method of articulating his ideas which could not easily be detected as such and at the same time be understood by those he wanted to address. Mao most probably chose this text in order to remind his entourage of the possibility of communicating

[55] Liu Xiuming 2000: 541–42.

on political issues by referring to the above-mentioned canon of anecdotes. At the same time, he strove to redefine the canon. His choice of 86 readings of ancient Chinese texts[56] shows that he intended to create a canon of references that supported his basic ideas about the necessity to overcome Confucian dominance in historical thinking, to praise the Legalist strain of thought and thus to propel the idea of continued renewal of the political system in the PRC. When Feng Xianzhi and Jin Chongji argue in their official biography of Mao Zedong that he had nothing to do with the Pi Lin Pi Kong Campaign they obviously try to distance Mao from a campaign that was regarded as a Second Cultural Revolution and spread a lot of fear. They also try to convince the reader that the relationship between Mao and Zhou Enlai was much better than others suggest. But they also point to a certain difference between Mao and the writers of those articles which stick to a structural approach as the basis of their historiography of allusions. As far as we can know from his writings of the 1970s, Mao did not put any stress on methodological questions. The reconciliation between Chinese history and world history did not seem to occupy any priority in his thoughts. He used the method of history in allusions pretty much in the same way as Wu Han, Deng Tuo, and Liao Mosha had used it, only with the important difference that he referred to a different set of anecdotes and a different way of interpreting them. By having the 86 readings distributed to his entourage he wanted to make sure that they knew their anecdotes and were therefore able to decipher the esoteric communications that were based on these materials. That is why the Han Feizi text plays an important role. Through this text, Mao made the secret known to those who were not born to know it. He also showed that this method was appropriate for those wanting to oppose the majority. He thereby hinted at another major difference between what was going on during the Pi Lin Pi Kong Campaign and what the authors of the "Evening Chats" had tried to do. The debate on issues of history during the Campaign to Criticize Lin Biao and Confucius was a "revolt" initiated from above and therefore legitimate. The "Evening Chats at Yanshan" were a "revolt" initiated from below by a minority of scholars communicating in esoteric language with each other. Also, the criticism of the 1960s was based on Confucianist ethic standards, much in contrast to Mao's criticism, which was based on the highly unconventional praise for Qin Shihuang. In any case, Mao must have realized by the end of his

[56] See Liu Xiuming 2000, for a meticulous listing of the texts.

life that to win over all "knowledgeable Chinese" was a precondition to avoiding restoration, and that restoration could not be avoided without allowing them to define themselves as a group in society privileged and different from the rest by their special form of communication. This communication was based on their shared knowledge of the past and their ability to use this knowledge to propel their ideas for the present and the future. At the beginning of the Cultural Revolution, these people were attacked as "academic authorities." The attack was aimed at getting rid of their influence and of their way of networking by esoteric communication. By the mid-1970s, those of them who had survived were reintegrated into the system. This is when Mao made his attempts at redefining the canon while allowing for the same methods of communication to continue that he had attacked in the early phase of the Cultural Revolution.

Wang Haiguang comes to the conclusion that the Campaign to Criticize Lin Biao and Confucius did not reach its aims.

> The criticism of Lin Biao and Confucius as well as the reassessment of Confucianism and Legalism was too distant from the level of knowledge as well as the cultural sophistication of the people. Even though major articles criticizing Confucius and Confucianism appeared without interruption in newspapers and journals, people had difficulty deciphering their meaning and read them without understanding what they were about. Nevertheless, articles appeared signed by authors of worker, peasant, or soldier background. However, these articles were actually designed by young scholars. Additionally, the method of "using the past to serve the present" created an overflow of abuses of history in the form of "historiography in allusions." The contradictions and incoherencies that occurred provoked the opposition of upright intellectuals.[57]

Conclusion

Wang Haiguang might be right with his argument that the so-called Second Cultural Revolution did not attract the "masses." What he forgot is that it did attract quite a number of intellectuals, some of them in influential positions at the time. The esoteric communication during the last years of the Cultural Revolution gave them the feeling that the dissonance between historical materialism and traditional Chinese thought could be overcome. However, this feeling did not survive Mao Zedong. Neither Yang Rongguo's attempt to integrate historical materialism and

57 Wang Haiguang 2000: 1115.

traditional Chinese historical thinking nor Mao's approach to using the past to criticize the present from a Legalist point of view has had a traceable impact on on-going debates in the PRC. However, the basic problem of how to reconcile Chinese history and world history, universality and particularity still haunts historians in the PRC.[58] Even in times when China is going through a period of undeniable progress the idea of the irreversibility of change has not taken root. The CCP leadership is trying to establish a new superstructure to its market economy with Chinese characteristics. The idea of a harmonious society integrating traditional Chinese thought with socialism stands at the center of this attempt. At the beginning of the twenty-first century, the Chinese political elite still seems to have found no alternative to what, according to Sima Qian, a scholar of the state of Qi said when asked about the possibility of the consolidation of Qin rule: "I have yet to hear of anything able to endure that is not based on ancient precedents."[59]

References Cited

Balazs, Etienne (1964), "History as a Guide to Bureaucratic Practice." In Wright, 1964: 129–49.

Bingdian 冰点 (2006), *Bingdian zhubian Li Datong fanbo zuopai xuezhe de gongji* 冰点主编李大同反驳左派学者的攻击 (The Editor in Chief of Bingdian Liu Datong Refutes the Criticism of Leftist Scholars). www.peacehall.com/news/gb/china/2006/02/200602281846.shtml.

Cheek, Timothy (1997), *Propaganda and Culture in Mao's China: Deng Tuo and the Intelligentsia.* Oxford: Clarendon.

Chen Yangfeng, Li Zilin, and Mei Qian (1973), "Lin Biao wei shenme zhouma Qin Shihuang?" 林彪为什么咒骂秦始皇 (Why is Lin Biao Cursing Qin Shihuang), in *Hongqi* 12 (1973): 57–62.

Chow, Kai-wing, Kevin M. Doak, and Poshek Fu [eds.] (2001), *Constructing Nationhood in Modern East Asia.* Ann Arbor: University of Michigan Press.

Chuang Hsin Cheng (1970), *Evening Chats at Yenshan, or the Case of Teng T'o.* Berkeley: University of California Press.

Dictionary of the History of Ideas (2003), *Analogy in Early Greek Thought.* http://xtf.lib.virginia.edu/xtf/view?docId=DicHist/uvaBook/tei/DicHist1.xml;chunk.id=dv1-09;toc.depth=1;toc.id=dv1-09;brand=default.

Dirlik, Arif (2003), "The Politics of the Cultural Revolution in Historical Perspective," in *Law* 2003: 158–83.

Duara, Prasenjit (2001), "The Regime of Authenticity: Timelessness, Gender, and National History in Modern China," in Chow, Doak, and Fu, 2001: 359–85.

Fan Shuzhi 樊树志 and Wang Qingyu 王庆余 (1977), " 'Gu wei bang yong' de elie biaoyan. Qingsuan Xuexi yu pipan liyong lishi fandang de zuixing." "古为帮用" 的恶劣表演— 清算《学习与批判》利用历史反党的罪行 (The Evil Performance of "Gu wei

[58] Weigelin-Schwiedrzik 2007.
[59] Sima Qian 1979: 177.

bang yong"—Exposing and Criticizing the Crimes of Using Hisotry to Oppose the Party Committed by "Study and Critique"), in *Lishi yanjiu* 3 (1977): 80–87.

Fang Ling (1973), "Qin Shihuang zhuan," in *Xuexi yu pipan* 4 (1973): 82–88.

Feng Xianzhi 逄先知 and Jin Chongji 金冲及 (2003), *Mao Zedong zhuan, 1949–1976* 毛泽东传 (*Biography of Mao Zedong, 1949–1976*). Beijing: Zhongyang wenxian chubanshe.

Gao Gao 高皋 and Yan Jiaqi 严家其 (1986), *"Wenhua da geming" shi nian shi* 文化大革命十年史 (The History of the 10 Years of Cultural Revolution). Tianjin: Tianjin renmin chubanshe.

Gao Mobo (2008), *The Battle for China's Past: Mao and the Cultural Revolution*. London: Pluto.

Goldman, Merle (1975), "China's Anti-Confucian Campaign," in *China Quarterly* 63 (1975): 435–62.

Gregor, A. James and Maria Hsia Chang (1979), "Anti-Confucianism: Mao's Last Campaign," in *Asian Survey* 19, 11 (1979): 1073–92.

Law, Kam-yee ed. (2003), *The Chinese Cultural Revolution Reconsidered: Beyond Purge and Holocaust*. Houndmills: Palgrave Macmillan.

Li Shi 李时 (1974), "Qin tongyi liu guo qi jueding zuoyong de shi shenme?" 秦统一六国起决定作用的是什么? (What's the Decisive Factor in the Unification of the Six Kingdoms by the Qin), in *Hongqi* 3 (1974): 15–21.

Liu Xiuming 刘修明 (2000), "Cong yinzhi 'da zi ben' guji kan Mao Zedong wan nian de sixiang he xintai," 从印制 "大字本" 古籍看毛泽东晚年的思想和心态 (Taking a Glimpse at Mao Zedong's Thought During his last Years Through the Printing of Ancient Books in "Big Characters"), in Zhang and Su 2000: vol. 1, 531–53.

Liu Yuanyan 刘元彦 and Zhu Geji 诸葛计 (1977), "Lin Biao fandang jituan gu ji de chong-yan—ping Li Shi de yi pian fandang luanjun heiwen," 林彪反党集团故伎的重演—评李时的一篇反党乱军黑文 (The Reenactment of Old Techniques by the Anti-CCP Clique of Lin Biao —A Critique of an Evil Text by Li Shi that is Opposing the Party and Causing Chaos in the Army), in *Lishi yanjiu* 6 (1977): 17–25.

Luo Siding罗思鼎 (1973), "Qin wangchao jianli guocheng zhong fubi yu fan fubi de dou-zheng" (秦王朝建立过程中复辟与反复辟的斗争) (The Struggle for and Against the restorationin the Course of the Founding of the Qin), in *Hongqi* 11 (1973): 30–40.

—— (1974), "Lun Qin Han zhi ji de jieji douzheng" 论秦汉之际的阶级斗争 (On Class Struggle During the Qin and Han [Dynasty]), in *Hongqi* 8 (1974): 16–26.

—— (1974b), "Lun Li Si" 论李斯 (On Li Si), in *Xuexi yu pipan* 8 (1974): 10–18.

San Mu 散木 (2007), " 'Pi Lin pi Kong' yundong zhong de 'hong ren'—Yang Rongguo." "批林批孔" 运动中的 "红人"—杨荣国 (The "Red Person" in the Movement "To Criticize Lin [Biao] and Confucius—Yang Guorong), http://museums.cnd.org/CR/ZK07/cr449.gb.html (accessed Aug. 26, 2008).

Sima Qian (1979), *Selections from Records of the Historian*. Trans. by Gladys Yang. Beijing: Waiwen chubanshe.

Teoh, Vivienne (1985), "The reassessment of Confucius and the Relationships among Concepts, Language, and Class in Chinese Marxism 1947–1966: A Study of the Thought of Feng Youlan and Yang Rongguo on the Scope of Benevolence," in *Modern China* 11, 3 (July 1985): 347–76.

Wang, Ban (2004), *Illuminations from the Past*. Stanford, CA: Stanford University Press.

Wang Gungwu (1975), "Juxtaposing Past and Present in China Today," in *China Quarterly* 61 (1975): 1–24.

Wang Haiguang 王海光 (2000), " 'Pi Lin pi Kong' yundong" 批林批孔运动 (Movement to Criticize Lin Biao and Confucius), in Zhang and Su 2000: vol. 2, 1097–1116.

Weigelin-Schwiedrzik, Susanne (2006), "In Search of a Master Narrative for 20th Century Chinese History", in *China Quarterly* 188 (Dec. 2006): 1070–91.

—— (2007), World History and Chinese History: 20th Century Chinese Historiography between Universality and Particularity, in *Global History and Maritime Asia Working and Discussion Paper Series*, Working Paper No. 5, Osaka University.

Wright, Arthur F. ed. (1964), *Chinese Civilization and Bureaucracy: Variations on a Theme*. New Haven, CT: Yale University Press.

Xiandai hanyu cidian (2002), *Xiandai hanyu cidian*, 2002 nian zengbuben. Beijing: Shangwu yinshuguan.

Yang Rongguo 杨国荣 (1954), *Zhongguo gudai sixiangshi* 中国古代思想史 (*A History of Ancient Chinese Thought*). Beijing: Renmin chubanshe.

—— (1973), "Kongzi—wangu de weihu nulizhi de sixiangjia" 孔子—顽固地维护奴隶制的思想家 (Confucius—A Thinker Stubbornly Defending the Slaveholder System), in *Renmin ribao* (Aug. 7, 1954): 1.

Ye Xiaoqun 邢小群 (2005), Guo Moruo de zun Kong yu fan Kong 郭沫若的尊孔与反孔 (Guo Moruo's Reverence for Confucius and His Opposition Against Confucius). http://book.sohu.com/20051214/n240965702.shtml (accessed Aug. 26, 2008).

Zhang Hua 张化 and Su Caiqing 苏采青 [eds.] (2000), *Huishou "wenge."* 回首文革 (Looking back at the Cultural Revolution) 2 vols. Beijing: Zhonggong dangshi chubanshe.

RECOLLECTION OF THE PAST AND THE POPULARIZATION OF HISTORY

CHAPTER SEVEN

POPULAR READINGS AND WARTIME HISTORICAL WRITINGS IN MODERN CHINA

Long-hsin Liu

Prologue

Over the past years, I have engaged in studying the development of modern Chinese learned societies. In my observation, the nature and activities of these societies varied widely. Some of them were academically oriented; others were devoted to public affairs. Among these societies, the Popular Reading Publishing House (*Tongsu duwu biankanshe*, 通俗讀物編刊社, PRPH hereafter), which emerged in 1930s in the context of nationalism and national salvation, is unique in its endeavor to promote popular readings. In my opinion, the emergence of the PRPH can certainly be examined in the context of nationalism after the Mukden Incident, but it can also be looked at from the perspective of the popularization of historical narrative and contrasted to the parallel development of professional and objective historiography during the same period. The latter ultimately sheds light on the multiple natures of modern Chinese history writings.

Until now, studies of the PRPH have focused mainly on Gu Jiegang's life-long activities. A typical work is Laurence A. Schneider's *Ku Chieh-kang and China's New History: Nationalism and the Quest for Alternative Traditions*, which discusses Gu's effort to spread popular readings and promote mass culture in the 1930s.[1] Schneider also gives a sketch of Gu's endeavor to make use of historical geography and popular readings to reflect the problem of how to shape Chinese nationalism—a practical and even urgent problem in wartime China. A trip to Northwest China, according to Schneider, reinforced Gu's belief in the importance of nationalism and mass culture for anti-Japanese propaganda. In short, Schneider has made an important contribution to studies of this field, yet due to limitations on the nature and accessibility of research materials, he could not paint a complete picture of Gu's activities. Another important work

[1] Schneider 1971: 279–300.

is Chang-tai Hung's *War and Popular Culture: Resistance in Modern China, 1937–1945.* Influenced by the theory of cultural history, Hung calls for a new perspective on the nature and function of popular culture in wartime China. In his book, Hung devotes an entire chapter to the concept of "new wine in old bottles" and issues related to debates over literature,[2] but he paid little attention to the relationship between popular readings and history writings. Among works in Chinese, Liu Qiyu's *An Account of Gu Jiegang's Life and Thought* devotes considerable attention to the origin and development of the PRPH, but most of his discussions are drawn from Gu's memoirs,[3] and he does not fully expound on the theme of history writings.

All the works mentioned above have clearly contributed to our understanding of the PRPH's promotion of popular culture, but none explores the relationship between popular readings and wartime history writings. There are two core reasons for this. The first is that the PRPH's publications are simply too voluminous to be completely reviewed and digested. From 1933 to 1940, the PRPH published various kinds of popular reading, including drum singing, drama scripts, fiction, New Year's pictures and cartoons, and other genres. It has been said that during its seven years of operation, the total number of the PRPH's publications—excluding various pictorials and reprinted readings—exceeded 600 kinds and 50 million volumes.[4] Dealing this enormous volume of materials undoubtedly presents a formidable challenge to any researcher. Moreover, some of the publications are difficult to locate and others cannot be found in libraries nowadays because to contemporary readers, many of these publications, such as textbook, were valueless and hence there was no thought of keeping and preserving them for later generations. Thus many publications have disappeared. No wonder that Japanese scholar Ogura Yoshihiko, who tried to collect the PRPH's publications in the 1980s, could write only a brief introductory article based on the books he located: a meager eleven volumes.[5] Fortunately, recent unearthed materials, especially the publication of Gu Jiegang's diary and the memoirs of members of the PRPH, allow us to re-sketch the features of the PRPH and to reconsider Gu's political stand and his attitude toward the PRPH and its promotion of popular readings. The present chapter draws heavily from the PRPH's periodicals and

[2] Hung 1994: 187–220.
[3] Liu Qiyu 1986: 235–38.
[4] Gu Chao 1993: 301.
[5] Ogura Yoshihiko 1983.

other related publications obtained from various libraries. I attempt to conduct a thorough examination of the relation and interaction between popular readings and popular culture; the aim, form, and methodology of the PRPH's reformation of history writings; and the PRPH's influence on reshaping modern concepts of nationality.

From the Three Clans Publishing House to the Popular Reading Publishing House

According to Gu Jiegang, the PRPH was founded on July 27, 1934, by twenty or more intellectuals, including Gu himself, Xu Bingchang, Fan Wenlan, and others.[6] Its predecessor was the Three Clans Publishing House (Sanhu shushe, 三戶書社, TCPH hereafter), which was established by the Anti-Japanese Association of the Faculty Members of Yenching University (Yanda jiaozhiyuan kangri hui, 燕大教職員抗日會, AJAYU hereafter) after the Mukden Incident. As propaganda secretary of the AJAYU, Gu Jiegang suggested offering monetary payment to solicit drum singing (drum-accompanied lyrics) and opera scripts (劇本) for anti-Japanese propaganda and promoting popular readings in these genres.[7]

About the same time as Gu put forward his proposal, a group of Yenching University students, returning from the battle zone after a morale-boosting trip in Rehe Province, related that the provincial voluntary army was fervently hoping that Yenching University would compile a textbook for them. Gu thereupon consulted professor Guo Shaoyu about opening a new workshop—Popular Literature Composition—in the Department of Chinese Literature with a view to training students to write folksong transcripts, play scripts, fiction, drum singing, and textbooks.[8] The TCPH was thus established, attached to the AJAYU for the publication of these works. The AJAYU also set up its own bookstore—Jinli Bookshop—for marketing and selling those publications.[9]

Aside from publishing the works of faculty members and students, the AJAYU also held essay competitions, encouraging creative writing from the public and publishing the winners' works.[10] In order to attract a broad

---

6 Gu Jiegang 2007: 3.216, Gu Chao 1993: 222.

7 Gu Chao 1993: 205.

8 Wang Xuhua 1990: 1036–37.

9 Gu Chao 1993: 210.

10 Zhao Buoyong won the first prize with his *Du Chuan sishou dujayu*, which was based on a real story about the volunteer's army. The second was *Cuihong guniang xunnanji*,

spectrum of readers, it replicated the style of contemporary bestsellers, such as a red-green chromatographic cover with some photographs of theatrical figures wearing fancy costumes.

Initially, the publisher did not anticipate a warm reception, and only 5,000 copies of each of its pamphlets were printed. However, to its surprise, most of them quickly sold out. Some booksellers even made pirated copies, reprinting these bestsellers. For instance, *General Song Zheyuan and the Fierce Battles at Xifengkou Pass* (Song Zheyuan dazhan Xifengkou, 宋哲元大戰喜峰口) was said to have been reprinted in pirated editions to the tune of over 70,000 copies.[11]

Presumably because of this rampant piracy, not only did the TCPH make no profit from the wide circulation of its publications, but its affiliated bookshop, Jinli, was even forced to close down in 1934.[12] The TCPH was fortunate in being spared from a similar fate after receiving subsidies from the Ministry of Education,[13] yet it had to rename the "Popular Reading Publishing House" upon the insistence of Wang Shijie, minister of education.[14] Thereafter, the TCPH broke away from the AJAYU and became an independent society for academic learning.

Thanks to Gu Jiegang's incessant efforts, the PRPH received various funds from the Ministry of Education, the Sun Yat-sen Institute for the Advancement of Culture and Education, and the headquarters of the KMT (Kuomintang/Guomindang).[15] With this help, the PRPH successfully tackled a series of financial problems and was ready to carry out the task of propagating anti-Japanese popular readings.

---

which mainly depicted the miserable scene after the surrender of Shenyang. The third and following prizes were awarded to, among others, the drama scripts *Songhuzhan*, *Paiziqu*, *Ku Chaoxian*, and twang singing *Yiyongjun Nüjiang Yao Reifang*, to name just a few. Gu Jiegang 2007: 51–52, Liu Qiyu 1986: 236.

11 Gu Jiegang 1988b: 365. There are various calculations of the total number of copies of *Song Zheyuan dazhan Xifengkou*. Gu Jiegang reported 70,000 volumes in "An Overview of Publishing Popular Literature." But during the war in a forum for reviewing "new wine in old bottles," the PRPH gave a new figure: about 700,000 volumes. This variant figure probably derives from the different time-point of the investigation. See "Guanyu jiuping zhuang xinjiu de chuangzuo fangfa zuotanhui jilu," in Gu Jiegang 1938a: 110.

12 Gu Chao 1993: 215.

13 Wang Shouzhen 1938b: 22.

14 Gu Chao 1993: 212, Gu Jiegang 2007: 3.92, 98.

15 Gu Jiegang 1988c: 336, Gu Chao 1993: 225.

New Wine in Old Bottles: Popularization and Secularization

Although the PRPH and the AJAYU went their separate ways in October 1933, many of their first members were from Yenching University. Most of them, such as Wu Shichang, Zheng Kanci, Yang Bin (Yang Jiwue, real name; also known by his pseudonym Yang Gang), were Gu's students. Famous writers like Wang Shouzhen (Wang Riwei, real name), Li Yifei, Zhao Jibin (also known by his pseudonym Xiang Linbing), Shao Hengqiu, Zhang Xiuya, and others came from abroad and joined in the PRPH in a later period. Once the PRPH became funded by the government and the KMT, two problems arose. The first was that it could no longer explicitly appeal to the public in the cause of the anti-Japanese movement. The second was that drum song composition was dropped, despite its popularity among readers, because writers from the academic world were not skilled in this genre. It was replaced with a series of adapted and complied popular stories.[16] Under Gu Jiegang's influence, the works of PRPH members were introduced and submitted to various newspapers and journals, such as the History and Geography Weekly of *Dagongbao* and *Middle School Student* (Zhong xuesheng, 中學生). From March 1936, the PRPH launched a special column, Popular Lecture Weekly (*Tongsu jiangzuo*, 通俗講座), in *Shenbao*, for which it received a monthly payment of $120.[17] Gu argued that the most suitable heroes of popular literature to cultivate a national spirit should be heroic figures from Chinese history. This idea was embodied in most of the works submitted to the newspapers mentioned above. Most of the stories were about people resisting alien invaders, such as Gou Jian, Zi Can, Yue Fei, Shi Kefa, Huang Zongxi, Gu Yenwu, and Zheng Chenggong.[18]

Apart from "searching for useful materials in history," the PRPH members also began to ponder upon what "useful knowledge" should be conveyed to the general public other than national consciousness. In the opening statement of the *Tongsu jiangzuo* in *Shenbao*, the editor advocated the idea of "awakening people":

> To "awaken people" is not only to call upon them to recognize the poverty of their life and the need for hard work and new knowledge, but also to make them realize that the nation faces a crisis and that it is their responsibility to

---

16 Gu Jiegang 2007: 3.330–32, Gao Chao 1993: 229–30.
17 Gu Jiegang 1988c: 366.
18 Gu Chao 1993: 230, He 1936: 5, Zhou 1936: 5, Li Jizhen 1936: 5.

> save their country from this crisis. In this transient epoch, everyone should be equipped with the necessary tools of knowledge needed by a modern citizen.[19]

The PRPH maintained that the Chinese people were not only poor in material life, but also intellectually deficient in knowledge and skills and therefore remained backward. From the traditions of the past thousands of years they had inherited dullness, pomposity, dependency, blind obedience, selfishness, fatalism, disunity, and indifference to politics, all of which disqualified them as modern citizens. The PRPH members believed that popular literature should be held responsible. Works of popular literature such as folksong transcripts, twang singing (tanci, 彈詞), and fiction contained too many obsolete feudalistic thoughts which dominated the public's living habits, thoughts and behaviors, forms of entertainment, and even funeral customs.[20]

Gu Jiegang and Wang Shouzhen believed that although the advocacy of vernacular literature and proletarian literature from the May Fourth Movement onward had popularized the written language to a certain extent, its influence was still limited to the urban and Western-educated elites,[21] whereas most peasants and workers were still beyond its reach.[22] Even though some popular works were considered as masterpieces in the New Culture Movement and said to have literary value, Gu still maintained that plots and contents were full of feudalistic thinking such as was found in reincarnation stories and tales of celestial fairies. All these, he contended, "obstruct social progress" and "violate the scientific spirit" and undoubtedly constituted the most tenacious obstacle to the introduction of new culture to the masses. In order to rectify these problems, the PRPH launched a new magazine entitled *Minzhong zhoubao*, and put forward the slogan "New wine in old bottles." It urged that popular literature employ old but familiar forms and styles of composition to imbue a

---

19 Tongsu duwu biankanshe 1936.

20 Tongsu duwu biankanshe 1936.

21 In the prewar era, most of the PRPH members, except for Gu Jiegang, wrote under pennames to protect themselves from Japanese investigations. For example, Wang Shouzhen (王守真), one of Gu's students, use Wang Shouzhen (王受真) as a pseudonym. Sometimes Gu revised students' compositions and then submitted them in his name. Thus, for example, the real author of "Why Should New Wine Fill Old Bottles?" was Wang Shouzhen, but used Gu Jiegang as author when it first appeared in *Minzhong zhoubao* 1, 5 (Oct. 30, 1936). Not until Shenghuo shudian published *Tongsu duwu runwenji* (Collected Writings on Popular Readings) did the editor replaced Wang Shouzhen as the author. Wang Shouzhen 1938a: 13–17.

22 Gu Jiegang 1936a: 15. This is also mentioned by Wang Shouzhen 1938a: 13.

new and modern national culture, including scientific knowledge for the masses, revolutionary concepts, a national consciousness and patriotic spirit, to replace superstition, absurdity, and obscenity.[23] This is, in a sense, an extension of the enlightenment spirit of the May Fourth Movement.[24] But as the PRPH members saw it, so-called secularized works had failed the test of "popularization" ever since the May Fourth Movement era. As Xiang Linbing put it, the mistake of the May Fourth intellectuals was to promote literature that "de-popularized secularization" and "un-secularized popularization",[25] i.e., literature that was either popular but obsolete or progressive but aloof.[26] The idea of "new wine in old bottles," therefore, was "secularization based on sound popularization, and popularization based on rational secularization."[27]

The concept of "new wine in old bottles" can be traced back to two origins: the theory of the popularization of literature and art advocated by the League of Left Wing Writers in the 1930s; and the debate in literary circles over "new or old forms" of literature. Basically, the attitude of the PRPH on the latter issue coincided with that of the Left Wing writers. Lu Xun, for example, argued that popular readings must adopt part, but not and never all, of old forms of literature. This partial adoption must consist of "exclusion and addition," which leads to a new form and style.[28] This is what Xiang Linbing meant by concepts like "inheritance as creation" and "submission as subjugation." Xiang insisted that the aim of adopting an old form is not catering to, or even surrendering to the backwardness of the masses. On the contrary, the affirmation of an old form is just the beginning of negation of oneself, in other words, a process from quantitative change to qualitative transformation.[29]

Actually, the 1930s debate in literary circles over new/old forms had a very significant implication, namely, the search for a cultural form of a new nation. The PRPH, therefore, embraced the idea of integrating popular culture with tradition, as seen in, for example, Gu Jiegang's folklore studies.[30] Gu attempted to find a nationally distinguished form of cultural expression in popular culture. Xiang Linbing and others associated

23 Tongsu duwu biankanshe 1936.
24 Gu Jiegang 1936a: 5, 16–17.
25 Xiang 1938b: 35–36.
26 Xiang 1938c: 60.
27 Xiang 1938a: 70–72.
28 Lu 2000: 16–18.
29 Xiang 1938c: 67–68.
30 Schneider 1971: 148–56.

with the PRPH argued that overall Westernization and opposition to the traditional heritage in the May Fourth era not only could not lead to an organic connection between Chinese and Western culture, but also failed to "incorporate [the cream of] traditional culture in a critical way" and China's "transcendental inheritance."[31] They insisted that in the context of war against Japan and the challenges of the nation building, a genuinely new form of cultural expression for a new nation could only be created by preserving positive elements in traditional mass culture.[32]

The compilation of popular literature and the formation of the theory "new wine in old bottles" were the result of a trial-and-error process. In order to understand the degree of public's reception of and the tastes in literature, in 1936 the PRPH dispatched Sun Xiangli to investigate the sales volume and marketing network of publications in Beijing.[33] According to Sun's report, the PRPH was astonished by the influential power of popular readings and their special marketing network.[34] If the PRPH could not effectively utilize this sales network, their publications would have little access to readers in the countryside. Because the readers in this network accepted nothing but the traditional "appearance" of a book, the strategy "new wine in old bottles," which meant new content with the old form of binding and layout, had to be applied here in order to take full advantage of this extensive commercial network.[35]

In order to adapt to the market demand and implement the notion of "new wine in old bottles," the PRPH adjusted the content of its publications several times. But who was really the "populace" to which the PRPH addressed itself? Was it the urban working class or rural villagers? In either case, the PRPH realized that to reach these illiterate lower-class readers it needed the help of "middlemen," i.e., traditional storytellers and drum singers. All new knowledge had to be fully comprehended and digested

---

[31] Xiang 1938a: 70–71.

[32] The issue of "the form of Chinese nation" was continually debated in the 1940s. Leftist writers, such Zhou Yang, Ai Siqi, He Chifang, and Xian Xinghai, held forums or took up the issue in their discussions in the press in Yan'an, in such publications as *Xin Zhonghua bao*, *Wenyi tuji*, and *Zhongguo wenhua*. The viewpoint of "new wine in old bottles," advocated by Xiang Linbing and the PRPH members, was seriously criticized by these leftists. The components of the debates varied widely, including literature, poetry, drama, music, and fine arts. See Tang 1992: 3.36–46, Hung 1994: 253–62.

[33] Sun wrote a report on the condition of the circulation of popular readings. See Wang Zhen 2006: 109–110.

[34] Wang Shouzhen 1938a: 14–15.

[35] Wang Shouzhen 1938b: 20–22.

by these "middlemen" before being passing on to the masses.[36] Therefore, most compositions and works in the PRPH's publications, including periodicals, chapter-style novels, drum singing, local drama and popular stories, and other forms of literature were written for these middlemen to understand, so that they could in turn reproduce the message by means of reciting or verbally relating it to the illiterate masses, and eventually transforming the thinking of the lower class and reforming society.[37]

Only after the question of the readership was established could publishers settle on the content and form of publications.[38] After several years' experience, the PRPH's writers became increasingly familiar with and even skilled in the "old form" of composition. In the beginning, the genres of drum singing, drama, and *tanci* were common. Later they discovered multiple and colorful forms in folk culture, such as singing melody (*changqu*), old drama, ditties (*xiaodiao*), folksongs, comic dialogue, *shuanghuang* (a two-man act with one singing or speaking while the other acts out the story), *layangpian* (a still-picture show viewed through a magnifying lens), cartoons and New Year's pictures, to name just a few.[39] All these new genres, born and adapted from various environments, represent indigenous characters and richness. While using these new forms, the PRPH also found that most of these chapbooks were created for performance rather than reading, in which singing weighed more than oral speech.[40] Hence, the PRPH began to study the expression structures and rhyming patterns of Beijing opera, drum singing, drama, melodies, and folksongs, etc., in order to compile more attractive works for the middlemen and the masses.[41] In 1937, the PRPH introduced a "training course for drum singing" and employed a famous blind drum singer, Zhai Shaoping, to lecture and to revise its publications.[42]

---

36 Wang Shouzhen 1938a: 13.

37 Gu Jiegang 1938c: 116.

38 The PRPH was never consistent about its targeted readership. The editor of the *Minzhong zhoubao*, after publishing the fourth issue, responded to reader's letters that the targeted readers were "teachers and middle school students in the countryside, urban peddlers and workers, plus the old literati who could read *Seven Heroes and Five Gallants* (*Qixia Wuyi*, 七俠五義)." Editorial Board 1936: 48–49.

39 Gu Jiegang 1938c: 116.

40 Most of the folk arts and literature, like drama, regional operas, drum singing, and even written novels, are spoken to an audience, not to individuals. This verbal form is inevitably combined with music, thus making it into a propagandistic composite art. See Xiang 1938c: 64–65.

41 Gu Jiegang 1938d: 2–7, Gu Jiegang 1938c: 116.

42 The nine singers, trained by Zai Shaoping, were invited to perform at a banquet that Xu Bingchang gave for Fu Zuoyi. Fu was delight by their performance and invited them

After several years of advocacy, the PRPH's guideline of "new wine in old bottles" was gradually accepted by writers and artists. Lao She (Shu Qingchun) may be one of the best examples. After the outbreak of the war, Lao She abandoned his attempt to write "New literature novels," and when he went to the rear in 1938, he began to adopt the idea of "new wine in old bottles" to compose drum songs. In order to write appropriate works with adequate rhythms and singing style, he asked Fu Shaofang and Don Lianzhi, drum artists who had escaped from the north, for advice on writing drum songs in the Beijing and Shandong ditty style. He also tried to learn the singing of Henan *zhuizi* (ballad songs) from an artist.[43] Basing on these techniques, Lao She composed various popular works, such as *Wang Xiao ganlu* (Wang Xiao Drives a Donkey), *Zhangzhong dingji* (Zhangzhong Sets the Plot), *Da xiaoRiben* (Fighting the Little Japanese) in drum-singing style; and *Xin cihu* (Killing the Tiger: A New Version), *Zhonglietu, Wangjiazhen* in the Peking opera genre, to name just a few. In his view, the "old form of literature," no matter what it is, would have the same effect on stimulating popular passion for the Anti-Japanese War.[44]

At the same time, Lao She and other leftist writers, such as Mao Dun, Guo Moruo, Feng Naichao, Xia Yan, and Hu Feng, organized the All-China Resistance Association of Writers and Artists (ACRAWA hereafter) in Hankou and then move their organization to Chongqing in August 1938.[45] There they received a subsidy from MOE and Central Propaganda Department of the KMT, and held various workshops and study groups on the composition of popular literature.[46] They invited members of the PRPH,

to perform *Bailing Temple* (*Bailing Miao*, 百靈廟) for his troops, to extol their fighting success and to boost their morale. Their vivid performances touch the heart of many audiences. Gu Jiegang 1988c: 367.

[43] Lao She 1982b: 33–36.

[44] Lao She 1938a: 42.

[45] The inaugural meeting was held on March 27, 1938, with hundreds of attendees, who elected Cai Yuanpei, Zhou Enlai, Romain Rolland, Agnes Smedley, and nine others to be the Honorary Presidium, and appointed Shao Lizi, Lao She, Feng Yuxiang, Guo Moruo, Chen Mingshu, Tian Han, Zhang Daofan and Hu Feng to be members of Presidium. They also recommended Zhou Enlai, Guo Morou, Ru Diyuan, Feng Yuxiang, Shao Lizi and Wu Zuxiang to co-draft "The Declaration for Anti-Japanese Movement from the Association of Literary and Art Circles." See Gan 1989: 92.

[46] The ACRAWA received a subsidy from the Central Board of Propaganda of the KMT and was commissioned to compose street theater, drum singing, children's readings, popular fiction, and instructions for popular entertainment. All these works were published by the Central Board of Propaganda. See "The Report of ACRAWA Committee Affairs: July 2," in *Kangzhan wenyi*, no. 11 (Oct. 15, 1938); "The Report of ACRAWA Committee Affairs: October 15," in *Kangzhan wenyi*, no. 18 (Oct. 15, 1938), cited from Lao She 1991a: 15.607–8, 613–15.

such as Xiang Linbing and Wang Zemin, to be the lecturers.[47] Although members of the PRPH and the ACRAWA did not share the same concept of "new wine in old bottles," occasionally they exchanged their experience in literary creation. When Laoxiang (Wang Xianchen, a member of the ACRAWA) composed a drama *Muhanyi* (Collecting Winter Clothing), he especially invited Gu Jiegang, Zhang Daofan, and Zhang Mosheng to the audition.[48] Gu was gratified with the ACRAWA's endeavor to promote the writing of drum songs and declared that "this work has [finally] been done diligently and broadly."[49]

In applying the idea of "new wine in old bottles," Lao She and the members of ACRAWA came to believe that what was needed was more than merely replicating the style of the "old forms," since the issues of importance to ordinary people, and their tastes, are important ingredients for attracting an audience as well. They noted that although some Chinese dramas, such as *Qiuhu Xiqi, Wujiapou,* and *Fenhewan,* use the same story structure, and in some respects even duplicate one another, people nonetheless liked all of them. This feature of immutable structure gave Lao She the idea that writers should not only replicate the same story structure but should also to follow and adopt a way of thinking similar to that reflected in the "old forms." Hence loyalty to the emperor in Chinese drama could be substituted by patriotism to the nation, and chivalry could be replaced by sacrifice for a righteous cause. In this way, the old structure and old thinking could made useful for arousing popular sentiment in the Anti-Japanese War. As Lao She remarked, to be effective as propaganda, "a story should be attractive in the sense that it resembles the old as well as reflects the new, and contains thinking that is both time-tested and fresh."[50]

In fact, Lao She's remark echoed what the PRPH claimed. After years of promoting "new wine in old bottles," the PRPH realized that people would not buy its chapbooks unless they had heard or seen the stories and unless the content was comprehensible. Authors had to position themselves in the readers' perspective and re-create subjects familiar to readers before replacing the old content. For example, it was said that

---

47 "The Report of ACRAWA Committee Affairs: November 5," in Lao She 1991a: 15.616–17.

48 "The Report of ACRAWA Committee Affairs: October 29." in Lao She 1991a: 15.615-16; Gan 1982: 111.

49 Gu Chao 1993: 289.

50 Lao She 1938c: 10, Lao She 1938b: 5.

if one attempts to introduce the history and geography of Xinjiang, one should not use these word "Xinjiang" as the title; instead, if one changes the title to *New Journey to the West* (*Xin Xiyouji*, 新西遊記),[51] for instance, people will be enticed to buy it. *Sister Wang Misses Her Husband* and *The Little widow Visits the Grave* are two other typical and successful examples. Sister Wang, a traditional obedient wife figure in Chinese literature, in the PRPH's revision, became a widow yearning for her husband who is fighting in the Anti-Japanese War, and the grave the widow visited is her husband's, who died in the war. After this revision these books became best sellers. Another way of implementing "new wine in old bottles" was to reprint well-received or old opera scripts with the addition of a new drum song on the last pages. In doing so, PRPH believed, intentional education goals could be fused into unconscious leisure activities.[52]

Overall, the protagonists in the PRPH's publications varied widely, from political leaders, to military commanders, soldiers, the wounded, refugees, students, women, guerilla fighters, traitors, members of the puppet army and even the enemy, to name just a few. The PRPH's publications were highly esteemed and well received everywhere. After the Sino-Japanese War broke out, Gu Jiegang unexpectedly found his name on a Japanese arrest list. He and PRPH members were forced to flee hastily to Suiyuan Province. Not until he reached there did Gu find that their publications were very well-received and widespread.[53] Publishers in Taiyuan and Lanzhou even reprinted a great many of their publications.[54] Local societies and people also copied their experiences in organizing similar groups and rewrote folksongs to promote a spirit of resisting the enemy.[55] In addition, the drum song "*The Fierce Battle at Pingxing Pass*" (Dazhan pingxing guan, 大戰平型關), which extolled Lin Biao, though irritated KMT headquarters and Military Council, was welcomed by the communist Eighth Route Army.[56] All this is evidence of the achievements of the PRPH's work in promoting popular readings in northwestern China.

---

[51] Wang Dazhen 1936: 36–41.

[52] Gu Jiegang 1988b: 368–69.

[53] Gu Chao 1993: 277.

[54] Gu Jiegang 1988b: 367.

[55] Youth in Lanzhou organized a similar society entitled The Folks (Laobaixing, 老百姓) and invited Gu Jiegang to be its director. It published a bi-weekly magazine. Gu Jiegang 1988c: 4.

[56] Gu Jiegang 1988c: 4.

The Construction of History Knowledge and Reshaping National Modernity in Wartime China

Before starting this section, let me briefly introduce the PRPH publications I obtained from various libraries. The PRPH published three major periodicals:, *Minzhong zhoubao*, of which the first issue was October 2, 1936; *Dazhong zhishi*, launched eighteen days later; and, two months later, *The Route to Survival* (Qiushengzhilu, 求生之路). Another main publication is *Tongsu jiangzuo*, a serial and special column in *Shenbao*; and the last is *The First and Second Series* (Jiazhong congshu 甲種叢書, Yizhong congshu 乙種叢書)—both highly influential publication but unfortunately difficult to find today. Although these periodicals were launched very close in time to one another, they were intentionally aimed at different readerships. Of the three, *Minzhong zhoubao* was the most encompassing in content. Gu Jiegang initially was the chief editor, but was replaced by Xu Bingchang after the second issue. The key figures on the editorial board remained Wang Shouzhen, Xiang Linbing, and Li Yifei. *Dazhong zhishi*, on the other hand, seemed more topically balanced and academic in content. Gu Jiegang was nominally the chief editor, but the person in charge was in fact Wu Shichang. Wu graduated from the Institute of Chinese Classics of Yenching University, and took on the post of editor at the Research Institute of Peiping. Other key figures under his editorship came from Yenching University and Fu Jen University, for example, Lian Shisheng, Yang Bing, Zhang Xiuya, Zheng Kanci, and others.[57] The readership of *Qiushengzhilu* is quite different from that of the other two. Intended for elementary school students, the journal's content and wording were simpler and easy to comprehend. The length of each issue—twenty some pages—was also a reasonable length for young pupils.

In an article entitled "The Compositional Methodology of New Wine in Old Bottles," Gu Jiegang argued that popular reading is suitable for the general masses, not students.[58] He considered that the former were the subject of the popular cultural movement, while the latter the youth movement. Hence the form and content of these two movements are differentiated. Based on this differentiation, *Minzhong zhoubao* and *Yizhong congshu* were actually periodicals for advocating the cultural movement;

---

[57] Gu Chao 1993: 259.

[58] Gu Jiegang 1938b: 10.

Dazhong zhishi and *Tongsu jiangzuo* in *Shenbao*, were oriented toward the student movement.[59]

Although promoting popular readings was a shared goal of these periodicals, there still existed some divisions of labor among them. This division manifested in a bi-trajectory route: one for the populace, and the other for young students and intellectuals. Gu Jiegang insisted that "reforming and rewriting the history of China should transform the masses' concepts of history."[60] It was with this aim in mind that he devoted himself to numerous publications in the prewar period. He told his friends more than once that his ambition is "let the upper class understand China through reading *The Chinese Historical Geography Semi-monthly Magazine* (Yu Gong, 禹貢半月刊), and the lower class identify themselves with Chinese nationalism through popular readings."[61]

To reform the masses' concepts of history, what Gu Jiegang attempted to convey is that the populace should be aware of their position and obligation in the process of nation building. Through rewriting historical biographies and events, the PRPH sought to prepare the masses to become "modern citizens" and to acquire the rudiments of knowledge. In a note inviting submissions, the PRPH stated its ultimate goals to be "enlightening the masses, inspiring their will for self-improvement, enhancing the national consciousness, and cultivating national self-confidence."[62] In confronting the national crisis, the PRPH took its responsibility to be to "awaken the people and call on them to shoulder a heavy burden in the name of 'salvation and striving for survival.'"[63]

Since "national salvation and striving for survival" were the objective, the most reoccurring protagonists in *Minzhong zhoubao, Dazhong zhishi,*

[59] *Minzhong zhoubao* ceased publication in February 1937. In May 1947, it was relaunched, under the title *Minzhong zhoukan*, by the publisher Minzhong duwu she (民眾讀物社, Popular Readings House). The editor emphasized that it is a popular magazine for "peasants, citizens, and the laboring masses." Unfortunately, this periodical only last for seven months and once again ceased publication. See "Suzhou guoli shehue jiaoyu xueyuan minzhong duwushe bian," *Minzhong zhoukan* 1: 1–1:29/30 (May 24, 1947–Dec. 14, 1947); Gu Chao 1993: 332. On the other hand, *Dazhong zhishi* only published twelve issues and ceased publication thereafter. After the war, Gu Jiegang organized the Shidi tongsu congshu bianjishe (Compilation House for Popular Series on History and Geography). Through this publisher, Gu revived the spirit and goal of *Dazhong zhishi* and published several well-received books, such as *Zhongguo minglen zhuan* and *Zhongguo lishi gushi xiaocongsu.* See Gu Chao 1993: 314, Gu Jiegang 1949: 7.

[60] Gu Chao 1993: 200.

[61] Gu Chao 1993: 237, Gu Jiegang 2007: 3.376.

[62] "Tongsu duwu biankan she zhenggao qishi," in *Minzhong zhoubao* 1, 1 (Oct. 2, 1936): 68.

[63] Bianzhe, "*Minzhong zhoubao* kaichangbai," in *Minzhong zhoubao* 1, 1 (Oct. 2, 1936): 2.

and *Tongsu jiangzuo* are generally national heroes such as Shi Kefa, Miao Kexiu, Qi Jiguang, and Lin Zexu. Heroines are also frequently present in the stories—for instance, Hoelun (Genghis Khan's mother) or the story about "Madam City" where the prefecture chief's mother commanded an array of women armies to protect the city from the attack of the Di 氐. Interestingly, as Shen Songqiao has suggested, during the late Qing the genealogy of historical heroes was constructed by the revolutionaries as "ethnic heroes" and by the reformists as "national heroes."[64] These heroes or heroines might have been in opposing camps, fighting against each other, in the actual historical context. Ambiguously, their hostile roles were switched in modern China so that they were portrayed as friends or even compatriots, in order to suit the changing boundaries of the Republican state, or more specifically, to suit the goals of the Anti-Japanese War. In these sorts of works, besides heroes and heroines, loyal courtiers and traitors—for instance Kou Zhun, Wang Anshi, Zhuge Liang, and Qu Yuan—were also portrayed. Gu Jiegang also depicted villains like Shi Jingtang, Zhao Yanshou, and especially molded Wang Qinruo into a "traitor condemned by all" in *The Son of Heaven Counterfeits the Book of Heaven* (Zhenlong tianzi Zaotianshu, 真龍天子造天書).[65] The contrast between loyalty and betrayal in these works not only highlights the value judgment concerning these figures, but more importantly, it had been re-shaped under the goal of constructing a homogeneous nation-state. Thus historical figures, no matter whether of Han or alien ethnicity, and no matter whether loyal subjects or traitors, were made to serve the current need of national integration and the rhetorical strategy of popular reading.

It goes without saying that when a nation is engaged in a life-and-death struggle, it is not necessarily men alone who may become heroes nor is it adults alone who may devote themselves to serving their country: women and children, as well as officials and even bandits, may answer the call of duty. "The Mother of Wang Sungu," written by Zheng Kanci, for example, is a story about the struggle between Yan and Qi in the Warring States period. The protagonist Wang Sungu, a fifteen-year-old boy, is the son of a senior official in Qi. He is encouraged by his mother to call on 400 loyalties, and successfully annihilates the army of Chu and thus saves

---

[64] Shen 2000: 107–43.

[65] Wang Naitang 1937b: 17–24.

the endangered Qi.[66] That Zheng Kanci chose as her protagonist not a famous historical figure but a child was obviously a rhetorical strategy. In her depiction, a bully could never fight against those who stand united, and thus people like women, children, the old and the disabled, who may seem weak, can turn out to be powerful if they join together and so create a general will.[67] The PRPH's authors vigorously advocated the spirit of perseverance and unification in bravely fighting the enemy in order to awakening their people to their responsibility to the nation.

If we compare the identical historical figures portrayed in these journals and publications with each other, we can easily perceive distinctive differentiated writing strategies and emphases. The portrayal of heroes in the special column "Historical Story" of *Dazhong zhishi* and *Tongsu jiangzuo*, for example, usually focuses on their integrity, nobility, and contribution to the nation; traitors, on the other hand, are characterized by their evil behavior and unpardonable vices. Typical of these stories is a free, easy, and comprehensible writing style interweaving dialogues to portray the protagonists as vivid characters. But this does not mean the stories were inauthentic. For instance, Zheng Kanci, whose works was praised by Gu Jiegang as "characterized by a popular tone but without the style of pupils,"[68] had once composed a tale about General Guo Ziyi. In her story, General Guo bravely broke into Uighurs' barracks, and single-handedly persuaded them to retreat. She wrote the story with a free style and easy diction, but made it as historically accurate as possible.[69]

Yang Liansheng (Lien-sheng Yang), who later became a famous professor of history at Harvard, wrote a story about Monk Xuanzang's pilgrimage to India in which he emphasized his amazing perseverance during the great pilgrimage and his lifelong devotion to the translation of sutras. Yang also took the opportunity to expound on Buddhist doctrine and its spread into China.[70] Another example can be seen in Wang Chongwu's much-lauded rendering of Hoelun, Genghis Khan's mother. Wang writes of Hoelun's sadness when forced to marry Yesugei, the hardship she endured by the Onon riverside, and her anger over the conflict among her sons. All this detailed depiction is encompassed within a five-page booklet, containing seven footnotes. The story, according to the author,

---

66 Zheng 1936: 5.
67 Jueren 1937: 3, Han Depu 1936: 5, Ji Zhen 1936: 5, Su Zihan 1936: 4.
68 Gu Chao 1993: 208.
69 Zheng 1937: 30–37.
70 Yang 1937: 20–26.

is strictly based on documented materials from *Yuan mishi*, *Yuanshi*, *Xin Yuanshi*, *Mengwu'er shiji*, and *Menggu Yuanliu*; the author even referred to textual criticism from "Notes on the History of Mongolia" by Wang Guowei.[71] Although a detailed and academic textual criticism is lacking in these stories, the authors never abandon the idea of delivering actual historical facts and information. The narration was fluent and always conformed to the historical evidence. The authors seldom made personal remarks and tried to be objective; if an inevitable circumstance arises, the author only insinuates and leaves room for readers to ponder or comprehend on their own. Overall, this type of literary narrative, though popular, was not written for illiterate peasants or workers.

However, the narrative style of portraying historical figures in *Minzhong zhoubao* and *Yizhong congshu* was quite different from that of *Dazhong zhishi* or *Tongsu jiangzuo*. Apart from frequently using drum singing and singing scripts, *bengbeng* drama, or *pihuang*, the narrative style and content are distinctively different from those of the other two magazines. Most of the popular folksongs, such as "Ten Cups of Tea" (Shibei cha, 十杯茶), "Ten Glasses of Wine" (Shibei juo, 十杯酒) or "Fengyang Flower Drum" (Fengyang huagu, 鳳陽花鼓), express nothing more profound than a clandestine affair or a wife longed for her husband's homecoming.[72] Their narration, akin to doggerel, deals with irrelevant historical figures for the purpose of achieving harmonious rhyme and easy comprehension. Take "Xinchu huagu ci" (新出花鼓詞), for instance:

> Jiang Taigong steadily sat on the fishing stand,
> Guangong was formerly a dofu dealer,
> Liu Bei was once a straw-sandal maker,
> Zhao Zilong came from Changshan,
> Xue Rengui crossed the sea and fought eastward,
> Li Jing invaded the Central Plain,
> Li Cunxiao was a muscle man,
> Prime Minister Xiao He met Zhang Liang, ...[73]

These incoherent lines, without any concern for reality or for temporal and historical order, were composed merely because they are easy to sing and understand. In order to rewrite these works, the PRPH writers applied the theory of new wine in old bottles to infuse current facts

---

[71] Wang Chongwu 1936: 26–32.

[72] Zaqiang changben: Shibeicha (n.d.), in Shuanghongtang xiqu 190. Zaqiang changben: Xinchu gailiang Shibeijuo (n.d.), in Shuanghongtang xiqu 190.

[73] Zaqiang changben: Xinchu Huaguci, Shiyinben (n.d.), in Shuanghongtang xiqu 190.

concerning the national crisis. The result was numerous works such as "Ten Cups of Tea for the War of Resistance" (Kang'Ri shibeicha, 抗日十杯茶), "The New Flower Drum" (Xin huagu, 新花鼓), "Twelve Months of National Calamity" (Guonan shi'er yue, 國難十二月), and "Folksongs of the War of Resistance" (Kangzhan geyao, 抗戰歌謠), to name just a few.[74] These works more or less apply the ten-stanza form of "Ten Cups of Tea" to convey the sentiment of protecting China and attacking traitors. Historians of popular history have suggested that this kind of seasonal folksong can be traced back to the Six Dynasties and became the forerunner of the "Twelve Months of Drum Singing" that emerged in Song times. The various and diverse folksongs were widely wielded as an instrument of enlightenment by intellectuals and gentry during the late Qing period.[75] However, the monthly sequence in these folksongs, though regularly ordered from January to December, derived its meaning from its function to indicate rhythmic breaks. By contrast, the temporal order in the folksongs created by the PRPH indicated an objective time sequence and constituted a reliable record of current events. For example, the "Twelve Months of National Calamity" adopted the vocal form of Henan zhuizi to redraw the history of the Republic of China. It briefly describes the history from the overthrowing of the Qing dynasty and founding of the Republic to the Xi'an Incident and the Japanese invasion, using the order of months as line-heading as, for example, in June, the battle between the Self Defense Army and Mongolian troops occurred; in August, the Battle of Shanghai began; in September Chahar was occupied by the Japanese army and the Eight Route Army fought in the Battle at Pingxing Pass; in December, Nanjing fell; to name just a few.[76] The sequence of months and incidents follows the timeline of actual events in the war. This linear temporal order is, by and large, quite different from the repetition and interchangeability in traditional folksongs.

This temporal arrangement implies that PRPH writers aimed to overturn the leisure tastes of the populace, and intentionally embedded a linear historical narration which conveyed temporal contextualization and causal reasoning. When people hummed or sang these songs, they were unconsciously aware of the national situation and recognized the responsibility of the nation's citizens.

---

74 Anonym 1938a: 7, Jiang 1938, Gu Jiegang 1988a: 68.

75 Li Xiaoti 1992: 184–91.

76 Li Ke 1938.

The portrayal of historical figures in *Minzhong zhoubao* or *Yizhong congshu* is of a different type. "The Records of Pacifying the Wokou" (Ping Wo ji, 平倭記), authored by Wang Zemin, is a story hinting at the anti-Japanese movement in which General Qi Jiguang leads people and resists the Wokou (Japanese pirates) during the reign of Emperor Jiajing. The author not only relates the brutality of the Wokou, but also presents a lengthy portrayal of the "domestic thief" Wang Zhi. A brief passage from the original text is as follow:

> A proverb says,
> Wild monsters wouldn't come unless seduced by domestic thieves.
> The chief of thieves named Wang
> Was born in Huizhou, Anhui province.
> Wang Zhi is his full name and he's a ruffian
> Leading a gang of bullies
> Sailing ships in the Eastern Sea
> First he only did business with aliens
> Then he brazenly became the king of pirates.
> Ganging up with the outlaw Xu Hai, and with Cheng Dong as his pawn
> Colluding with the bedbug Ōjima Tada, the chief of the Wokou,
> Seizing Fuzen, Fungo, Tsumei, the thirty-six isles
> Building vessels and warships called dalianfang.[77]

Under the premise of patriotism, the portrayal of the character of protagonists is usually stereotypical. Traitors must confront heroes, and resisters never coexist with collaborators. Characters must be clear-cut in order for readers to identify them without ambiguity. Wang Zhi in the above cited excerpt cannot at the same time be a merchant and a leader of the coastal brigands trading in foreign lands, not to mention his role as a colonizer of foreign lands. He is presented only as a ruffian. Under the government's restriction of activities that threatened cordial diplomatic relations with Japan, the drum minstrel cleverly uses the word Wokou as a metaphor and "the bedbug Ōjima Tada" as a way to taunt the Japanese without employing any terms or phrases connoting anti-Japanese sentiments. In this type of writing, intended for the illiterate majority, it seems unnecessary for authors to observe any strict and accurate historical facts.

Obviously, the commander-in-chief is not the only activist in stories of the War of Resistance. The masses play a prominent as well. A story where the mass are absent cannot be considered as expressing the people's determination to resist. In "The Chanyuan Treaty," authored by Wang

77 Wang Zemin 1937: 24.

Naitang, Kou Zun insists that the emperor command his troops in person, but the emperor alone cannot accomplish the task successfully. In the story, the status of the masses is elevated to a more significant place. For example:

> The people shout "A long life!" to the emperor by the side of the road
> Crowds of people shout to the sky
> One says the emperor is nothing but a worthless louse
> But he surprises all by leading troops to the frontier in person.
> The other says no wonder the king is worthless
> Scared whenever hearing that fighting has erupted
> ....
> One says whether the emperor resists or not
> Our people still come out to fight for the nation's honor
> Unite the people to fight the Tartars
> Die so that the nation can survive
> The other says since the government demands resistance
> We people organize an army of volunteers
> Unite the army and the people
> Slaughter the enemy, and march bravely.[78]

Only through the phrase of "one says" and "the other says" is the sublime determination of the people to resist manifested. People can themselves be motivated to organize army of volunteers whether the government fights or not. The narrative of drum singing or singing scripts sends a clear message to the readers that resistance does not rely on the government alone. The author points out clearly that people can also be emancipated from their repressed role if they remain strong-willed and determined.

However, the effectiveness of famous historical figures as models for the reader to admire and imitate could not be as influential as the real story experienced by common people during the war. Thus the contemporary protagonists created by PRPH writers were usually unknown common people such as Miao Kexiu, who joined the Self Defense Army after the Mukden Incident, or Zhao Jinbiao, a fictive hero who kill innumerable of the enemy as recounted in the story "The Bloody Battle at the Marco Polo Bridge."[79] The unsung heroes, such as vagrants, rickshaw coolies, or students, no matter whether fictive or authentic,[80] embodied a combative

78 Wang Naitang 1937a: 29–32.

79 Anonym 1937b, Lao Bai 1938, Anonym 1937c, Anonym 1937a, Yemin 1936.

80 As Shen Songqiao has suggested, the glorious deeds of heroes, no matter whether fictitious or true, usually play an important role in reinforcing the unity of the national community. This characteristic can also be found in the works of the PRPH. The

spirit and the will to die for their country. Sung by drum singers, these stories aimed to console the soldiers' heart and soul, stimulate people to identify themselves with the masses and thus achieve the goal of war mobilization.

But people do not just awaken barehanded and go fight without any weapons. They acquire powerful weaponry from an accumulation of knowledge, national consciousness, and a historical sense of being citizens with a contemporary mentality. The PRPH writers continually adjusted their works to prevent stereotyped writing and to address the call for patriotic war.[81] They timely add various modern skills and common sense to enhance people's capability to resist the enemy. The story of Wu Xun became a paradigm of the PRPH's advocacy. Gu Jiegang was the first author to compose drum songs lauding Wu Xun's outstanding merits. The story urges that people not ignore the problem of illiteracy and value the tremendous impact of literacy on people.[82] Wu lived on leftover food for thirty years and saved every penny to run a school and to realize his dream of promoting education for the poor and eradicating illiteracy. Literacy became the most important thing for national survival and self-strengthening. The story was revised by Mi Baoshun, who added a more precise account and details, and was published in *Minzhong zhoukan* after the war.[83]

Furthermore, *Minzhong zhoubao* published a drum singing serial that is an adaptation of the story of a famous patriotic prostitute Sai Jinhua. With a view toward accurately reviving the story, a PRPH editor interviewed the heroine, Fu Caiyun, herself.[84] The author Zhao Zhizhong took this well-known story and transformed it from an obscene legend into a historical story with rich, modern ideas. The plot in which Fu Caiyun is forced into prostitution was singled out to criticize contemporary social injustice and to urge readers to redress this injustice.[85]

At the beginning of chapter four, Hong Jun, Fu Caiyun's first husband, brings her to Beijing, where he is serving. On the way there, he explains to his wife about world geography and the idea of race evolutionism: "races

---

protagonists were magnified in order to stimulate national consciousness. See Shen 2000: 85, 87–88.

81 Xiang 1938d: 83–84.

82 Gu Jiegang 1936c: 31–40.

83 Mi 1947: 6–9.

84 Wang Xuhua 1990: 1038, Gu Jiegang 2007: 3.185.

85 Zhao 1936a: 45.

need to be aggressive in order to survive."[86] Fu asks her husband, if China is the Heavenly Dynasty, how and when can it be bullied? Hong then tells her the story of modern China. He recounts the events of ceding territories after the Opium War, how tariff rights were surrendered to foreign powers, and how China gradually lost of its subject states. The paragraphs concerning the Tongwen Press (Tongwenguan), training the navy, and the Empress Dowager Cixi peculating military funds in order to construct the imperial Yihe Yuan garden, transmit a meta-historical concept of modernization.[87]

After eight issues the story by Sai Jinhua finally ceased printing. Probably the story is too long to come to an end, and no offprint version survives. Zhao Zhizhong adapted this well-known legend and colored it with new content. In this new version, world geography, Sino-foreign relations, and modern Chinese history are interwoven with a series of national disgraces. He makes use of every opportunity to educate his readers in new moral values and national thought. The paragraph describing the tension between Fu Caiyun and Hong Jun's first wife, for example, is actually intended to criticize orthodox Confucianism.[88] All these elements are what PRPH members believed to be the promotion of new characters for a modern citizen, which was the objective of popular readings.

Epilogue: Internal and External Predicaments

From its founding in 1933 onward, the PRPH was continually beset by financial shortfalls and subject to political intervention. Its publications received a lukewarm response from the very beginning. The slogan "awaken the people" thus ironically revealed that people were passive and indifferent to what the PRPH advocated. The result was poor sales, which reduced the willingness of publishers in Beijing and Tianjin to print for the PRPH, and forced it to seek the aid of Shanghai publishers.[89] Gu Jiegang once lamented that he was "really helpless in the face of people's unconcern for knowledge."[90] In this circumstance, the PRPH

86 Zhao 1936b: 48–49.

87 Zhao 1936c: 50–51.

88 Zhao 1936b: 45–46.

89 The Shanghai Yadong tushuguan promised Gu Jiegang that it could publish 200,000 words of popular readings and 80,000 words of bi-weekly periodicals per month. Gu Jiegang 2007: 513.

90 Gu Jiegang 2007: 3.125.

could not but seek financial aid in the form of subsidies from the government and the KMT or private donations.[91] But Gu complained more than once that the Ministry of Education obviously had submitted a budget that contained funds for the PRPH but was reluctant to allot the subsidy.[92] Moreover, under suspicion of harboring communist members in its organization, the PRPH was nearly shut down in 1936.[93] Its incessant promotion of anti-Japanese sentiment also caused it all along to be harassed by the Japanese and warned by Chinese government. In this circumstance, the *Minzhong zhoubao* was forced to cease publication because it presented "a hindrance to diplomatic relations" Furthermore, the operation of *Dazhong zhishi* began running into difficulties from the sixth issue, and it barely managed to print six more issues before it inevitably ceased publication. After its periodicals folded, the PRPH could not sustain itself by publishing popular readings. Although after the outbreak of the war the government no longer discouraged advocating anti-Japanese sentiment, wartime inflation and the scattering of the members of the PRPH forced it to disband in 1940. Though subsidies from the Ministry of Education and Executive Yuan were provided until 1944, by then the PRPH existed in name only.[94]

In my view, that the PRPH dissolved in the mid-war period can be attributed to both internal and external factors. Externally, its financial difficulties were merely a superficial cause. The root of its downfall lies in the active advocacy of anti-Japanese sentiment and a leftist tendency which KMT government could not tolerate. This involves an issue worthy of further discussion: the political stance of the members of the PRPH and their relationship with Gu Jiegang. Thanks to the publication of Gu's diary, we now have a clearer understanding of this issue.

As we have seen, the movement of compiling and composing popular readings and literature was closely attached to the stimulus provided by the Mukden Incident, but the national crisis that ensued, disrupted people's lives as well. Gu Jiegang, for instance, admitted in his diary that writing popular readings was somewhat related to his earlier research theme, but organizing and managing the PRPH was only a "social service" and indeed not part of the academic career he had in mind. The

91 Ibid., 3.380.
92 Ibid., 3.442, 501, 504.
93 Gu Jiegang 1993: 247.
94 Ibid., 301.

national calamity forced intellectuals out of their ivory tower.[95] To be the backbone of the country and to advocate resisting the enemy are natural reactions of intellectuals. However, the advocacy of the anti-Japanese war was unfortunately caught up in party interests and the political struggle between KMT and CCP. Although at the beginning the PRPH won the government's support and subsidies, its praise of Song Zheyuan and Fu Zuoyi,[96] whom Chiang Kai-shek could not tolerate after the Central Plains War, make it harder to gain further financial support from the government and the KMT. Conversely, its sympathy for Fu Zuoyi and Song Zheyuan made it bedfellows with the CCP and hence led the PRPH to become the battleground in which the CCP tried to expand its propagandistic manipulation.

Gu Jiegang endured pressure from all sides during this period. In the central government, Gu was widely rumored to be in cahoots with communists or even to have become a leftist. Even Fu Sinian told Gu privately that he heard many vicious remarks about Gu and warned him not to encourage the Ji-Cha local government to condemn the central government.[97] A more astonishing rumor was that Gu had instigated the Sian Incident. Gu was extremely resentful of this kind of malicious gossip that could almost lead to a death sentence. In Gu's mind, publishing popular readings was a responsibility for all who were unwilling to see the nation fall.[98] However, Gu privately visited Zhang Xueliang,[99] accepted a subsidy from Feng Yuxiang and Sun Ke, and admired Fu Zuoyi.[100] But all this, while at the same time accepting a subsidy from the government without any attempt to help the government's propaganda, was, in the eyes of the central government, undeniably evidence of treachery.

Additionally, no matter how hard Gu tried to clarify his non-leftist position, the fact that the PRPH's editorial policy was dominated by leftist or pro-leftist members of the PRPH, such as Wang Shouzhen, Li Yifei, and Xiang Linbing, led the KMT to suspect the rumors to be true. According to Wang Shouzhen's memoir, Gu gave free rein to the members of the PRPH and hence seldom concerned himself with editorial affairs or even

95 Gu Jiegang 2007: 3.204.

96 The PRPH published many works on Feng Yuxiang's army, such as *General Song Zheyuan and the Fierce Battles at Xifengkou Pass* mentioned above, and "The Bloody Battle at the Marco Polo Bridge," *The Great Battle at Tianjinwei*, to name just a few.

97 Gu Jiegang 2007: 3.577, 599.

98 Ibid., 3.518.

99 Ibid., 3.560.

100 Ibid., 3. 579–81, 664.

personnel matters. The loose criterion for membership in the PRPH also gave a chance for many leftist or pro-leftist youth to join.[101] Their gathering at the PRPH was mainly a reaction to the KMT's prohibition on the expression of resistance to the Japanese and thus a projection of their sympathy for Marxism and its theory of class revolution, which, of course, was advocated by the CCP but nonetheless anathema to the KMT. Even if expressions of anti-Japanese sentiment were no longer prohibited, the PRPH's writers' lauding of the valor of the 29th Route army and even of the Eighth Route Army was definitely something the KMT government could not tolerate, and hence eventually resulted in the cutting off of the budget to PRPH. Without this financial aid, the disbanding of the PRPH was only a matter of time.

Gu Jiegang was not totally oblivious to this situation. In fact, he was clearly aware that the dilemma was a consequence of his personality and great reputation. As he deeply reflected on the journey fleeing from Beijing to Suiyuan in 1937:

> Youths would like to gather around me and be attached to me owing to the fact that I always treasure the talented. This is where I have succeeded and paradoxically also where I have failed. As for maintaining their livelihood, I have to contact various parties and seek jobs for them. However, people who misunderstand and suspect that I am trying to build up my influence deride me as "Boss Gu" or "Master Gu." But honestly speaking, sometimes I feel I am a puppet surrounded by mediocre students who intend to use my reputation for their own private purposes. If they succeed, they get what they want,, but unfortunately I have to bear their curses if I fail. That's why it is widely rumored that "Gu has become a leftist" or "Gu has been captured by CCP."[102]

Gu believed educating people was one of his social responsibilities. However, since he took on too many responsibilities and was too busy to effectively oversee the publishing activities in which he was involved, he had no choice but to rely on his associates and to take a laissez-faire attitude. His confession of being a puppet more or less reflected his position and attitude. At that time, Gu was worried about the state breaking up and what would happen in the cultural sphere on the one hand, but paradoxically, on the other hand he was relieved he was not being chased by the Japanese and was free from having to deal with administrative details. What he did not anticipate, however, is what would transpire

101 Wang Zhen 2006: 104–6.
102 Gu Jiegang 2007: 3.673.

once he reached Lanzhou. As soon as he got to Lanzhou, a messenger from the Eight Route Army contacted him. Although Gu recognized that the CCP only shared his commitment to patriotic propagate for the common people, he ironically worried about unjust rumors arising once more.[103] Eventually, he broke away from this situation and decided to be done with the PRPH. In April 1938, Gu resigned as the head of the PRPH when he was designated as a member of the Committee for Education Project sponsored by the Directorate of the Boxer Protocol Indemnity. He finally ended his seven-year-long life as a "puppet" and thereafter was again never involved in the affairs of the PRPH.[104]

The publication of Gu's diary helps us to reconstruct his relationship with the PRPH, to clarify his complex psychological conflicts, and to draw a clearer profile of the editorial policies of the PRPH. No matter how deeply the PRPH bore the imprint of Gu's ideas and will, his experience was not unique, but rather epitomized that of intellectuals at that time. When someone pointed to "The Fierce Battle at Pingxing Pass" as evidence of Gu's pro-leftist inclinations, he indignantly retorted that:

> The KMT and the CCP have cooperated with each other since the outbreak of the war, and of course the Eighth Route Army is dominated by the CCP but it also belongs to the country. Thus the victory at Pingxing Pass is undoubtedly an achievement of our country. People who are dissatisfied with my praise of this victory should not condemn me as a leftist if they want me to die; straightforwardly accusing me of being a traitor to China would be more effective![105]

Lao She was also attacked by KMT for his advocacy of the theory of new wine in old bottles, and he angrily fought back that those accusations were nothing more than a plot of "disrupt the War of Resistance camp."[106] Many intellectuals who were patriotic but rewarded by the KMT by being labeled renegades thus were pushed to join the CCP camp.

If the struggle between the KMT and the CCP over which would lead the War of Resistance was the external dilemma of the PRPH, then the debate over the old-new form of literature, which referred to the universality or particularity of national character, was its internal dilemma. The PRPH unavoidably faced a predicament in decided on its editorial guidelines and practice with regard to popular literature.

---

[103] Ibid., 3.748, Gu Chao 1993: 281.
[104] Gu Jiegang 2007: 4.59, 282, 345.
[105] Gu Jiegang 1992: 283–84.
[106] Gan 1989: 130.

As we mentioned above, the idea of "new wine in old bottles" was a reflection of the PRPH's project of building a bridge between popular culture and tradition. Unlike their May Fourth predecessors, who overall repudiated tradition, the PRPH writers believed popular culture possessed vigorous and creative energy from which intellectuals could and should draw in order to construct a new culture with a more national flavor. However, their advocacy was widely criticized by the left wing camp during the debate over "Questions on the national form," which began in 1939 and last for two years. The debate spread widely—reaching such places as Chongqing, Guilin, Kunming, Chengdu, Hong Kong and Shanghai—and involved many writers and artists who reflected on and discussed the issue of "the popularization of literature and art." Some criticized the PRPH's idea of "new wine in old bottles" as overemphasizing "folk forms" and assuming that "old forms" were the only path toward popularization—a path that led to arbitrary applications in literature. On the other hand, the PRPH's emphasis on pouring a new national consciousness into the old forms familiar to people was also criticized as merely intellectuals' condescending gesture to the masses, a deficient perspective on popularity.[107]

Xiang Linbing, the main theorist of the PRPH, became a target for all in this debate. In 1940 he published a series of essays concerning the "origin of the national form" as a defense of and justification for the theory of "new wine in old bottles." He divided the discussions on "the national form" into two camps: one consisting of writers who argued that the national form should be based on the May Fourth spirit; and the other, of writers who embraced the notion that the national form must start from forms that are attractive to and loved by the populace.[108] Xiang himself argued that "the national form" is not something "already in existence" but merely a burgeoning process that, at that moment in the 1940s, possess "residues of feudal society" but nonetheless stands in opposition to it and eventually will become a tool for undercutting feudalism:

> The new always emerges as an embryo from the old and through a process of self-negation becomes a new and independent entity. [According to this proposition] the creation of the national form should not be molded on the Chinese Literature Movement, but on the contrary, it must be shaped by the process of denying the existing literary forms.[109]

---

107 Xiang 1940a, Wang Aisong 1996: 27–28, Liu Zhong 2004: 137–43.
108 Xiang 1940a.
109 Ibid.

The main concern in Xiang's dialectical rhetoric is that the new national form must not follow the May Fourth route, a deformed product of urban culture only catering to the tastes of the bourgeoisie and deficient in verbal form.[110] Xiang argued that "existence determines consciousness," and thus the reason popular literature was attractive to and loved by the masses was based on habituation: that is, it was something that the common people saw and heard every day. The form of the folk literature, to which the common people were accustomed and which was appropriate to the oral and dialogic form, was thus the key to creating a new national form.

The outcome of the debate on new-old literature in the 1930s and on national form in 1940s has been variously evaluated. I do not intend to re-evaluate them here. What I am concerned with is the very inner issue in this debate: the controversy over the universality or particularity of nationality.[111] Under the discourse of nationalism, the PRPH's theory of "new wine in old bottles" not only confronts May Fourth Western-oriented literature, but also reaffirms the representation of the "old forms" of popular literature and recombines them with traditional culture. Here "tradition" is attached with a whole new and different connotation which was once attacked by the May Fourth intellectuals as "forest literature" (*shanlin wenxue*, 山林文學) or of outmoded moral value. The definition of tradition is always floating concomitantly with the varied connotations of "modern," as well as "the nation." At the time of the War of Resistance, tradition not only represented the enormous energy of popular culture, but also was entrusted with the task of reconstructing the nation. As Xiang Linbing put it: "The debate on 'the national form' symbolizes the discovery of dynamic social reform in China mirrored in literature, ... and the people are the very agent by which social dynamism is implanted".[112] The social status of popular culture was thus lifted up and replaced the universal value of Westernization and modernization that May Fourth intellectuals had embraced. The war created an urgent need for a clear-cut identification between the enemy and allies; only an easily identifiable appeal could qualifyed as an agent of nationalism. Thus the

110 Ibid.

111 Guo Moruo argued that the issue of "national form" was influenced by the debate in Soviet Russia, which emphasized "adopting the universal form to discard national particularity." However, the core issue for Chinese was interweaving "Chinesization" and "popularization," which involved attempting to "apply national particularity to enhance universal content." Guo 1992: 19.31–47.

112 Xiang 1940a, see also Xiang 1940b: 4.

"New Literature," with its modern literary style which could only be read and comprehended by a few intellectuals, was replaced by literature from the masses in the name of war mobilization.

In this nationalistic context, it was more important to seek national particularity than modernistic universality. The theory and practice of "new wine in old bottles" highlighted by the PRPH tried to satisfy the need to construct a new national order which could preserve the old form of popular culture as well as instantiate a new national consciousness. However, the process of transforming from form to content paradoxically implied the interweaving of national particularity and modernistic universality. This dilemma reflects a taken-for-granted hypothesis of the PRPH's members when they tried to bridge the gap between form and content in literature. As Xiang Linbing said:

> Popular [literary] form is counterrevolutionary only when it is associated with feudal content, such as outmoded popular literature; or combined with the imperialism, akin to the propaganda circulated in the Japanese-occupied area. It will be a powerful and valuable weapon if merged with revolutionary thought.[113]

Xiang believed that since content determines form, correct content will transform form.[114] This assumption still marks "old bottles" (old form) and "new wine" (new content) as polarized and counter to each other. Moreover, the content is to be determined by intellectuals who consider "old bottles" as only a tool which always passively receives "new wine," hence there is no difficulty in the transformation between form and content. However, Lao She, who had actual experience with applying this theory, thought Xian's assumption was too optimistic.

Lao She felt that composing in the style of "new wine in old bottles" was not only a painful but impossible mission. In 1941 he announced that he was abandoning this theory, and gravely declared that "new is new, and can by no means be merged with old. To compromise is to surrender!" He admitted that he once believed that applying the old forms for composition was as easy as winking, but he came to realize its difficulty in practice. Old drama and drum singing, though old enough, were still alive and beloved by the people. Writers should learn these lively forms and deliberately and harmoniously add new elements. Moreover, the most difficult work for Lao She was composing works that were instructive

113 Xiang 1940a.
114 Xiang 1940b: 4.

and served the purpose of the War of Resistance while at the same time applied the given form and language but nonetheless were polished writing. Shouldering this task, Lao She complained, caused him to "breathe heavily." This misfit, in short, was the problem of intellectuals who were deficient in real experience and had scant empathy for the real life and sense of the people. Unless writers became involved, they could not write works that would resonate with the masses. Lao She thus confessed that his works on old drama and drum singing were merely a product learned from urban dwellers, thus "still far away from truly popular literature." He argued that genuine and vivid popular literature could only be composed by writers who "live closely with street performers and could compose and revise impromptu"; those who "live in urban areas and work behind closed doors could never reach the people."[115]

Though Lao She's experience and confession were very different from those of Xiang Linbing, he never directly responded to the form–content and "national form" debates. However, he underscored that he had already replied to these debates when he decided to discard the "old forms."[116] He also argued that "it seems natural that new [literature] is bred from the old, but do not overlook the difficulties arising from impetuousness. The expectation that our compatriots can swiftly become a modern nation is simply a wishful thinking."[117] Lao She's change of heart showed that the theory of "new wine in old bottles" was all along an interweaving of national particularity and modernistic universality and hence underestimated the gap between intellectuals and the people. Moreover, it oversimplified the clash between form and content. To some extent, by rejecting the modern literary theory of the May Fourth generation, the PRPH also denied its value to modern Chinese literary history and hence it was attacked by some intellectuals. Ironically, the PRPH's insistence on the theory of "new wine in old bottles" overlooked the subjective activity of the people and therefore in the end remained aloof from the common people. Wavering between the populace and intellectuals, the PRPH thought it could raise the status of popular culture, but ironically all along it merely intended "to stand with," not "become" a part of the populace. This inner paradox is, by and large, the major and inescapable cause of the dissolution of the PRPH.

---

115 Lao She 1982b: 33–36.
116 Lao She 1991b: 439–41.
117 Lao She 1938b: 5.

References Cited

Anonym (1937a), "Xiezhan Lugou qiao" 血戰蘆溝橋 (The Bloody Battle at the Marco Polo Bridge), in *Yizhong congshu*, Xi'an: PRPH, 1937.
—— (1937b), "Kangri yingxiong Miao Kexiu" 抗日英雄苗可秀 (The War Hero Miao Kexiu), in *Yizhong congshu*, no. 1, 1937.
—— (1937c), "Shou Shangdu," 收商都 (The Regain of Shang's Capital), in *Yizhong congshu*, new series no. 4, Xi'an: PRPH, 1937.
—— (1938a), "Xinkan kangzhan guci mulu" 新刊抗戰鼓詞目錄 (Republication of the Bibliography of Drum Singing during the Anti-Japanese War), in Anonym, *Kangzhan geyao*, compiled in *Yizhong congshu*, xinkan di 11 zhong. Wuchang: Tongsu duwu biankanshe.
—— (1938b), "Minguo ershiliu nian" 民國二十六年 (The Year 1937), in *Yizhong congshu*, xinkan di 14 zhong. Xi'an: Tongsu duwu biankanshe.
Editorial Board (1936), Letter Column, *Minzhong zhoubao* 1, 4 (Oct. 23).
Gan Hairan ed. 甘海嵐 (1989), *Lao She nianpu* 老舍年譜 (*Chronological Biography of Lao She*). Beijing: Shumu wenxian chubanshe.
Gu Chao 顧潮 [Gu Jiegang] (1993), *Gu Jiegang nianpu* 顧頡剛年譜 (*Chronological Biography of Gu Jiegang*). Beijing: Zhongguo shehui kexueyuan chubanshe.
Gu Jiegang 顧頡剛 (1936a), "Tongsu duwu de shidai shiming yu chuangzuo fangfa" 通俗讀物的時代使命與創作方法 (The Epochal Mission and Methods of Composition of Popular Readings), in *Minzhong zhoubao* 1:1 (2 Oct.).
—— (1936b), "Wu Xun taofan xingxue" 武訓討飯興學 (Wu Xun Promotes Education through Begging), in *Minzhong zhoubao* 2, 3 (Jan. 15).
—— (1938a), *Tongsu duwu lunwenji* 通俗讀物論文集 (*Collected Writings on Popular Readings*). Hanko: Shenghuo shudian.
—— (1938b), "Jiuping zhuang xinjiu de chuangzuo fangfa lun" 舊瓶裝新酒的創作方法論 (The Composition Methodology of New Wine in Old Bottles), in Gu Jiegang 1938a: 10.
—— (1938c), "Women zenyang xiezuo tongsu duwu" 我們怎樣寫作通俗讀物 (How We Compose Popular Readings), in *Kangzhan wenyi* 2, 8 (Oct. 29).
—— (1938d), "Daguci zenyang zuofa" 大鼓詞怎樣作法 (Composition Methods of Drum Singing), in *Minzhong zhoubao* 3, 3 (Apr. 16).
—— (1949), "Bianji Zhongguo mingrenzhuan shuomingshu" 編輯中國名人傳說明書 (Explanations of How to Compile the Biographies of Chinese Celebreties). *Shangyou ji*. Shanghai: Hezhong tushuguan.
—— (1988a), *Xibei kaocha riji* 西北考察日記 (Diary of [the Trip] to Investigate Northwest China). Zhongguo renmin zhengzhi xieshanghui Gansusheng weiyuanhui wenshi ziliao yanjiu weiyuanhui, ed., *Ganqing wenjianji*. Lanzhou: Gansu renmin chubanshe.
—— (1988b), "Bianyin tongsu duwu de jingguo" 編印通俗讀物的經過 (The Proces of Compiling and Publishing Popular Readings), in Gu Jiegang 1988a.
—— (1988c), "Xibei kaocha riji xu" 西北考察日記序 (Foreword to the Diary of [the Trip] to Investigate Northwest China), in Gu Jiegang 1988a.
—— (2007), *Gu Jiegang riji* 顧頡剛日記 (The Diary of Gu Jiegang). Vol. 3 (1933–1937); Vol. 4 (1938–1942). Taipei: Lianjing.
Guo Moruo 郭沫若 (1992), "Minzu xingshi shangdui" 民族形式商兌 (Discussion on "the Form of Nation") in *Chongqing Dagongbao*, weekly review supplement, June 9, 1940, sec. 2; June 10, 1940, sec. 3. This article is collected in Guo Moruo, *Guo Moruo quanji, wenxue pian*, vol. 19. Beijing: Renmin wenxue chubanshe, 31–47.
Han Depu 韓德溥 (1936), "Xingqi xingxue de wuxun" 行乞興學的武訓 (Wu Xun Promotes Education Through Begging), in *Shenbao tongsu jiangzuo* no. 25 (Sept. 3, 1936).
He Guang 賀光 (1936), "Zheng Chenggong" 鄭成功 (Zheng Chenggong), in *Shenbao tongsu jianzuo* no. 2 (Mar. 26).

Hung, Chang-tai (1994), *War and Popular Culture: Resistance in Modern China, 1937–1945.* Berkeley: University of California Press.

Jiang Mi 江枚 (1938), *Kangzhan shiling geyao* 抗戰時令歌謠 (Season Ballad for Combating War). Two volumes. Hanko: Shenghuo shudian.

Jueren 孓人 (1937), "Li Kan qi: Yige yong yu kangdi de furen" 李侃妻:一個勇於抗戰的婦人 (A Brave Woman Resisting the Enemy), in *Shenbao tongsu jiangzuo* no. 44 (Jan. 14).

Ji Zhen 季珍 (1936), "Jin gongren" 晉弓人 (Bowmen of Jin), in *Shenbao tongsu jiangzuo* no. 36 (Nov. 19).

Lao Bai 老白 (1938), "Da'nao Wangjia zhuang" 大鬧王家庄 (Turmoil in Village Wang), in *Yizhong congshu,* new series no. 28, Wuchang: PRPH, 1938.

Lao She 老舍 (1938a), "Zhonglietu" 忠烈圖 (*Portrays of the Martyrs*), in idem, *San si yi.* Chongqing: Duri chubanshe.

—— (1938b), "Tan tongsu wenyi" 談通俗文藝 (On Popular Drama and Literature), in *Ziyou Zhongguo* no. 2 (May 10). Also in Lao She 1982a.

—— (1938c), "Zhizuo tongsu wenyi di kutong" 製作通俗文藝的苦痛 (The Distress of Composing Popular Literature), in *Kangzhan wenyi* 2, 6 (Oct. 15), in Lao She 1982a.

—— (1982a), *Lao She quyi wenxuan* 老舍曲藝文選 (Anthology of Lao She's Works on Folksongs and Literature). Beijing: Zhongguo quyi chubanshe.

—— (1982b), "Wo zenyang xie tongsu wenyi" 我怎樣寫通俗文藝 (How I Compose Popular Literature), in Lao She 1982a.

—— (1991a), *Lao She wenji* 老舍文集 (*Collected Writings of Lao She*). Beijing: Renmin chubanshe.

—— (1991b), "1941 nian wenxue quxian de zhanwang" 1941 年文學趨向的展望 (The Prospect of the Literary Trends of 1941), in Lao She 1991a, vol. 15.

Li Jizhen 李季珍 (1936), "Yue Fei di shaonian shidai" 岳飛的少年時代 (The Youth of Yue Fei), in *Shenbao tongsu jianzuo* no. 39 (Dec. 10).

Li Ke ed. 李克 (1938), "Guonan shi'eryue" 國難十二月 (Twelve Months of National Calamity), in *Yizhong congshu,* new series no. 14. Wuchang: Tongsu duwu biankanshe.

Li Xiaoti 李孝悌 (1992), Qingmo de xiacheng shehui qimeng yundong, 1901–1911 清末的下層社會啟蒙運動 (Lower Class Enlightenment in the Late Qing Period). Taipei: Institute of Modern History, Academia Sinica.

Liu Qiyu 劉起釪 (1986), *Gu Jiegang xiansheng xueshu* 顧頡剛先生學述 (*An Account of Gu Jiegang's Life and Thought*). Beijing: Zhonghua shuju.

Liu Zhong 劉忠 (2004), "Zhongguo xiandai wenxue hua yu xingshi de sanci lunzhen" 中國現代文學話語形式的三次論爭 (Three Debates on the Discursive Formation of Modern Chinese Literature], in *Shehue kexue yanjiu* (June).

Lu Xun 魯迅 (2000), "Lun jiuxingshi de caiyong" 論 "舊形式的採用" (On the "Adoption of Old Form"), in *Qiejieting zawenji.* Beijing: Renmin wenxue chubanshe.

Mi Baoshun 米寶順 (1947), "Xingqi xingxue de Wu Xun" 行乞興學的武訓 (Wu Xun Promotes Education Through Begging) in *Minzhong zhoukan* 1, 2 (May 31).

Ogura Yoshihiko 小倉芳彥 (1983), "Tsûzoku dokubutsu henkansha no shubanbutsu —1930 nendai no kôni pamphlet 11 shu" 通俗讀物編刊社の出版物—1930年代の抗日ペンフレット11種 (The Publications of the Popular Readings Publishing House: Eleven Pamphlets Advocating anti-Japanese War during the 1930's), in *Gakushūin daigaku tōyō bunka kenkyūjo chyōsa kenkyū hōkoku* 17 (Mar.): 61–112.

Schneider, Laurence A. (1971), *Ku Chieh-kang and China's New History: Nationalism and the Quest for Alternative Traditions.* Berkeley: University of California Press.

Shen Songqiao 沈松僑 (2000), Zhen Dahan zhi tiansheng—minzu yingxiong xipu yu wang Qing de guozu xiangxiang 振大漢之天聲: 民族英雄系譜與晚清的國族想像 (Imagine China: Construction of the Genealogies of Chinese 'National Heroes' in Late Qing), in *Bulletin of the Institute of Modern History* no. 33 (June): 77–158.

Su Zihan 蘇子涵 (1936), "Suzhou wuyishi" 蘇州五義士 (The Five Brave in Suzhou), in *Shenbao tongsu jiangzuo* no. 8 (May 7).

Tang Tao 唐弢 (1992), *Zhongguo xiandai wenxueshi* 中國現代文學史 (*The History of Chinese Modern Literature*), vol. 3. Yan Jiayan, ed. Beijing: Renmin wenxue chubanshe.

Tongsu duwu biankanshe 通俗讀物編刊社 (1936), "Tongsu jiangzuo fakanci" 通俗講座發刊詞 (Foreword for the First Issue of Popular Lectures), in *Shenbao*, March 19.

Wang Aisong 王愛松 (1996), " 'Dazonghua' yu 'Huadazhong': sanshi niandai yige wenxue huayu di fansi" 大眾化與化大眾: 三十年代一個文學話語的反思 (Popularization and/or Cultivating People: Re-assessment of the Literary Discourse during the 1930's), in *Nanjing daxue xuebao* no. 2.

Wang Chongwu 王崇武 (1936), "He'elun" 訶額崙 (He'elun), in *Dazhong zhishi* 1, 3 (Nov. 20): 26–32.

Wang Dazhen 王大真 (1936), "Xin Xiyouji" 新西遊記 (The New Journey to the West), in *Minzhong zhoubao* 1, 1 (Oct. 2).

Wang Naitang 王乃堂 (1937a), "Chanyuan meng" 澶淵盟 (The Chanyuan Treaty), in *Minzhong zhoubao* 2, 5 (Jan. 29): 29–32.

—— (1937b), "Zhenlong tianzi zaotianshu" 真龍天子造天書 (The Son of Heaven Counterfeits the Heavenly Book), in *Minzhong zhoubao* 2, 6 (Feb. 5): 17–24.

Wang Shouzhen 王受(守)真 (1938a), "Weishenme yaoba xinjiu zhuangzai jiuping li" 為什麼要把新酒放在舊瓶裏 (Why We Need to Fill New Wine into Old Bottles), in Gu Jiegang 1938a.

—— (1938b), "Zailun weishenme yaoba xinjiu zhuangzai jiuping li" 再論為什麼要把新酒裝在舊瓶裏 (Re-discussion on New Wine in Old Bottles), in Gu Jiegang 1938a.

Wang Xuhua 王煦華 (1990), "Gu Jiegang xiansheng xueshu jinian" 顧頡剛先生學術紀年 (Academic Chronicle of Gu Jiegang), in Yin Da and Zhang Zhenglang eds., *Jinian Gu Jiegang xueshu lunwenji*. Vol. 2. Chengdu: Bashu shushe.

—— (2006), ed., *Gu Jiegang xiansheng xuexinglu* 顧頡剛先生學行錄 (Academic Biography of Gu Jiegang). Beijing: Zhonghua shuju.

Wang Zemin 王澤民 (1937), "Pingwoji" 平倭記 (Account of Pacifying the Japanese), in *Minzhong zhoubao* 2, 5 (Jan. 29).

Wang Zhen 王真 (2006), "Ji Gu Jiegang xiansheng lingdaoxia de Tongsu duwu biankanshe" 記顧頡剛先生領導下的通俗讀物編刊社 (Commemorating the Popular Reading Publishing House under the Leadership of Mr. Gu Jiegang), in Wang Xuhua, ed, Gu Jiegang xiansheng xuexinglu. Beijing: Zhonghua shuju.

Xiang Linbing 向林冰 (1938a), "Jiuping zhuang xinjiu zai wenhua fazhanshi shang de renwu" 舊瓶裝新酒在文化發展史上的任務 (The Cultural Historical Mission of "New Wine in Old Bottles"), in Gu Jiegang 1938a.

—— (1938b), "Jiuping zhuang xinjiu shiyi" 舊瓶裝新酒釋義 (Definition of "New Wine in Old Bottles"), in Gu Jiegang 1938a.

—— (1938c), "Jiu xingshi de xin pingjia" 舊形式的新評價 (New Evaluation of Old Forms), in Gu Jiegang 1938a.

—— (1938d), "Tongsu duwu biankanshe de ziwo piping" 通俗讀物編刊社的自我批評 (The Self-critique of the Popular Readings Publishing House), in Gu Jiegang 1938a.

—— (1940a), "Lun minzu xingshi de zhongxin yuanquan" 論民族形式的中心源泉 (On the Central Source of the National Form), in *Chongqing Dagongbao, Zhanxian* (Mar. 24).

—— (1940b), "Fengjian shehui de guiluxing yu minjian wenyi di zairenshi: zairun minzhu xingshi de zhongxin yuanquan zhiyi" 封建社會的規律性與民間文藝的再認識: 再論民族形式的中心源泉之一 (The Regularity of the Feudal Society and a New Appreciation of Popular Literature: Again on one Central Source of National Form), in *Chongqing Xinshubao, supplement shudao*, no. 101 (April 21).

Yemin 野民 (1936), "Bailing miao" 百靈廟 (The Bailing Temple), in *Minzhong zhoubao* 1, 12 (Dec. 18).

Yang Liansheng 楊蓮生 (1937), "Xuanzang fashi" 玄奘法師 (The Monk Xuanzang), in *Dazhong zhishi* 1, 6 (Jan. 5): 20–26.

Zaqiang changben: Shibeicha 雜腔唱本: 十杯茶 (n.d.), in *Shuanghongtang xiqu 190*; Full-text Database of Chinese Rare Books, compiled by the Institute of Oriental Culture, Tokyo University.

Zaqiang changben: Xinchu gailiang Shibeijiu 雜腔唱本: 新出改良十杯酒 (n.d.), in *Shuanghongtang xiqu 190*. Full-text Database of Chinese Rare Books, compiled by the Institute of Oriental Culture, Tokyo University.

Zaqiang changben: Xinchu Huaguci, Shiyinben (n.d.), in *Shuanghongtang xiqu 190*; Full-text Database of Chinese Rare Books, compiled by the Institute of Oriental Culture, Tokyo University.

Zhao Zhizhong 趙之中 (1936a), "Sai Jinhua" II 賽金花 II (Sai Jinhua II), in *Minzhong zhoubao* 1, 3 (Oct. 16).

—— (1936b), "Sai Jinhua" IV 賽金花 IV (Sai Jinhua IV), in *Minzhong zhoubao* 1, 5 (Oct. 30).

—— (1936c), "Sai Jinhua" V 賽金花 V (Sai JinhuaV), in *Minzhong zhoubao* 1, 7 (Nov. 13).

Zheng Kanci 鄭侃嬨 (1936), "Wang Sungu di muqin: Haizi dongdi ziqiang, Qiguo buwangle" 王孫賈的母親——孩子懂得自強, 齊國不亡了 (The Mother of Wang Sungu: The Qi Could not be Subjugated, because their Children Understood [the Meaning of] Self-strengthening), in *Shenbao tongsu jiangzho* 15 (June 25).

—— (1937), "Guo Ziyi" 郭子儀 (Guo Ziyi), in *Dazhong zhishi* 1, 9 (Feb. 20): 30–37.

Zhou Yuan (1936), "Shi Kefa" 史可法 (Shi Kefa), in *Shenbao tongsu jianzuo* no. 38 (3 Dec.).

CHAPTER EIGHT

FIGURING HISTORY AND HORROR IN A PROVINCIAL MUSEUM: THE WATER DUNGEON, THE RENT COLLECTION COURTYARD, AND THE SOCIALIST UNDEAD*

Haiyan Lee

Introduction

In the late 1950s and early 1960s, the fallout from the Hundred Flowers and the Great Leap Forward Campaign convinced Mao Zedong (1893–1976) that the Chinese Communist Party (CCP) had prematurely placed class struggle on the backburner during the decade of socialist transformation which had essentially eliminated the economic foundation of class exploitation. In 1962, Mao alerted his comrades to the dual nature of class struggle in a socialist society: against both the remnants of "feudalism," who hankered for a "change of sky" (*biantian*), *and* the new bourgeois elements bred by an entrenched bureaucracy. As such the struggle would be "protracted, complex, and sometimes acute."[1] A draft resolution dealing with problems in rural work commonly known as "The First Ten Points" (May 1963) cites the following directive from Mao:

> Class struggle, production struggle, and scientific experiment are the three great revolutionary movements that build up a powerful socialist nation. They are a guarantee for the Communists to do away with bureaucratism, to avoid revisionism and dogmatism, and to stand eternally invincible. They are an assurance for the proletariat to be able to unite with the massive labor populace in order to realize democratic dictatorship of the proletariat. Otherwise, the landlords, the rich peasants, the counter-revolutionaries, the bad elements and the evils [*niugui sheshen*] would all come out; our comrades would do nothing about it, and many people would even resort to collusion without distinguishing friend from foe, and thereby allow the enemy to erode and invade, to divide and dissolve, to abduct and penetrate. Many

* I would like to thank Prasenjit Duara, Michael Dutton, Henrietta Harrison, Viren Murthy, David Wang, and Yiching Wu for reading and commenting on the earlier drafts. Thanks also go to Axel Schneider and Viren Murthy for inviting me to present this paper at the 2007 Leiden workshop on the writing of history in East Asia.

[1] Mao 1996: 25.

> workers, peasants, and intellectuals would then become prey to enemy tactics of coupling force with inducement. If things were allowed to go on this way, the day would not be too far off—few years, over ten years, or few decades at the most—when the resurgence of a nation-wide counter-revolution becomes inevitable. It would then become a certainty that the Party of Marxism and Leninism would turn into a party of revisionism, of Fascism. The whole of China would then change color. Let all fellow comrades give it a thought! Isn't that a most dangerous situation![2]

In his speech at the 10th Plenum of the 8th Central Committee (September 1962), Mao urged the nation to talk about class struggle every year, every month, every day, at congresses, at party delegate conferences, at plenums, and at every meeting.[3] Soon, Mao's exhortation "never forget class struggle" (*Qianwan bu yao wangji jieji douzheng*) would be emblazoned across the nation and new campaigns such as the Four Cleanups and Socialist Education would be launched, preparing the stage for the Cultural Revolution.

In rural western Sichuan, a museum responded to Mao's urgent resurrection of class struggle by mounting a new exhibit called the Rent Collection Courtyard (*Shouzuyuan*) featuring an extraordinary collection of over a hundred life-size clay sculptures that depicted scenes of class exploitation and oppression. The sculptures would catapult the museum to national and international fame and would become the most sensational specimens of socialist art aside from the eight model plays.

The museum itself had been established not long before this. In 1958, the Ministry of Culture convened a museum conference and called upon all counties and villages to set up local museums (*bowuguan*) and exhibition halls (*zhanlanshi*) and to promote mass participation in the preservation of historical sites and cultural artifacts.[4] Thereupon the Sichuan provincial authorities decreed that one of the manor houses (*zhuangyuan*) of Liu Wencai, a wealthy landlord in Anren township, Dayi county (50 kilometers west of Chengdu), be preserved and converted to a *chenlieguan*, or display site, in order to showcase the cruel exploitation and oppression of peasants by the feudal landlord class.[5] Local officials immediately went to

---

[2] Baum and Teiwes 1968: 70.

[3] Schram 1974: 190.

[4] Denton 2005, Flath 2002.

[5] The choice of *chenlieguan* instead of the usual *bowuguan* seems deliberate. For lack of a better term, I will refer to the site as a museum in this chapter. James Flath (2002) notes that the museum initiatives of the 1950s and early 1960s "served the function of preserving a historical building, despite the frequent religious and capitalist 'bad class background' of

work, evicting the troops stationed in the main compound, taking stock of the artifacts, recovering dispersed items, and researching the Liu family history. In merely two months' time, the museum opened informally to flocks of curious local visitors. The response was overwhelming. It became an overnight sensation in Sichuan and caught the attention of national media as well. More artifacts were recovered and the museum opened officially later in the year with the calligraphy of Marshal Chen Yi (1901–1972) gracing the front gate placard: "Daiyi Landlord Mansion Museum."

Liu Wencai (1887–1949) was the fifth of six sons born into a well-to-do peasant family in Anren. Before he became the arch symbol of the landlord class in the 1960s and 70s, he was dwarfed in notoriety by his younger brother Liu Wenhui (1895–1976) and cousin Liu Xiang (1890–1938), both swashbuckling warlords who ruled Sichuan for much of the early part of the twentieth century. Liu Wencai started out as the owner of 2 hectares of farmland and a liquor distillery, but managed to amass vast amounts of wealth during the ten years (1922–1932) in which he served in various lucrative official posts in charge of tax collection under Liu Wenhui's wing in Yibin, a port city on the Yangzi River. After the latter's military career suffered a fatal setback at the hands of Liu Xiang and retreated to Xikang (Kham) in western Sichuan, Liu Wencai took his riches back to Anren and proceeded to acquire land, build mansions and villas, procure wives and mistresses, consort with the Elder Brothers' Society, raise militias, eliminate rivals and put down CCP-led insurrections, set up shops and factories, open banks, monopolize the narcotics trade, and sponsor charities and public welfare projects (including the famed Wencai Middle School).[6] He died of tuberculosis on October 17, 1949, in the very month that Mao proclaimed the founding of the People's Republic. He had been largely

the host property" (48). The *chenlieguan* model seemed to be a solution to the difficulties posed to "an ideologically populist state" by "the contradictions inherent in the nationalization and management of a body of largely elitist relics" (57). In Li Zhensheng's photographic documentary of the Cultural Revolution in Heilongjiang, a rich peasant in Acheng county named Yuan Fengxiang was denounced during the Socialist Education campaign and had his "mansion" seized and converted to a temporary museum. One photograph shows the peasants lined up outside his "landlord mansion" and another—an interior shot—shows the "illicit possessions" (*zangwu*) on display inside the museum: electric lights, a Swiss watch, a deerskin jacket, a silk dress and stockings, etc. (Li 2003: 52, 53). I thank Karl Gerth for alerting me to Li's work.

[6] A *China Daily* report gives the following inventory of Liu's wealth: "[Liu] owned 28 houses and mansions around the province, a bank with 22 branches in South China, five pawn shops, 27 warehouses, seven factories, two power plants, four boats on the Yangtze River, some 1,500 shop fronts and 1,000 hectares of farmland." See Chen 2001.

forgotten for nearly a decade until in 1958 when his fortress-like tomb was blasted open by a group of steel workers and his bones were tossed onto the river bank. The brutal disinterment would inaugurate his metamorphosis into what journalist Xiaoshu aptly calls "a political monster" (*zhengzhi konglong*) perched over the "Jurassic Park" of his manor-house-turned-museum.[7] His name would become a household name, his crimes would be written into school textbooks and shown on screens across the nation, and his best-known victim would travel the region to tell her story of being imprisoned in his secret dungeon. He would become the personification of an unjust economic system, a long dark age, and absolute evil. The museum that put his crimes on display brought the dark past into the present and enabled millions of Chinese to relive the horrors of class exploitation and oppression.

The museum was established on the site of the Old Mansion (*lao gongguan*) which, built between 1932 and 1942, was the residence of Liu Wencai and his wives and servants. Covering some 7,000 square meters, it

---

7 Xiaoshu is the penname of Chen Ming (b. 1962), a graduate of Sun Yat-sen University and a freelance journalist who maintains a lively profile on the Chinese blogosphere. His revisionist reportage *Liu Wencai zhenxiang* (The true Liu Wencai 1999) attracted considerable attention among Chinese readers for whom Liu Wencai is a larger-than-life villain. Its digital version was widely circulated on the internet and stimulated a storm of responses, both pro and con. One blogger wrote: "Not so long ago, at the mere mention of this man's [Liu Wencai] name, a host of words would bubble up in our mind: tyrant-landlord, exploitation, oppression, the Old Society, the Water Dungeon, the Rent Collection Courtyard.... Some of us would even react physiologically, becoming nauseous and upset in the stomach and our blood pressure climbing up" (Jiang Hai, "Ping *Liu Wencai zhenxiang*" [Commenting on *The True Liu Wencai*], www.blogms.com/blog/CommList.aspx?BlogLogCode=1001545346 [last accessed August 8, 2006, URL no longer active]). The Hong Kong-based Phoenix Satellite Television Station produced a mini-series that drew heavily from the book and thereby further widened the book's impact. Xiaoshu is mostly concerned with reconstructing Liu's personal and clan history against the panoramic backdrop of warlord-era Sichuan, but he also devotes a lengthy prelude (about a quarter of the book) to the transformations of the museum. This part of the book has been most useful to me and I have borrowed extensively from its sources and insights. I have compared Xiaoshu's book with one that defends the orthodox image of Liu Wencai by the prolific reportage writer Yingquan and found the former a better example of scholarship than the latter. Xiaoshu documents his sources fairly carefully and gives a selected bibliography; Yingquan, by contrast, writes in a fictional mode and does not bother with such scholarly trappings as citations (Yingquan, 2000). However, I do not intend to take sides on specific issues that are under dispute, for my arguments concerning the Liu Wencai myth rest primarily on the material and visual cultural texts that once were or are still on display in the museum, and much less on archival materials or oral histories. In other words, I deal with the politico-aesthetic phantasm of Liu Wencai as "an evil landlord" and objective enemy, not with the flesh and blood Liu Wencai or whether he was indeed so wicked.

comprised 27 courtyards, three gardens, and about 180 halls and rooms.[8] In the initial phase, the museum converted 19 rooms into display spaces and grouped them into three exhibition blocks: extravagant lifestyle; the pentad rule of officials, tyrants, warlords, bandits, and secret societies; and instruments of exploitation and violence. The contents of the exhibitions were artifact-centered. Thus, the Lius' lavish lifestyle was presented directly in the restored living quarters: the audience halls, bedrooms, prayer chambers, tobacco rooms, wardrobes, china and jewelry collections, art and antique collections, etc. Their unholy alliance with other "feudal" powers was documented in correspondence, birthday gifts, funeral presents, and various congratulatory plaques, couplets, and calligraphic and art works. The instruments of exploitation included the two-sized pecks (*dou*), tenant rosters, and account books; torture instruments included chains, cangues, handcuffs, pistols, whips, and rods (plus bloody shirts). In 1960, the museum added several new features. First there were the 17 wax tableaux (*laxiang*) and 87 mini-dioramas (*baizuitu*) re-creating the many episodes of extortion, usurpation, abuse, and torture perpetrated by the Lius and their "running dogs," as well as a few instances of peasant resistance. Although it has been revealed that not a few of the stories were exaggerated, misattributed, taken out of context, or outright fabricated, at the time these were believed to be based on peasant denunciations and verified and substantiated by research. The museum also purchased and requisitioned a substantial number of artifacts, regardless of their origins, to make up for those lost in the intervening years. Soon afterward, eight large plaster tableaux were erected in a rear garden, all enacting allegedly true stories of peasant resistance. But the best-known addition in this period was the (re)created Water Dungeon (*shuilao*) in an opium storage room that launched the political career of Leng Yueying (1911–1984), supposedly the dungeon's sole survivor.

The museum reached its zenith of fame in 1965 with the Rent Collection Courtyard sculptures, which, unlike the above-mentioned features, were no longer pegged to concrete individuals or incidents. Then in the mid-1970s, the New Year goods storage room next to the prayer chamber was

---

8 Liu Wencai allegedly annexed, by force and by manipulation, more than ten peasant homesteads for the Old Mansion to achieve this gargantuan size. The New Mansion nearby was built for Liu Wenhui, but he stayed in Khams and never moved back to Anren. In the late 1940s, Liu Wenhui secretly defected to the CCP and facilitated the "liberation" of Sichuan; he later served as a minister of forestry and lived out the rest of his life in Beijing. When he is mentioned in the vast propaganda literature on Liu Wencai, it is usually only in his former role as a warlord.

converted into a "torture instruments room" (*xingjushi*) and the four adjacent china storage rooms were set up as "torture chambers" (*xingxingshi*). This would be the final burst of creativity on the part of the museum staff. In the early 1980s, spurred by rumors and complaints, investigation teams began to probe the veracity of the Water Dungeon. In 1988, the authorities declared it phony and ordered the site dismantled and relabeled to reflect its original use as an opium storage room. The torture chambers were also declared an error and the hair-raising instruments were quietly removed. But the Rent Collection Courtyard stayed and continued to attract attention, even if much diminished and un-orchestrated.[9] In 1996, the museum was renamed the Dayi Liu Clan Manor Museum (*Dayi Liushi zhuangyuan bowuguan*), dropping the word "landlord" and substituting the more neutral *bowuguan* for the ominous *chenlieguan*. It was also added to the government's cultural relics protection list. Nowadays, an eerie quietude hangs over the compound and a few scattered visitors have replaced the erstwhile thronging crowds. The museum has had to adjust to the market economy by touting its architecture and artifacts as the crown jewels of western Sichuan's cultural heritage. It has also installed a permanent exhibition of folk wedding customs in the New Mansion built for Liu Wenhui but never occupied. The luxury lifestyle of the Liu clan now stokes fascination with high life and surrounding restaurants unabashedly offer inventive dishes named after Liu Wencai, the political monster turned consumerist icon.

This chapter approaches the museum, in its *chenlieguan* phase, as a material and visual cultural text caught up in the vortex of the Maoist permanent revolution. It is not intended to be an institutional history of the museum or an archive-based study aiming to set the record straight on Liu Wencai as a historical figure. Rather, I focus on the Water Dungeon narratives and the sculptures of the Rent Collection Courtyard as instances of socialist art that exemplified the Maoist discourse of class. I borrow insights from history, psychoanalysis, and political philosophy in order to shed light on the process in which a local tyrant was turned into the archetype of the landlord class and enemy of the socialist nation. In particular, I seek to understand how the category of "objective enemy"

[9] In June 1999, Cai Guoqiang, a Chinese artist residing in the United States, was awarded the highest honor at the Venice Biennale for orchestrating a replication of the Rent Collection Courtyard sculptures (using the hired labor of 11 sculptors). Controversy ensued over the nature of Cai's prize-winning "action art": was it an ingenious enactment of postmodern aesthetics or a crass act of plagiarism? See Daozi 2003.

was necessitated by a linear, progressive conception of history, and how the museum mobilized the fetishism of land rent to foreclose a genuine Marxist critique of the emergent class formations in post-1949 Chinese society. Toward this end, I emphasize the *nationalist* axis of Chinese communism and how class was racialized and thereby depoliticized in Mao's China.

Class as Identity and the Logic of Displacement

Soon after the founding of the People's Republic, the CCP used the so-called *chengfen* system to affix a "class designation" to each person on the basis of his or her (past) relation to the means of production and the degree of economic exploitation this relation enabled. But as the new socialist nation proceeded to nationalize the economy and liquidate private property, the *chengfen* system became increasingly anomalous. How could anyone be a "landlord" or "capitalist" if he or she no longer *owned* land or machinery on such a scale as to allow him or her to live on the surplus value of others' labor? At the same time, socialist state-building introduced a new system of stratification that was tied neither to individual ownership of property nor to market mechanisms, but to the socialist political economy and its bureaucratic apparatuses. This new social hierarchy structured the everyday lifeworld in the form of the work-grade system, the urban-rural apartheid, and the theory and practice of *guanxi* (back-channel redistribution and mobility networks based on access and patronage).[10] This was a class system in which party officials and technocrats became what Wallerstein calls the "administrative bourgeoisie"[11] with privileged access to material and cultural resources that were mostly denied to average workers and especially to peasants.[12] But the ideological

[10] See Yang 1994.

[11] Wallerstein 1991: 141.

[12] Claude Lefort faults the socialist Left for failing to recognize the separation of administrative, regulatory, and policing power of the modern state from society and the consequent formation of a bureaucratic class, a fact also willfully denied in CCP class ideology. Lefort could very well be describing the PRC when he writes: "in such a society, the state alone appears to all and represents itself to itself as the sole instituting principle, as the great actor that possesses the means of social transformation and the knowledge of all things. It is the emergence of this 'point of view of the state'—of a state potentially at the centre of power and knowledge—that makes possible the formidable expansion of bureaucracies, whose members may cultivate their own interests and derive as much power and advantage as possible from it" (Lefort, 1986, 280). The strength of the bureaucracy as a class, however, is in sharp contrast to "the fragility of the position of its members,

visibility and shrillness associated with the *chengfen* system, which disenfranchised members of the former elites and their descendants, obfuscated recognition of the new class structure and obviated a genuine class critique for three decades. Thus in the very name of class struggle, class was for all intents and purposes displaced from Maoist politics. Richard Kraus writes:

> As originally constituted, the system of class designations provided the Party simultaneously with two sorts of information. It offered a view of the stratification of property and exploitation, and it concurrently served as a guide for distinguishing potential enemies and friends. But the failure to update the designations [to reflect the socialist transformation of economic relations] meant that they could not longer indicate ongoing economic positions of individual Chinese, and the system's identification of enemies and friends was useful only as a historical point of reference.[13]

But the invariable recourse to the economically defunct class categories in political campaigns proves that the *chengfen* system was never meant to be a neutral historical reference point. Indeed, Kraus acknowledges that the hierarchy of property-based class relations eventually degenerated into that of caste, or "a static ordering of revolutionary honor and shame."[14]

For class to become caste-like identity, it must go beyond the purely economic. Kraus shows that the CCP's calculus of class position had from the start incorporated political and attitudinal considerations, so that rectification and thought reform could be the means by which a person of a particular class origin realigns him or herself with a different class standpoint.[15] If early on this was necessary to resolve the intellectual problem of accommodating revolutionaries of nonproletarian origins, in the post-1949 period, it served to make class a more efficient tool of caste-formation. The behavioral supplement of class was the system of labeling or identifying politically deviant behavior for which the figurative term of "hat" (*maozi*) was deployed. While class designations were quasi-hereditary and universally applied, political labels could be assigned and removed on an ad hoc basis, so that even a person with a "good" *chengfen* could still be made to wear a counterrevolutionary "hat" for a discrete crime or misstep.

---

who were constantly threatened with annihilation, whatever their rank and authority, on account of their subjection to political power" (293).

[13] Kraus 1981: 57.

[14] Ibid., 51.

[15] Ibid., 21.

Crucially, the two systems worked in tandem, as evidenced in the Chinese acronym for the "five bad elements" of the 1957 anti-rightist campaign: *difufanhuaiyou*, standing for landlords, rich peasants, counterrevolutionaries, hooligans, and rightists.[16] Of the five categories, the first two were class designations and the remaining three were political labels, but the acronym was used simply as "a shorthand conceptualization of the class enemy."[17] While the old property-based class enemies—the landlords, rich peasants, and capitalists—could always be counted on to serve as the usual suspects in each campaign, everyone else was not therefore immune from being "revealed" as new behavior-based class enemies. The specter of "hats" kept everyone on his or her toes. Labeling thus prevented the *chengfen* system from becoming not only economically imprecise but also political obsolete by destabilizing the boundary between "red" and "black" and by turning everyday conduct into telltale signs of class identity.

Essentially, the two systems interpenetrated to render the entire population a vast reservoir of what Hannah Arendt calls "objective enemies" in referring to the Jews in Nazi Germany and the descendants of the former ruling classes in Soviet Russia. Rather than being really suspected of any hostile action, they were declared "objective" enemies of the regime in accordance with the ideology of Nazism or scientific Marxism.[18] Arendt cites a Nazi jurist who regarded his opponents "not as individuals but as carriers of tendencies endangering the state and therefore beyond the pale of the national community."[19] In communist ideology likewise, the legality of individual action is largely irrelevant in the prosecution of class struggle and in determining the boundaries of the national community. Class enemies are objective enemies because they are, from the "objective" perspective of scientific Marxism, on the wrong side of History. History needs to be capitalized because the Enlightenment conception of linear, progressive history as systematized by Hegel and dogmatized by scientific Marxism has come to possess the same inexorable truth value as "race" does in Nazi ideology, hence the Marxist cant of the Law of History.[20] The extirpation of objective enemies is thus the necessary precondition for the inexorable motion of historical progress. Whether a particular person is guilty of a specific crime or not is not meaningful. The motion of

16 Ibid. 58–61, 92–93.
17 Ibid., 60.
18 Arendt 1985: 121.
19 Ibid., 122n96.
20 See also Duara 1995.

History requires the continued identification and destruction of objective enemies. “As soon as one category is liquidated, war may be declared on another.”[21] Or else History would grind to a halt.

This is precisely the theoretical basis of Mao’s notion of permanent revolution and the practice of setting quotas for determining who the enemy du jour is. Kraus quotes Mao as saying: “in accordance with materialist dialectics, contradiction and struggle are perpetual; otherwise, there would be no world.”[22] He also suggests that Mao developed the theory of the two kinds of contradictions in the 1950s to resolve the fundamental problem of historical progress: “if class struggle is the engine of history, what is to be the motor for progress after the socialist revolution brings about the ultimate end of classes? Mao’s answer is to retain a notion of struggle detached from its connection with class.”[23] But Mao subsequently jettisoned the idea that non-antagonistic contractions could be the main motive force pushing Chinese society forward. Instead, in urging his countrymen never to forget class struggle, he identified antagonistic contradictions as the centerpiece of the Law of History, and welcomed the violence unleashed in class struggle because it was inevitable and because it “cleared things up.”[24]

The notion of contradiction within the people (*renmin neibu maodun*) was poorly theorized and haphazardly invoked because it sat uneasily with the denial of internal division that, according to Claude Lefort, was fundamental to the ideological “phantasy” of the “People-as-One.” The only legitimate division was to be external, between the People-as-One and the Other—the forces of the Old Society and the emissaries of foreign imperialism. “The constitution of the People-as-One requires the incessant production of enemies. It is not only necessary to convert, at the level of phantasy, real adversaries of the regime or real opponents into the figures of the evil Other: it is also necessary to invent them.”[25]

The figure of class enemy as an ideological invention, a phantasy, is a familiar one in post-Mao literature. A passage in the novel *Furong zhen* (Hibiscus town, by Gu Hua, 1981) describes the ideological necessity of objective enemy this way:

21 Arendt 1985: 123.
22 Kraus 1981: 67.
23 Ibid., 45–46.
24 Wakeman 1975: 295.
25 Lefort 1986: 298.

> [Crazy Qin] was labeled Iron-Hat Rightist, to show that his label would go to his grave with him.... Fortunately he had no wife and therefore no children to inherit this political legacy. And he knew that revolution needs a target. Unless each village and town kept a few "dead tigers" and living targets, how could mass movements and struggles be mobilized? Each time the higher-ups urged them to grasp class struggle, the local cadres called meetings, paraded and denounced the Five Categories, then reported the number of class enemies struggled against and recalled their past bitterness to educate the masses. In teams where the Five Categories had died out, their children took their place. Otherwise, how to convince people that in the historical period of socialism there would always be classes, class contradictions and struggles?... After Land Reform there had been a number of movements to redefine class status. As the means of production had been collectivized, putting an end to private property, the criterion for this was each one's political behavior, which leaves much room for play. But children could still inherit their parents' class status....[26]

Once an identity becomes quasi-hereditary, it is not far from slipping into a racist category. Indeed, both ideologically and institutionally, the primacy of class struggle as socialist identity politics amounted to a virulent form of "class racism." Etienne Balibar insists that racism is not just a prejudice or a form of false consciousness, but is instead a genuine mode of thought, an insistent desire for knowledge, and a way of defining who we are.[27] He proposes a broad definition of racism to take into account the "necessary polymorphism of racism, its overarching function, its connections with the whole set of practices of social normalization and exclusion."[28]

According to Balibar, racism has always been linked to class differentiation and is a much broader category than ethnic racism accompanied by biological theories. Historically, various disadvantaged and minority social groups—women, sexual deviants, the mentally ill, and so on—have been racialized and subjected to naturalizing practices of exclusion and depreciation. In his genealogical study of the "race myth," Balibar shows that it was in regard to the "race of laborers" that the modern form of racism first emerged. With the birth of the proletariat, race became detached from its historical origins in the aristocratic representation of the nobility as a superior "race" and in the slave-owners' representation of slaves as an inferior "race," and entered the "field of equivalences between sociology, psychology, imaginary biology and the pathology of the 'social body'."[29]

---

[26] Gu 1983: 141–42, 172; translation slightly modified.
[27] Balibar 1994: 200.
[28] Balibar 1991b: 49.
[29] Balibar 1991a: 207–9.

The proletariat, in its dual status as the exploited and the politically threatening, was identified as a racial Other at the very moment of its discursive constitution:

> For the first time those aspects typical of every procedure of racialization of a social group right down to our own day are condensed in single discourse: material and spiritual poverty, criminality, congenital vice (alcoholism, drugs), physical and moral defects, dirtiness, sexual promiscuity and the specific diseases which threaten humanity with 'degeneracy'... Through these themes, there forms the phantasmatic equation of 'labouring classes' with 'dangerous classes', the fusion of a socioeconomic category with an anthropological and moral category.[30]

Balibar concludes that the modern notion of race did not initially have a national or ethnic origin, but a class or caste signification, and that it was only retrospectively ethnicized so as to be incorporated into nationalist politics.[31] As a mechanism that naturalizes social differences, class racism works both ways. Indeed, in communist discourse, it is the bourgeoisie itself that is the target of racializing rhetoric. Michel Foucault points out that it was the socialist Left, in its nationalist phase, which first made political use of the "scientific ideology" of racism and propagated the idea that "the rotten, decadent class was that of the people at the top, and that a socialist society would have to be clean and healthy."[32]

The obsession with purity is where racism finds partnership with nationalism. Balibar maintains that racism and nationalism are never far apart, if only in the form of disavowal, and that the seeds of racism lie at the heart of all nationalist politics. This is because no nation has a natural ethnic basis and the idea of nationhood is always fictive. Racism produces fictive ethnicity through a naturalizing discourse of similarities and differences and a set of constructed ties of affect and genealogical evidence. Racism is thus historically articulated to nationalism not as an expression but as an internal supplement, always in excess of nationalism and hence repudiated in theory, but always indispensable to its constitution.[33] This is how Balibar accounts for the paradox of nationalism:

> [Racism] is already that internal excess of nationalism which betrays, in both senses of the word, its articulation to the class struggle. Hence the ever recurring paradox of nationalism: the regressive imagining of a nation-state

30 Ibid., 209.
31 Ibid., 207.
32 Foucault 1980: 223.
33 Balibar 1991b: 54.

> where the individuals would by their nature be 'at home', because they would be 'among their own' (their own kind), and the rendering of that state uninhabitable; the endeavor to produce a unified community in the face of 'external' enemies and the endless rediscovery that the enemy is 'within', identifiable by signs which are merely the phantasmatic elaboration of its divisions.[34]

The supplementarity of racism and nationalism, I believe, goes a long way toward explaining some of the puzzling aspects of Chinese communism. Indeed, characterizing the CCP discourse of class as a form of racism refocuses our attention on the undertheorized nationalist face of communism, which has exercised a fatal attraction on the independent-spirited but fiercely patriotic intelligentsia. It also helps us understand why the new nation proved "uninhabitable" for many even as it claimed to be the emancipated homeland for all "Chinese." Much like the myth of Aryanism, whose initial referent was class not nation, communist class racism simultaneously shrinks and enlarges the socialist nation, rendering class enemies as false nationals to be ejected from the body politic and the proletariat of the world as class brethren to be united and liberated. The racist obsession with purity and with rooting out the false, exogenous, and cross-bred elements finds expression in the elastic and exuberant typology of class enemies based on past property relations and present political behavior.[35] It posits a radical dichotomy between the exploited and the exploiter, each with its own genealogy, political agenda, moral values, affective regime, lifestyle and pseudo-biological characteristics. Class becomes an identity fortified by a quasi-biological theory of "political bloodline" (*zhengzhi xueyuan*), and the idea of "class origin" is institutionalized in the *chengfen* system.

Discussing the theory of political bloodline and its incorporation into the governmental technology of the personnel file, Michael Dutton presents a pyramid diagram that divides Chinese society into four blocks or stratifications (*jieceng*): at the top of the pyramid are CCP officials and members (*dang guan yuan*); next are advanced masses (*xianjin renmin*), or what Arendt would call "sympathizers" who people the "front organizations" of a totalitarian movement (62), followed by ordinary masses

---

34 Balibar 1991a: 215.

35 In addition to the "five black elements," common terms included spy, traitor, opportunist, renegade, capitalist roader, counterrevolutionary, conspirator, putschist, enemy agent, etc. Demonizing epithets were also in plentiful supply: ox demon, snake spirit, freak, blood sucker, black claw, vermin, scum, bastard, slippery fish, diabolical wind, and so forth.

(*putong renmin*); and at the bottom are the class enemies (*jieji diren*). Crucially, a shaded zone marks off the class enemies from the rest of the pyramid. The dark horizontal line above the shaded zone is identified as the "political bloodline" (*zhengzhi xueyuan xian*).[36] Clearly, this line is qualitatively different from the ones that divide the other three *jieceng*: to be a class enemy is to be of an alien "race" whereby one enters the zone of biological fatality and immutability. As the diagram indicates, the class enemies are literally in the gutter. They are enemy aliens (*jieji yiji fenzi*) who do not to belong to the people-nation. Hu Ping unflinchingly calls the class enemies the "Jews" of China: "shall we say that, in China's anti-'Semitic' campaigns, the landlords and rich peasants—these inferior 'Jews' of class rather than racial origins—have been 'purified' from social life?"[37] The descendants of the "five black elements," moreover, are denigrated for their political "black skin" (*zhengzhi fuse shang de heiren*) much in the same way that the descendants of African slaves have been treated in American history.[38]

The socialist nation is to be the nation of the laboring classes—the exploited, oppressed, innocent, upright, compassionate, indomitable, and robust—the People-as-One. The exploitative classes are the enemy within and fatally associated with the external enemy of the nation. The internal enemy is therefore always already the traitor, collaborator, or spy who identifies and colludes with the imperialists. A typical example of the essential unity of the internal and external enemy is the redoubling of Huang Shiren as landlord *and* traitor in the evolution of the archetypal narrative of socialist China, *The White-Haired Girl* (*Baimaonü*). In the early oral legend, the "White-Haired Goddess" is said to have been a peasant girl abducted by Huang Shiren the wicked landlord around 1930 and liberated by the Red Army a few years later. The opera version (1945) and the film version (1950) both move the abduction to 1935 in order to make the date of her liberation coincide with the outbreak of the Sino-Japanese War in 1937. In the opera, Huang is alluded to as the ringleader of the local militia (*tuanzong*) whose in-laws' family is wiped out by invading Japanese troops.[39] The ballet film version (1972), however, pushes the abduction even later, to 1937, and immediately establishes Huang as a traitor by showing him consorting with a Japanese officer who appears

36 Dutton 2004.
37 Hu 2001: 438.
38 Ibid., 450.
39 He et al. 1954.

to be inviting him to serve on the Pacification Committee (*weichihui / ijikai*).

If this appears counter-factual, it is faithful to the logic of class racism, which posits an inexorable tie between the enemy within and the enemy without. Such poetic license would be exercised again and again in socialist realism. Indeed, the two thousand years of Chinese history is renarrativized to satisfy the genealogical drive to establish transhistorical class filiations. In this scheme, the capitalists and the landlords of the twentieth century are bound to the kings, emperors, feudal lords, scholar-officials, landowners, and merchants as their class "ancestors." The workers and the peasants, on the other hand, are the progeny of generations of exploited and oppressed peasants and laborers. Past riots, rebellions, and uprisings come to make up a long and glorious history of "revolutions" culminating in the communist victory of 1949.

Because class racism, like most forms of racism, operates in an inverted fashion, the socialist imaginary is centrally structured around the quasi-hallucinatory visibility of the class enemy, the false Chinese. Following the logic of excess, a host of real-imaginary traits accumulate around class enemies who figure prominently in socialist gothic as the "undead." Economically dispossessed and politically disenfranchised, they sustain their "undying hearts" (*xin bu si*) with the hope of a swift and vengeful restoration of the ancien regime with its system of wanton exploitation and grotesque inequality. They are usually associated with fetishized objects such as the secretly buried land title deed or account book (dubbed "restoration ledger" [*biantianzhang*]), the cleverly camouflaged radio transmitter or camera. In high socialist realist narratives, the climax is usually when vigilant villagers catch the unreconstructed (former) landlord furtively unearthing his stash of documents and hungrily poring over them while fantasizing about reclaiming his properties when the "sky" changes color. Alternatively, it is the counterrevolutionary who infiltrates the revolutionary ranks and spies for his foreign paymasters—typically Chiang Kai-shek and his American patrons. The enemy agent either uses his gadget to transmit espionage information, or attempts to blow up a power plant or dam with his time bomb, conspiring for a counterrevolutionary offensive against the newly-minted socialist nation. The confiscation or burning of these fetishized objects thereby stages the triumphant destruction of the enemy within and the symbolic purification of the body politic.

The entire conception of the Dayi landlord museum can be said to exemplify the inverted logic of class racism whereby the People-as-One is only conceivable by inferring from the visibility of the class enemy.

Museumfying the native habitat, genealogy, and criminal history of the landlord class is tantamount to proclaiming its status as the alien, the stranger who is not one of "us." A museum of the Other, however, invariably courts exoticism and other unintended consequences. In some ways, the landlord museum was also a socialist haunted house that offered visitors the pleasure of horror and abjection in the tradition of the folk rendition of the 18 layers of purgatory in temple murals. Its displays and artworks encouraged aesthetic, affective, and bodily engagement even as they emphatically invited the moral and epistemological gaze. As a ritual site where audiences were incited to surrender themselves to a subjunctive world of "as if,"[40] it could not foreclose the inherent ambiguity and instability of the ritual process. So, predictably, it failed to pass the scrutiny of Cultural Revolution Red Guards armed with the "telescope and microscope of Mao Zedong Thought."[41] In the remaining parts of this chapter, I examine the aesthetic and affective politics of the museum's two most famous loci and conclude by reflecting on the nature and implications of the Maoist political.

The Water Dungeon and Socialist Horror

Throughout the 1960s and 1970s, the Dayi landlord museum had a basement room exhibit known as the Water Dungeon. The caption ran like this:

> A water dungeon was secretly built next to Liu Wencai's prayer chamber. It is said that the construction workers were slain immediately upon the completion of the dungeon. No one knows for certain how many people have perished in it. This hell on earth was filled with icy cold water; rotting corpses were piled up in it, emanating a rancid odor. There was an iron cage with spikes and sharp nails on all sides. One could neither stand nor sit in it, so it was nearly impossible to get out of it alive. Leng Yueying, a National People's Congress representative, model worker, and Communist Party member, is the sole survivor of the water dungeon.[42]

Leng Yueying was once a tenant farmer for Liu Bohua, a cousin of Liu Wencai. Around 1951, she began to tell a story about how, in 1943, merely three days after she had given birth, she was taken to the Liu manor house

---

40 Seligman et al. 2008.
41 MacFarquhar and Schoenhals 2006: 47.
42 Xiaoshu 1999: 29.

with a blindfold and was imprisoned in a water dungeon for owing Liu Bohua five pecks of rent grain. Dayi county officials soon seized upon the story and commissioned a mini-diorama re-creation of the water dungeon accompanied by live narration in 1954 as part of a larger exhibition to contrast the Old and New Society and to celebrate rural collectivization. In 1958, the year the landlord museum was launched, planners filled the basement room adjacent to the prayer chamber with water, installed a replica cage, and simulated bloody smudges on the walls—all according to the diorama model. Leng Yueying soon became the most famous peasant woman in China—affectionately known as Mother Leng—and was highly sought after in the *yikusitian* (remembering past bitterness and cherishing present sweetness) lecture circuit. According to Xiaoshu's estimate, in the space of two decades, she gave nearly a thousand lectures in front of a total audience a million strong.[43]

The enormous popularity of Leng Yueying's Water Dungeon story was undoubtedly the result of the party's propaganda machine going into overdrive. But what Leng offered was first and foremost a tale of suffering, and a surprisingly graphic tale to boot. Her recounting of the bloody water, spiked cage, rotten flesh, and killer stench, tied in with the macabre pit re-created in the basement, made available a rare chance to experience gothic horror in socialist China. It seems logical that the story would become more and more sensationalized as her audience grew bigger and bigger, essentially evolving into a socialist gothic tale. As such the Water Dungeon was a primary material-textual site for the ritual process of abjection that delivered the birth of the socialist subject. Let us take a closer look at Leng's story, particularly the ghastly details of her dungeon experience.

[43] Xiaoshu 1999: 42. Xiaoshu does not question Leng's claim that she was imprisoned, with a blindfold, by her landlord, Liu Bohua, in a room that had some water built up in it. Shortly before Leng went public with the water dungeon story in 1951, rumor had it that two militiamen had stumbled upon a hidden cellar in the Liu manor house and found it filled with thigh-deep, stinking water with human skulls and bones in it (Xiaoshu 1999: 47). The rumor was never verified. But it seems to have prompted Leng into imagining that she was jailed in "a water dungeon," allowing her tale of horror to take wing. In the early 1980s when the authorities began to look into the veracity of the water dungeon, Leng began to shun public appearances, invoking failing memory and poor health. Reportedly, only once when pressed by a reporter did she blurt out that it was the county officials, not she, who invented the whole story and that the reporter should be going after them instead (49). My object here is not to further investigate the historical authenticity of the water dungeon. I am more interested in the fictive truth that it embodies.

In the 1967 version recorded by Red Guards and preserved in the museum's archives, Leng dates her imprisonment to 1937. She was then Liu Wencai's tenant farmer and a draught-induced poor harvest caused her to be 5 pecks short in her rent payment. Liu's lackeys ransacked her thatched hut, beat up her children (her husband had gone into hiding), tied her up and dragged her to the manor house (without a blindfold):

> After turning this way and that way, we descended the steps of a stone stairway and immediately a foul stench assailed me. I passed out. A thug thrust me down into the water. When I stood up, the water reached my waist. And if you can even call it water! Icy to the bone and nauseatingly pungent. I had just given birth and blood was still trickling out of me. The icy bloody water made my insides churn as if a knife had been plunged into my abdomen. Not being able to steady myself, I fell down and the foul bloody water rushed into my mouth. I struggled to my feet, but immediately keeled over again, one hand clutching a bone and one foot stepping on a skull.[44]

Leng was shoved into a cage and then brought out three times to be interrogated. Each time she was returned to the dungeon, the horrific experience of wading through a gooey mass of disintegrated body parts was repeated. Seven days and seven nights later, her husband managed to redeem her freedom and carried her insensate body home. She learned afterwards that her blood-soaked clothes had hardened like ox hide and petrified human flesh was stuck to her legs. It took over ten days of soaking and scraping to separate the fetid clumps of rotten flesh and clotted blood from her own flesh.

Unsurprisingly, Leng's tale elicited wrenching emotional responses from her listeners. Reportedly, in 1966, when she spoke in front of some three hundred cadets of a PLA air force unit, many of them wept themselves into a fainting spell and had to be revived and propped up by museum workers.[45] A documentary made in the same year by CCTV recorded audience members being moved to torrential tears and thunderous sloganeering: "Never forget the bitterness of class oppression! Forever remember the debt of blood and tears!"[46] But Leng's story is no mere tear-jerker. Rather, it provides a narrative space in which a psychological mechanism called "abjection" can be enacted. According to Julia Kristeva, abjection refers to the body's desire to rid itself of waste: phlegm, pus, menstrual

---

[44] Cited in Xiaoshu 1999: 38.
[45] Xiaoshu 1999: 36.
[46] Ibid.

blood, and excrement. "What is abject . . . is radically excluded."[47] It is a process of ejecting and keeping at bay that which threatens the autonomy and meaning of the bodily self. There is also a temporal dimension to abjection, which refers to the body's attempt to jettison the memory of the mother's womb and to stall the fate of becoming a decomposing corpse—what one permanently thrusts aside in order to live (3). What is of interest to us is Kristeva's point that "it is not lack of cleanliness or health that causes abjection but what disturbs identity, system, order" (4). In this vein, abjection is as much a social mechanism of boundary making as a psychological process of ego-formation.[48] Enacted on a collective and historical scale, abjection becomes recognizable as racism, as in anti-Semitism wherein the Jews are the shady, sinister, and scheming abject that must be exposed and cast out in order for the body politic to become clean, pure, and integral.[49]

Kristeva believes that abjection is accompanied by narcissistic pleasure because it is a process in which the self becomes a subject. Subjectivation is the obverse of abjection, and subjectivity is premised on the rejection and denigration of the Other. In other words, the Other cannot be a neutral entity, but must necessarily be as undesirable, indeed revolting, as excrement. She calls the strange combination of disgust and fascination that accompanies abjection the "powers of horror." In this light, the Water Dungeon is a primal site of abjection where individuals become clean and inviolable socialist subjects by symbolically immersing themselves in the abject and then peeling it away. The fetid pit of flesh and bones serves as a synecdoche of the dark, murky, and horrific age of suffering. With her vivid loquacity, Leng takes her audience on a harrowing journey of merging with the most abject, most repulsive substances. In her surviving the dungeon, the socialist subject experiences the pleasure of abjection and subjectivation.

However, the abject cannot be repudiated once and for all. It stubbornly returns, not least because the subject needs to enact, in a ritualistically repetitive manner, the process of abjection in order to consolidate its identity and revalidate its autonomy. "Abject . . . is something rejected

---

[47] Kristeva 1982: 2.

[48] Kristeva's theory of abjection has a good deal in common with the anthropological theory of pollution whereby what is considered polluting in many cultures is also what disturbs or transgresses bodily or social boundaries. See Douglas 2002.

[49] A ghostwriter for Mao once non-too-subtly compared the CCP purge routine to defecation: "Societies are like that, [like people] they also have to defecate, which is not a bad thing" (quoted in MacFarquhar and Schoenhals 2006: 253).

from which one does not part."[50] In other words, the self cannot be a self without the Other, summoned hither only in order to be turned out, ad infinitum. Or, as Martin Jay puts it, abjection is "a necessary impulse in an economy of subjective constitution, which feeds off precisely what it seems to reject in a kind of endless game of fort-da."[51] The game of fort-da was famously played by Sigmund Freud's grandson, who, as observed by Freud, repeatedly threw a spool out of his crib only to retrieve it by yanking the attached string, all the while spiritedly uttering "fort… da!" (gone… back). Freud theorizes this game as the little boy's attempt to cope with his mother's intermittent absence. The spool is imagined to be the mother, and in making it disappear and reappear, he symbolically takes control of the situation and learns to regard himself as a separate being, on his way to be a subject. If the Old Society is the ugly mother from whose decaying womb the socialist subject is born, then it must be perpetually going and returning, in order for the socialist subject to reconstitute itself. Hence Leng Yueying must tell her story over and over again and the dungeon must be re-created so that the socialist game of fort-da could be played endlessly. To be sure, the Water Dungeon was not the only site in the museum that incited fear and horror. A nearby ground-level room fitted with cages and torture instruments was called the Earth Dungeon (*dilao*). But it never achieved the notoriety of the Water Dungeon. No living victim was identified to keep its horrors alive. At least part of the reason, in my view, is the Earth Dungeon's relative lack of abject substances, that is, those viscous and foul-smelling things that threaten to engulf the self and whose cleansing is the precondition of the birth of the subject.

In the 1967 version of her story, Leng Yueying was taken out of the dungeon three times to be interrogated, with Liu Wencai appearing in person during the last session. One almost suspects that the interrogations were added to her story just so that the act of being thrown into the revolting pit and the experience of abjection could be narratively repeated for maximum effect. The scripted nature of the narrative is also most blatant in the interrogation passages. The 1974 version gives an elaborate account of two consecutive interrogations (with Liu Wencai appearing in the first one) in which she was cajoled, whipped, and forced to confess in whose house she was hiding her grain. The scene bears a remarkable resemblance

[50] Kristeva 1982: 4.
[51] Jay 1998: 156.

to literary and cinematic moments in which brutish and huffy Japanese or Nationalist (KMT) officers interrogate steely Communist heroes but can never get anything out of them except dagger looks and fiery slogans. Leng tells us that she chose to be whipped rather than implicate her class brothers. She also resolved to survive the dungeon because the town herbalist and scribe had once told her about the Red Army that would one day come and smash the wolfish cabal of Liu Wencai and liberate the poor. She had at that time tattooed a red star between her thumb and forefinger. Now, thrown back into the dungeon and abruptly brought to her senses by the icy cold water, "I raised my hand with difficulty and gazed upon the star on my palm. In a daze, I seemed to see octagonal capped Red Army soldiers carrying a red flag and walking toward me. I was immediately filled with strength. Thus rallied, I crawled, crawled, and crawled until I reached one corner of the pit and leaned my head against the wall. I fainted again."[52]

Leng's 1974 narrative goes on to tell of her political activism in the 1960s and 1970s. In particular, she relates how she identified and captured a former steward of Liu Wencai who attempted to go underground, how she foiled the sinister plot of another Liu Wencai lackey and confiscated his time bomb, and how she endeavored to catch other saboteurs red-handedly. But above all, she affirms her primary duty of recounting the dungeon story: "for the sake of educating our young, and for the sake of preventing our country from changing color, I will go on telling my story for as long as I live!"[53] The Water Dungeon and its narrative are therefore a safeguard against the conspiracies of the internal and external enemy and a mechanism of purging the false nationals along with their miasmic theory of "the extinction of class struggle" (*jieji douzheng ximie lun*)—precisely by constantly and tirelessly keeping them (or their specters) in sight and by inducing the cathartic experience of fear, horror, and disgust. Abjection is thus the hysterical underpinning of the socialist order.

The Rent Collection Courtyard and the Law of History

Kristeva distinguishes the abject from both the subject and the object. If the abject is the repellent waste matter that the subject cannot bear to behold or contemplate, the object confronts the subject and demands

52 Chinese Department 1974: 16.
53 Ibid., 21–22.

precisely to be looked at and pondered. In its reliance on architecture and artifacts, the museum on the whole cannot avoid turning the class enemy into the object of a contemplative gaze. The magnificent manor house itself, with its innovative fusion of rural Sichuan and Shanghai/European architectural styles, is certain to inspire awe, envy, and perhaps even admiration. The exquisite furniture, china, objets d'art, jewelry, artworks, and antiques call attention to themselves as appurtenances of refined taste. Even the litany of Liu's vast wealth in the museum's accounts of clan history tends to conjure up an image of Liu as a very "capable" (*nenggan*) individual. The Red Guards, for all their obtuse fanaticism, keenly picked up on the inherent ambiguity of material culture in their attacks on the section of the museum called "The Restored Sites of Liu Wencai's Criminal Life" (*Liu Wencai zui'e shenghuo xianchang fuyuan guan*):

> From the Ford sedan to the study, everything is faithfully arrayed according to the recollection of Liu's chief of staff. A "memorial" it has indeed become! In our view, to put on display all the rotten and utterly reactionary things exactly the way they were, without any critical intervention or with merely a few lines of routine condemnation, is tantamount to disseminating these poisonous things and corrupting the revolutionary masses, particularly the revolutionary youth. Isn't it proof enough that the utterly reactionary plaques, couplets, poems, and paintings are displayed without any accompanying critique? Although there are a few notes of denunciation next to Liu Wencai's dog lair, Wang Yuqing's [Liu's youngest concubine] fox hole, and the Chinese-style and Western-style reception halls, the revolutionary masses can still tell from a single glance that these are meant to flaunt Liu's lavish and thoroughly corrupt lifestyle. In fact, these things have already done their damage to our revolutionary youth. According to what we have observed ourselves as well as the reports of the museum's revolutionary cadres, some comrades have taken to lingering in front of Liu's dog window in deep appreciation and have endless praise for Wang's fox hole. Even worse, one comrade has been heard saying: "Liu Wencai's bed looks so comfortable! I would die just to lie in it for a spell."[54]

Although the Red Guards did not descend on the museum until after the key exhibits of the Water Dungeon and the Rent Collection Courtyard had already been installed, they still found problematic the basic conception of the museum, particularly the artifactual contents of its displays. The Red Guards were not so wide of mark in their apprehension that too much attention to the material remains of a lavish lifestyle can paradoxically fuel fantasy and spur desire. There was already a lore about the Lius'

---

[54] Xiaoshu 1999: 82–83.

many private-life excesses, and it seemed to be driven primarily by the popular appetite for the bizarre and yet oh-so-fascinating habits of the rich. For example, Liu Wencai was known to have kept a bevy of wet nurses in his inner quarters to provide a steady supply of human milk to nourish his sickly body. "Pumping milk" (*jinai*), therefore, became a standard motif in narratives of suffering told by peasant women. Indeed, three of the five stories featuring a female protagonist in the "peasant family histories" collection produced by the museum (in collaboration with Sichuan University) tell of their ordeals in the inner recesses of the manor house.[55] It was the only occasion in which the female anatomy was named unflinchingly and in which class violence took on a sexual, even kinky, flavor—as when Liu's steward plunged a dagger into a wet nurse's breast or when Liu sank his teeth into the nipple of another wet nurse who was forced to suckle him directly through a hole in the mosquito net.[56] Liu's concubine Wang Yuqing was said to have a peculiar fondness for the webbed feet of ducks. In order to satisfy her exotic craving, 30 ducks had to be slaughtered daily to come up with enough webbed feet for one meal.[57]

Moreover, a 1965 brochure published by the Antiques and Relics Press (Wenwu chubanshe) devotes an entire chapter to the Lius' "corrupt lifestyle," which, condemnatory rhetoric aside, is essentially a detailed catalogue of the museum's holdings. It includes a dozen photographs featuring various luxury items such as the Ford sedan, ivory dining sets, and a gilded four-poster bed.[58] It is difficult to rule out the possibility of fascination overrunning hate. Criticisms similar to those of the Red Guards had in fact been aired early on, and the wax tableaux, the mini-dioramas, and the Water Dungeon were created largely in response to such criticisms. These exhibits directed the audience's attention to the boundless suffering of the laboring masses, treating suffering as a badge of honor conferring on the latter moral and political rectitude and guaranteeing their class standing. Narrativization of suffering was often supplemented by the ritual process of abjection, so that "speaking bitterness" (*suku*) became the most ubiquitous technology of self in socialist China. The celebrated metaphor of *fanshen* (turning over) owed its power to the reclaiming of experience indexed to emaciated and bruised bodies. It was always already the

---

55 Chinese Department 1974.
56 Wang Zhi'an 2001: 229.
57 Dayi Landlord Mansion Museum 1965: 38.
58 Daiyi Landlord Mansion Museum 1965.

experience of bodies in pain, bodies that were real for having known hunger, cold, fear, and incarceration, and bodies that strained to turn over in order to be free, healthy, and whole. The Maoist cult of experience thus rested on a vast repertoire of *fanshen* narratives that spoke of liberation as the shedding of pain and fear and yet thrived on the powers of horror.

However, the cult of experience necessarily operated at the phenomenological level, with its avowed adherence to the real and the concrete. The episodic tales of *fanshen* did not easily add up to a coherent narrative of history. The need for an embodied narrative that nonetheless conveyed the abstract Law of History was the impetus behind the creation of the Rent Collection Courtyard exhibit. It departed from the documentary mode (however faulty) of earlier exhibits with its generically sculptured and narratively sequenced clay statues that were meant to illuminate the structural exploitation built into the institution of tenant farming (*diannongzhi*). It was much more interested in visualizing class positionality and the mechanism of class exploitation than brute violence. It was a visual/narrative illustration of class struggle as the Law of History in which "rent" was the fetishized agricultural equivalent of industrial "surplus value." In breaking away from the (nominal) constraints of the real, it was able to deflect the kind of criticisms leveled at the artifact-based exhibits.

The Old Mansion had a large and small courtyard in the northwestern wing. Every year during the harvest season, hundreds of tenant farmers were summoned by Liu's "rent call" (*cuizuling*) to come hither to have their rent grain inspected, winnowed, measured, and retained/rejected. In the heightened political climate of the Socialist Education campaign, the propaganda chief of Wenjiang Precinct, Ma Li, came up with the idea of re-creating the rent paying scenes in the courtyards. Interestingly, his inspiration came from popular religion: "In the Old Society, there were murals of blade mountains and oil cauldrons in every King Yama temple or city god temple. Those images were very vivid and powerful. Why don't we take advantage of Liu Wencai's rent collection courtyards to make some clay sculptures? They hardly cost anything and can be used to educate the masses about class."[59] But unlike the Water Dungeon, the sculptures would not go the route of the macabre and the abject, because the goal was not to tell a literal, empirical truth (however adulterated), but a higher Truth.

---

59 Cited in Wang Zhi'an 2001: 157.

A group of museum workers and art school teachers and students were then assembled in Anren and began brainstorming. They visited and interviewed Liu's former tenants (including Leng Yueying), went to markets to observe rural customs and material culture, visited local temples to study folk sculpture, and produced four successive narrative drafts outlining the ambitious project. Beginning with the second draft, a group of Sichuan opera actors were recruited to model the various scenarios and to be photographed. Finally, in June 1965, the fourth draft and the accompanying photographic blueprints won approval from the authorities. The artists immediately went into frenzied production and completed the project in four months, in time for the National Day (October 1) celebration.

The opening statement of the fourth draft declares:

> Our goal is to re-create the scenes of rent delivering, inspection, collection, and seizure in order fundamentally to disclose the wickedness and brutality of the landlord class and the criminal essence of its cruel exploitation of the peasantry. "The landlord's rent collection courtyard is the poor man's gate of hell." The sculptures will graphically illustrate how Liu Wencai's lascivious and rotten lifestyle was founded on the squeezing dry of the peasants' blood, sweat, and life, and will inspire the audience's hatred of the landlord class.[60]

The draft outlines four sections, each highlighting a dominant affect: 1) delivering rent—bitterness (*ku*); 2) inspecting rent—wickedness (*jian*); 3) settling accounts and extorting rent—viciousness (*e*); 4) coda—hatred (*hen*). The final product has a fifth section: struggle. The official narrative guide begins by asking the audience to reflect on what they have presumably just seen—the "Liu Wencai's Corrupt Lifestyle" exhibits that one must pass through to reach the courtyards in the rear. Why was it that, even though the landlords never lifted a finger all year round, they lived in such obscene luxury? And why did the laboring masses never have enough to eat or wear even though they toiled year in and year out? The answer, the guide intones, is in the rent collection courtyard. This is followed by a quotation from Mao: "The primary method of landlord exploitation was rent collection. Other methods included lending loans, hiring farmhands, and running shops and factories. But extracting rents from peasants was the primary means of class exploitation."[61]

---

60 Ibid., 231.
61 Ibid., 296.

The prologue thus unambiguously identifies an economic relationship—"exploitation"—as the overriding theme of the exhibit. But how does one visualize an abstract economic relationship? Let us closely examine the artworks and narrative guide. The first section, "Delivering Rent," features a line of ragged peasants assuming postures of pain and exhaustion while carrying, pushing, or pulling rent grain into the courtyard. The guide calls our attention to the childless aged widow leaning on her cane, her face gaunt and eyes set deeply and sorrowful. The basket next to her has a hen in it. We learn that the hen has been brought along to supplement the insufficient rent grain. We see a widowed mother with an infant on her back and a skeletal boy at her side. She is clutching a half-filled sack that holds their entire yield. The guide ventriloquizes in the child's voice: "Mama, I'm hungry. Mama, I want to eat." It then offers the following commentary:

> In the cannibalistic Old Society, there was not just one helpless mother, and there was not just one old widow; the suffering of this sick man, this gray-haired grandpa, and this small child is not singular or isolated. No! This is not about one family or one household. This is a microcosm of the Old Society, a true portrayal of the miserable life of millions of laboring masses![62]

The truth claim here is grounded in the theory of "typicality" first expounded by Soviet literary theorists. The idea is that socialist art should not slavishly "mirror" reality in the deluded fashion of bourgeois realism or naturalism. Rather, it should transcend life's distracting multiplicity in order to arrive at a higher truth. The quest for a higher truth allows one to go beyond the phenomenological surface of things and discern their essence, or the Law of History. Literary or artistic characters should not be hamstrung by the requirements of verisimilitude. They should be "typical" instead of merely plausible or life-like.

The next group of sculptures portrays the exacting procedures to which the rent grain is subjected: visual inspection (*yan*), winnowing (*feng*), measuring (*guodou*), and account settling (*suanzhang*). The highlight in this section is the wind-winnowing device (*fengguji*) that can eliminate over 50 percent of the grain as subpar. When a child tries to take home some of the rejected grain, a steel whip is aimed at him; to the side, his grandfather is beseeching a lackey to accept their grain. The measuring procedure features the Lius' infamously oversized peck (*dadou*) which,

---

62 Ibid., 298.

being significantly larger than the standard ones used in the marketplace, used to "send shivers down the spines of all who beheld it" (*wang zhe zhe zhi dou, hunshen dou san dou*). (They also had an equally infamous small peck which was used when loaning grain out to peasants.)

In the account settling section, Liu Wencai appears on the scene. The guide states: "With words of benevolence and morality on his lips but thievery and debauchery inside him, with his fingers on prayer beads but his heart more ruthless than that of a wolf, Liu Wencai did not come to the rent collection courtyard everyday. He left it to his stewards and lackeys. Occasionally, though, he came here to take a look for the pleasure of it [*xinshang*]."[63] One scene shows that when a father and son are accused of falling short on rent, Liu instructs his steward to draw up a contract mortgaging the son to the KMT army.[64] The section called "Extorting Rent" includes the most iconic tableau: a blind old man being led by his orphaned granddaughter to the courtyard. We are told that he has toiled all year with his granddaughter's assistance and heaven knows how many falls he has sustained and how much blood and sweat he has shed. To get through the Gate of Hell, he has to turn over every single kernel of the blood- and sweat-soaked harvest to the landlord. And what does he receive for all this in the end? His granddaughter's indentures that read: "I am unable to pay the three pecks of rent grain owed to the Lius. I hereby willingly give my granddaughter in payment of the rent . . ."[65]

Another tableau shows a young mother being dragged away from her infant. Here the guide informs us that Liu Wencai's habit of imbibing human milk caused the deaths of many infants who were deprived of their mothers' milk. As an example, it relates Luo Erniang's story of having to suckle Liu through a mosquito net. "When Liu sucked a few times and no milk came forth, he savagely bit into her nipple. Blood gushed out and Luo Erniang rolled on the floor in pain. Even today there is still a scar on her nipple, testifying to the deep hatred of the working people."[66] The guide invites the audience to imagine their own mothers in the situation:

---

[63] Ibid., 301.

[64] The official narratives do not make any distinction between the (in-fighting) warlord forces in Sichuan and Chiang Kai-chek's KMT regime to which the former owed nominal allegiance. In retaliation against Liu Wenhui and his son for defecting to the CCP, Chiang instructed his general Hu Zongnan to wipe out the Liu clan and pillage its estates on the eve of the KMT retreat to Taiwan. The clan members went into hiding and were mostly unharmed, but the manor house was ransacked by Hu's troops.

[65] Ibid., 303.

[66] Ibid., 304.

"How would you feel? What would you think?" In case their feelings and thoughts might be distracted by the invocation of female anatomy, the guide follows up with another incantation of the poetics of typicality:

> In the evil Old Society, there was never just one Luo Erniang, nor was there just one blind old man. How many more fell by the wayside and how many more were driven to the brink of death for not being able to pay their rents? How many parents were forced to abandon their children and turn to mendicancy? How many families had their fathers conscripted by the warden (*baozhang*) and their houses turned upside down? Millions and millions! Countless! In that society, who did not have a blood debt and which family did not have a history soaked in blood and tears?[67]

The coda section emphasizes resentment, anger, and the resolve to rebel. One muscular young man is shown standing beside his empty baskets, his head inclined forward and his hands clenched into tight fists behind his back. Two youths are grabbing their shoulder poles apparently ready to strike back at the lackeys. The guide announces that the peasants have resolved to take the revolutionary path by joining the guerilla forces and dedicating themselves to armed struggle under the leadership of the CCP. The narrative voice has now reached a crescendo of revolutionary rhetoric, urging the audience to remember class suffering, fight to prevent history from replaying itself, carry the revolution to the end, and ultimately liberate all oppressed peoples of the world.

The five sections are arrayed in the circular space of the courtyard sequentially, creating a monument-in-motion effect. The 114 figures are clustered into twenty tableaux evenly spaced out so as to convey the large number of tenant farmers subjected to Liu's tyranny without overcrowding the space and obscuring the typicality of the individual tableaux. Of the 114, there are eighteen "negative characters" (*fanmian renwu*) consisting of Liu's stewards plus an accountant, a KMT soldier, a *baozhang* (warden), as well as Liu himself. Even though it is commonly known that Liu rarely appeared in the dusty, noisy, and crowded courtyard during rent collection times, he is made to be present so that the class contradiction can be personified and anchored.

During the sculpting process, nearby peasants often gathered at the site and counseled the sculptors on ethnographical details, such as the preponderance of women delivering rents because men feared forced conscription, or local women's habit of keeping a single instead of double

---

[67] Ibid.

braid. When the team was working on the winnowing scene, one onlooker remarked that the *fengguji* did not give the impression of generating "wind." In response, the artists adjusted the peasant figure standing in front of the device so that his gait and the billows of his shabby garment became suggestive of blowing wind. But the folk artist who was in charge of sculpturing clothes ended up making the figure look rather like a wondering immortal because the folds were too delicate and wavy. So efforts were made to coarsen the texture.[68]

In anecdotes like this, we catch a glimpse of the innovations that went into the project and the artistic appeal of these works beyond ideology. High-level officials and intellectuals, including Jiang Qing (1914–1991), Kang Sheng (1898–1975), Peng Zhen (1902–1997), Deng Tuo (1912–1966), Lu Dingyi (1906–1996), Zhou Yang (1908–1989), Guo Moruo (1892–1978), and Hua Junwu (b. 1915), responded to the exhibit with hyperbolic enthusiasm, praising it as the supreme achievement of socialist art nourished by folk traditions and sublimated by communist ideology. Zhou Yang, for example, exulted at the prospect that these sculptures would serve as sketching models for art school students: "Yes! They won't have to keep drawing Greek noses anymore!"[69] National media coverage of the exhibit was orchestrated by Hua Junwu, a *People's Daily* editor. It did not take long before the entire nation learned about "the great revolution in sculpturing art" (*diaosujie de dageming*); images from the Rent Collection Courtyard soon became ubiquitous in newspapers, magazines, pamphlets, portfolios, academic journals, and children's comic books. The Central Newsreel Studios shot a documentary which was screened across the nation. The Foreign Languages Press produced a high-quality color photo collection. Pro-CCP international media also joined in the chorus and foreign visitors trickled in to marvel at this proud achievement of socialist art.

Unlike earlier exhibits, the Rent Collection Courtyard was not primarily about brute physical violence (the focus of the Hundred Crimes dioramas) or the abject horrors of the Old Society (the focus of the Water Dungeon). It was intent on showing tenant farming as an exploitative system analogous to the wage labor system, thereby mapping the Marxist Law of History onto rural Chinese society. The rent grain was identified as the agrarian surplus value usurped by the landowning class to sustain its shameless lifestyle and to buttress the ruling machine. In an industrial

---

68 Xiaoshu 1999: 57.
69 Ibid., 62.

society, the factory and machine tools bring the capitalists (or their representatives) and the workers together on the primal site of production, a fact that conduces to the representation of antagonistic class relations. At least, the workers can be shown making something that is then taken away from them, not to mention the visually arresting contrast between clanking and churning machinery and the mute and fragile bodies that it not infrequently chews up and spits out maimed or dead (think of Charlie Chaplin's *Modern Times*). Tenant farming, however, does not afford such possibilities of dramatic visualization. Farming is a prosaic and drawn-out process and the peasants are on their own, working according to their own schedule and oftentimes using their own implements. There are usually no ferocious-looking overseers breathing down their necks as they hoe the soil. The land may wear out the toiling body, but it does not literally swallow it up. The peasant is what Arendt would call *animal laborans* whereas the industrial worker is the *homo faber*.[70] What the peasant produces in his or her "metabolism with nature" leaves nothing permanent or monumental behind. The peasant's *labor* makes possible other forms of *work* (such as manufacturing) that more immediately contribute to civilizational progress. The challenge is then how to glorify peasant labor and how to visualize the peasant's relation to the landowner as one of acute exploitation and oppression, contrary to all superficial manifestations.

The exhibit meets the challenge by fetishizing the rent grain. Although it is destined to be consumed, the rent grain is still the closest thing to the industrial product endowed with a use value and exchange value. The grain surrendered to the landlord as rent payment becomes a commodity and acquires an exchange value when the landlord, who cannot possibly eat it all, trades it for goods and services. Like the capitalist, the landlord is presumed to want to squeeze as much rent out of the peasants as possible, so as to acquire more land and live even more decadently. This is where an economic relationship morphs into a moral and political one. The narrative guide at one point claims that Liu Wencai's "iron board rent" (*tiebanzu*, meaning fixed or flat rent) often exceeded the average annual harvest and never budged in times of natural or human disasters.[71] At another point the guide asserts that the rent constituted 40 to 80 percent of a harvest. The discrepancy might be a result of the Red Guards' denunciation of the museum's introductory statement not-

70 Arendt 1958.
71 Wang Zhi'an 2001: 297.

ing that the average peasant household could have about 600 *jin* of grain left after paying rent. What has the blind old man got left, the Red Guards demanded to know, after he paid rent? "Nothing but a piece of paper indenturing his granddaughter to Liu Wencai!"[72] Most of the tableaux show the peasants literally turning their entire harvests over to the landlord and having absolutely nothing left to eat; or worse still, they sank deeply in debt because they invariably fell short of what was demanded—either because the rent exceeded the harvest, or because the grain was so excessively vetted in the inspection process that much of it was outright seized from the peasants (it is repeatedly emphasized the rejected grain was not allowed to be taken back by the peasants).

One should not be blamed for wondering how Liu Wencai could keep any tenants farming his land for more than one or two years before they were starved to death or sold into bondage or conscripted into the KMT army. If the typical capitalist is willing to pay a subsistence wage only in order to keep the worker returning to the factory the next day, the next week, and the next month, Liu seemed not to care at all whether his tenants could return to the land and cultivate the next year's crops. The tales of suffering almost always end with the tenant family losing their lease, in addition to being driven to utter destitution. There seemed to be an endless supply of tenant farmers continually throwing themselves into Liu's merciless jaws. If the Rent Collection Courtyard set out to represent the landlord-tenant relation as an unjust economic relation, it ended up going much further to cast it as ritual cannibalism. The peasants did not just farm Liu's land; they offered up themselves, flesh and bones, to his rapacious appetite. The tenant farming system was the fulcrum of "feudal" society and would go on perpetuating itself until the to-be-eaten rose up in rebellion and revolution, which is emphatically portrayed in the final section. Here, history did not merely take a step forward from feudal to socialist relations of production to keep apace with the advancing forces of production; it was also a moment of redemption, of saving history itself from being engulfed and cannibalized by the waste products of a decaying civilization.

It was in the redemptive mode that class racism had its fullest articulation. The landlords might look human, but they were "wolves and jackals" (*cailang*) at heart. The avoidance of grotesqueries in the Rent Collection Courtyard was compensated by the fetishization of the rent grain.

72 Xiaoshu 1999: 81.

To fetishize something is to take it as a substitute for a desired or dreaded thing/condition that one is either too ashamed or too frightened to acknowledge. What the rent grain revealed and concealed at the same time was the reification of the landlord as the objective enemy, or the enemy within that must be exposed, excluded, and exterminated. In using the rent grain to figure the Law of History, the exhibit constructed socialist redemption as a precarious enterprise to which the class enemy was an indispensable supplement. Rent collection brought the antagonistic classes face to face to enact the Law of History, kernel by kernel. When the grain was whittled down next to nothing, the socialist subject was born. He or she would return it to its wholesome use value as nourishment for the human body, instead of an insidious instrument through which its usurpers cannibalize its producers.

Conclusion

I have argued in this chapter that the ideology of class in socialist China is best approached through the prism of the racism-nationalism nexus. Class racism was the internal supplement of CCP nationalism, not its ugly antithesis. In racializing social classes, it supplied a fictive ethnic basis for the socialist nation. Its mechanisms of exclusion and denigration resembled those of the more common strand of racism accompanied by biological theories. In particular, it invented a phantasmagoric imaginary of the class enemy that powerfully elicited the aesthetic-affective experience of horror and abjection. It was a quasi-hallucinatory imaginary because the pseudo-ethnicity called "class enemies" was little more than an empty signifier designating more an ideological fantasy than a socioeconomic reality. Not only was China's bourgeoisie, in Meng Yue words, "a sham character, an absent enemy,"[73] but even the historicity of the putatively omnipresent and omnipotent landlord class was open to question. Social historians have shown that large-scale landowners made up at most a minor presence in the Chinese countryside and that the majority of peasants tilled their own land. In rural north China especially, in Prasenjit Duara's view, "the relations between landlord and tenant did not constitute a major contradiction."[74]

[73] Meng 1993: 119.
[74] Duara 1988: 252.

During the fateful modernization reforms of the late Qing and Republican periods, the primary source of local grievances was the conflict-ridden state-society relations. The reform measures taken by the state—bureaucratizing subadministrative staff and devolving tax-collecting and other governmental functions to the village level—in fact spurred the proliferation of "entrepreneurial brokers" who preyed on local communities instead of protecting local interests from the extracting state. Duara calls this process "state involution."[75] The entrepreneurial brokers, also known as "local bullies" (*tuhao, lieshen, eba, wulai*), were in fact the principal agents in local strife. Even where landlordism did become dominant, the prevalent form of absentee landlordism made it difficult to speak of class oppression on a day-to-day basis and to mobilize villagers along class lines.[76] William Hinton's remarkable documentary of land reform demonstrates this difficulty in heartrending detail.[77] Wang Hui argues that the disparity between the "objective reality" of class and the "expressive reality" of class discourse both justifies the class revolution in rural China and explains the injury and tragedy caused by an ideology spilling over the boundaries of lived reality.[78]

The collectivization of land in socialist China made the landlord a veritable anachronism, who nevertheless had to continue to live out his or her landlord identity and even pass it on to his offspring thanks to the patrilineal *chengfen* system. In Wang Hui's view, the transmutation of class as identity is at the crux of the depoliticization of the Chinese revolution. If, traditionally, class has always been a moral-political category, not a hereditary one in China, under socialism class paradoxically became a fixed, essentialized category that served to choke off the political agency it was supposed to animate. Hence the tragedy of class politics is not only a matter of ideology overtaking reality, but also of class as identity suppressing the possibility of political agency.[79] The fetishisms of class racism served to conceal the continuing state-society conflict as the socialist state set out to industrialize the nation by squeezing the peasants and demanding unconditional self-sacrifice from the population in general.

Arif Dirlik points out that self-reliance and delayed gratification at the behest of a developmentally committed party-state was no more than a

---

75 Ibid., 73–77.
76 Ibid., 76.
77 Hinton 1966.
78 Wang Hui 2006a: 18.
79 Ibid., 21; see also Wang Hui 2006b.

cover for exploiting the labor of the people: "The claim that the state of the people cannot exploit the people is not very convincing, based more on a 'social imaginary' of the state, than on actual relations between state and society."[80] Nationalism—supplemented by class racism—allowed the state-society fault-line to remain deeply buried beneath a social imaginary fixated on the life-and-death struggle between the socialist nation and its internal and external enemies. Even though Mao spoke of the dual nature of class struggle—against the remnants of the former exploiting classes and against the new power-holders—the *chengfen* and *maozi* systems were so entangled operationally that a genuine class critique was never developed. The behavior-based class designations degenerated into a handy tool for factional fights and power struggles. As Yiching Wu astutely notes, even though late Maoism pointedly targeted the new bureaucratic ruling class with the state as its central institution, when an official was denounced as a revisionist or capitalist roader, a thoroughgoing critique of the party as a new ruling elite or of the party-state system as the source of new power-based class relations usually gave way to an entirely individualized rhetoric that would "expose" the culprit as a traitor, enemy agent, or member of a conspiratorial cabal that had infiltrated the party.[81] He or she would then have to be "dragged out" (*jiu chulai*) of the revolutionary ranks so that purity could be restored to the revolution. Thus class struggle became a mere pretext for abjection. Wu writes that although the Cultural Revolution began with a call to attack the party "power-holders," it ended with a campaign to "purify class ranks" that emphasized *chengfen* class designations, a development that he considers "a significant retreat on the issue of defining class in postrevolutionary China."[82] For Wang Hui, the retreat was inevitable because Chinese socialism embedded its own seeds of depoliticization from the very start: "Depoliticization was not a tendency external to twentieth-century revolutionary politics; rather, it was contained within the very concepts of class and class struggle that drove the revolutionary process."[83]

---

80 Dirlik 2003: 165.

81 Wu n.d.: 8, 14. Wu also points out that Mao himself tended to refer to the "new bourgeois elements" as "cliques" instead of as a stratum or class. When Mao did use the term class (*jieji*), it was usually not in the rigorously Marxist sense: "Rather than referring to a structurally defined notion of class centering on the control of political power and means of production, Mao's use of the term referred instead to only a certain segment of the bureaucracy—those who were ideologically vulnerable" (Wu n.d.: 9).

82 Ibid., 16.

83 Wang Hui 2006a: 21.

I would combine this point with Duara's insight that socialism, despite its internationalist spirit, had to be undertaken by a sovereign state that was invariably committed to state- and nation-building,[84] and suggest that it is the nationalist core of Chinese communism that helped smother its Marxist urge to forge class subjectivity and solidarity, to foster public criticisms and mass debates, to incite popular distrust of state apparatuswa, and to launch relentless campaigns to counter the bourgeoisification of the bureaucratic stratum. Depoliticization—the abandonment or, indeed, the thorough negation of mass democracy and the retrenchment of the party-state system—was inevitable in part because, in the world system that was consolidated in the post-World War II era, the nation-state was the only legitimate authority to shore up sovereignty claims and to gain advantage in the global competition for "wealth, value, respect, and... recognition."[85] As Duara puts it, "the construction of socialism around the nation-state and its discursive commitment to a linear history and progressive future entailed a competitive understanding of the world which would, under particular historical circumstances, enable both unconditional nationalism and capitalism to prevail."[86] One may well argue that both unconditional nationalism and capitalism (capital accumulation and industrialization at the expense of the peasantry) prevailed in the name of building the socialist fatherland.

The Maoist class struggle was thus a kind of red herring that deflected recognition of the depoliticizing process whereby "the proletarian dictatorship was transformed from a political system that was to be a radical departure from all previous state formations and endowed with a genuine autonomous civil society and participatory democracy to an authoritarian state structure that monopolized violence."[87] The Liu Wencai landlord museum was a visually and narratively compelling contribution to this ideological misrecognition. Both the typical and peculiar circumstances of Liu Wencai's life made him an ideal candidate for the phantasmagoria of class racism. Although his father was only a prosperous farmer (*zigengnong*) and liquor distiller, he and his brothers and cousins rose to the heights of provincial power precisely during the chaotic years of state involution. He had indubitably become a local tyrant *and* big landlord who actually lived on his estate, even though he was wealthy enough to

---

[84] Duara 2009: 62.
[85] Duara 2009: 62.
[86] Ibid.
[87] Wang Hui 2006a: 7.

own several mansions in Chengdu, where two of his wives in fact lived for nearly two decades to escape the boredom of rural life. His physical presence lent force to the visibility of the class enemy and put a human face on what was otherwise an abstract relationship. But his rootedness in his lifetime did not prevent his name from becoming a free-floating signifier, a most versatile "political puppet."[88] He was the undead haunting the socialist house of horrors. He was not only a stranger to the socialist nation, but also morally execrable and aesthetically hideous—*pace* Schmitt's definition of the political enemy.[89] His physical demise on the threshold of the New Society was no obstacle to his political utility: his phantom would be struggled against for all actual and objective crimes throughout the Mao years as the posthumous accomplice of Liu Shaoqi, Lin Biao, Confucius, Deng Xiaoping, and even the Gang of Four.[90]

In visualizing the racist logic of the ruling ideology, the landlord museum helped obscure the emerging structures of inequality in socialist China. The fetishism of the rent grain made it possible to disavow the fact that segments of the population—both fixed (the peasants) and fluid (the class enemies)—were permanently marked down for expropriation and disenfranchisement. The racialization of the landlord class, on the other hand, maintained a social outcast group—the socialist *homo sacer*—which even the most powerless peasant or a mere wisp of a student can despise and assail with impunity. The Rent Collection Courtyard was a material metaphor for the politically willed "state of exception" in which those designated as class enemies by birth or by creed were subjected to ideologically sanctioned ritual violence. Giorgio Agamben characterizes the state of exception as "a zone of anomie in which all legal determinations—and above all the very distinction between public and private—are

[88] Xiaoshu 1999: 96.

[89] "The political enemy need not be morally evil or aesthetically ugly.... But he is, nevertheless, the other, the stranger; and it is sufficient for his nature that he is, in a specially intense way, existentially something different and alien" (Schmitt 1996: 27).

[90] Xiaoshu 1999: 92–96. In 1976, the following paragraph was added to the narrative guide of the Rent Collection Courtyard: "Comrades, in the Old Society, there were tens of thousands of such vile rent collection courtyards. Indeed the Old Society itself was the biggest rent collection courtyard. The Gang of Four maliciously opposed the revolutionary line of Chairman Mao and attempted to restore capitalism, in order to reenact the crimes of the Rent Collection Courtyard and make us laboring masses suffer the pain of paying rent all over again. We shall not allow it to happen!" (Xiaoshu 1999: 95). The charges so vehemently thrown at the Gang of Four were not essentially different from those habitually leveled at the many "class enemies" who were supposedly the victims of the Gang. This is not as ironic as it might appear only because the logic at work was still that of class racism.

deactivated."[91] He makes a forceful link between totalitarianism and the state of exception: "Modern totalitarianism can be defined as the establishment, by means of the state of exception, of a legal civil war that allows for the physical elimination not only of political adversaries but of entire categories of citizens who for some reason cannot be integrated into the political system."[92] Although the extent to which Mao's China was a totalitarian society is debatable, the state of "legal civil war" certainly obtained at the high points of various political campaigns wherein the targets were not just Mao's political adversaries, but also entire categories of the population marked as lives undeserving to live. The violence enacted in the state of exception is what Walter Benjamin calls "pure violence" because it neither makes nor preserves law (mythico-juridical violence), but rather disposes of it, thereby instituting the zone of anomie.[93]

In the state of exception, class designations (*chengfen*) and political labels (*maozi*) worked in tandem to render all lives precarious. Having participated in the ritual violence of previous campaigns not only did not immunize one against being discovered as the newest objective enemy, but rather augmented the anomic force of the class racist logic. If racism accompanied by biological theories has more or less stable (though specious) referents, then class racism as a state ideology was an entirely phantasmatic affair not least because "class" had become an alibi. The class racist subject must be able to play both the executioner and the victim. At the height of the Cultural Revolution, museum officials and artists who were the chief architects of the Liu Wencai myth were denounced as "living demons" (*huogui*) by the Red Guards. The museum itself, with the exception of the Water Dungeon and the Rent Collection Courtyard, was forced to close down between 1967 and 1971. Xiaoshu notes with irony that as these "living demons" faced their accusers—a ferocious, bloodthirsty mob of youngsters—they should not have felt so wronged, for were not the youngsters putting into practice precisely what the accused had imparted to them?[94]

Today, the museum is no longer a heterotopia of estrangement that reifies the class enemy as the stranger. Scraping by on the margins of an accelerating market economy, it has had to, like most former revolutionary sites, make the crucial transition that Dutton identifies in discussing

---

91 Agamben 2005: 50.
92 Ibid., 2.
93 Ibid., 53–64.
94 Xiaoshu 1999: 88.

the transformation of the Mao Cult to the Mao Craze: from a model of politics that "intensifies political conviction and 'throws' the subject" to one that " 'grounds' the subject and keeps him or her grounded through instrumental reason and endless consumer distractions."[95] Kirk Denton, too, points out that museums have edged away from narratives of class struggle to representations that "legitimize the contemporary ideology of commerce, entrepreneurship, and market reform."[96] Wang Hui laments that just as class differentiation in the original Marxist sense (as pertains to property relations) is reemerging in reform-era China, the discourse of class has all but disappeared as a result of the widening and deepening depoliticization process globally.[97] The exhaustion from the Maoist class politics that devolved into hideous power struggles and mass violence is so pervasive that anyone who invokes class risks being accused of wishing to bring back the dark age of the Cultural Revolution. If in those days "class" was an ideological alibi, a state-orchestrated bloody distraction, then today it is something from which distraction is sought with the tacit encouragement of the state and the eager assistance of a culture industry to which the museum increasingly belongs.

Distraction has indeed become the raison d'être of the museum. The occasional tourists are here not to swoon, to partake of ritual abjection, or walk out with clenched fists, but to consume a bit of local cultural heritage and to be amused and diverted. The world of the *bowuguan* may no longer afford ideological transcendence, but it has surely become a local source of enchantment. Since the late 1990s, the Anren government has made it mandatory that all buildings on the town's main drags feature façades in the western Sichuan architectural style as exemplified by the Liu manor house. And if this was not enough to give the town an amusement park look, ideas were floated from the enterprising sections of the municipal government that all townsfolk would soon be required to dress up in period attire and revert to traditional means of transportation—such as the wheelbarrow made famous by the Rent Collection Courtyard sculptures. Dutton argues that the kitsch forms of commodity culture that plague post-Mao China are the most determined gravediggers of Mao's "Siren politics."[98] If the townsfolk could indeed be persuaded to don long gowns and push wheelbarrows, it would not be because of any Siren songs coming from a crazed visionary, but because they were hoping to share

95 Dutton 2005a: 151.
96 Denton 2005: 567.
97 Wang Hui 2006a: 14.
98 Dutton 2005a: 155.

in the spoils of a tourist economy, or because they simply enjoyed the playfulness of it all. But more likely than not, it would pass, like all fads in a commercial culture, without any class enemy being offered up as a sacrificial lamb.

References Cited

Agamben, Giorgio (1998), *Homo Sacer: Sovereign Power and Bare Life*. Trans. Daniel Heller-Roazen. Stanford, CA: Stanford University Press.
—— (2005), *State of Exception*. Trans. Kevin Attell. Chicago: University of Chicago Press.
Arendt, Hannah (1958), *The Human Condition*. Chicago: University of Chicago Press.
—— (1985), *Totalitarianism*. San Diego: Harcourt Brace Jovanovich.
Balibar, Etienne (1991a), "Class Racism," in Balibar and Wallerstein, 1991: 204–16.
—— (1991b), "Racism and Nationalism," in Balibar and Wallerstein, 1991: 37–67.
—— (1994), Masses, Classes, Ideas: Studies on Politics and Philosophy Before and After Marx. Trans. James Swenson. New York: Routledge.
——, and Immanuel Wallerstein (1991), *Race, Nation, Class: Ambiguous Identities*. Trans. Chris Turner. London: Verso.
Baum, Richard and Frederick C. Teiwes (1968), *Ssu-ch'ing: The Socialist Education Movement of 1962–1966*. Center for Chinese Studies, University of California, Berkeley.
Chen Liang (2001), "Sichuan Landlord Mansions Depict Past," in *China Daily*, April 7.
Chinese Department, Sichuan University, and Sichuan Daiyi Class Education Exhibition Hall (1974), *Xuehai shenchou shouzuyuan: Sichuan sheng Dayi pinxiazhongnong jiashi* 血海深仇收租院: 四川省大邑贫下中农家史 (A sea of blood and hatred at the Rent Collection Courtyard: Family histories of poor and lower middle peasants in Dayi, Sichuan). Reprinted, Nongcun duwu chubanshe, ed. Chengdu: Sichuan renmin chubanshe.
Daiyi Landlord Mansion Museum, Sichuan (1965), *Wan'e de dizhu zhuangyuan* 万恶的地主庄园 (The iniquitous landlord mansions). Beijing: Wenwu chubanshe.
Daozi (2003), "Shouzuyuan de fuzhi yu hou xiandai zhuyi"《收租院》的复制与后现代主义 (Replicating the Rent Collection Courtyard and postmodernism). http://www.culstudies.com/plus/view.php?aid=1474 (accessed Aug. 8, 2006).
Denton, Kirk A. (2005), "Museums, Memorial Sites and Exhibitionary Culture in the People's Republic of China," in *China Quarterly* 183: 565–86.
Dirlik, Arif (2003), "The Politics of the Cultural Revolution in Historical Perspective." In Kam Yee-Law, ed., *The Chinese Cultural Revolution Reconsidered: Beyond Purge and Holocaust*. Basingstoke, UK: Palgrave Macmillan, 158–83.
Douglas, Mary (2002), *Purity and Danger: An Analysis of Concept of Pollution and Taboo*. London: Routledge.
Duara, Prasenjit (1988), *Culture, Power, and the State: Rural North China, 1900–1942*. Stanford, CA: Stanford University Press.
—— (1995), *Rescuing History from the Nation-State: Questioning Narratives of Modern China*. Chicago: University of Chicago Press.
—— (2009), *The Global and Regional in China's Nation-Formation*. London; New York: Routledge.
Dutton, Michael (2004), "Mango Mao: Infections of the Sacred," in *Public Culture* 16, 2: 161–87.
—— (2005a), "From Culture Industry to Mao Industry: A Greek Tragedy," in *boundary 2* 32, 2: 151–67.
—— (2005b), *Policing Chinese Politics: A History*. Durham, NC: Duke University Press.
Flath, James A. (2002), "Managing Historical Capital in Shandong: Museum, Monument, and Memory in Provincial China," in *Public Historian* 24, 2: 41–59.

Foucault, Michel (1980), *Power/Knowledge: Selected Interviews and Other Writings, 1972–1977*. Trans. Colin Gordon. New York: Pantheon.

Gu Hua (1983), *A Small Town Called Hibiscus*. Trans. Gladys Yang. Beijing: Panda Books.

He Jingzhi, Ding Yi, and Ma Ke (1954), *The White-Haired Girl: An Opera in Five Acts*. Peking: Foreign Languages Press.

Hinton, William (1966), *Fanshen: A Documentary of Revolution in a Chinese Village*. New York: Vintage.

Hu Ping 胡平 (2001), *Zhanzheng zhuangtai* 战争状态 (The state of war). Wuhan: Changjiang wenyi chubanshe.

Jay, Martin (1998), *Cultural Semantics: Keywords of Our Time*. Amherst: University of Massachusetts Press.

Kraus, Richard Curt (1981), *Class Conflict in Chinese Socialism*. New York: Columbia University Press.

Kristeva, Julia (1982), *Powers of Horror: An Essay on Abjection*. New York: Columbia University Press.

Lee, Haiyan (2010), "'Nowhere in the World Does There Exist Love or Hatred without Reason'," in Ban Wang, ed., *Words and Their Stories: The Language of the Chinese Revolution*. Leiden: Brill, 149–170.

Lefort, Claude (1986), *The Political Forms of Modern Society: Bureaucracy, Democracy, Totalitarianism*. Ed. John B. Thompson. Cambridge, MA: MIT Press.

Li Zhensheng (2003), *Red-color News Soldier: A Chinese Photographer's Odyssey through the Cultural Revolution*. Ed. Robert Pledge. London: Phaidon.

MacFarquhar, Roderick and Michael Schoenhals (2006), *Mao's Last Revolution*. Cambridge. MA: Belknap Press of Harvard University Press.

Mao Zedong 毛泽东 (1996), *Jianguo yilai Mao Zedong wen'gao* 建国以来毛泽东文稿 (Mao Zedong's Post-1949 Manuscripts), vol. 10. Beijing: Zhongyang wenxian chubanshe.

Meng Yue (1993), "Female Images and National Myth," in Tani E. Barlow, ed., *Gender Politics in Modern China*. Durham, NC: Duke University Press, 118–36.

Schmitt, Carl (1996), *The Concept of the Political*. Chicago: University of Chicago Press.

Schram, Stuart ed. (1974), *Chairman Mao Talks to the People: Talks and Letters, 1956–1971*. New York: Pantheon.

Seligman, Adam B., Weller, Robert P., Puett Michal J., and Simon, Bennet eds. (2008), *Ritual and Its Consequences: An Essay on the Limits of Sincerity*. Oxford: Oxford University Press.

Wakeman, Frederic E. (1975), *History and Will: Philosophical Perspectives of Mao Tse-tung's Thought*. Berkeley: University of California Press.

Wallerstein, Immanuel (1991), "The Bourgeois(ie) as Concept and Reality," in Balibar and Wallerstein, 1991: 135–52.

Wang Hui 汪晖 (2006a), "Qu zhengzhi hua de zhengzhi, baquan de duochong goucheng yu liushi niandai de xiaoshi" 去政治化的政治, 霸权的多重构成与六十年代的消失 (Depoliticized politics: The multiple structures of hegemony and the disappearance of the 1960s). Unpublished ms.

—— (2006b), "End of Politics? From East to West," in *New Left Review* 41 (Sept.–Oct.): 29–45.

Wang Zhi'an 王治安 (2001), *Hongtian juechang: Shouzuyuan nisu qiguan* 轰天绝唱: 收租院泥塑奇观 (Songs that reached the heavens: The Legendary Rent Collection Courtyard Clay Sculptures). Chengdu: Tiandi chubanshe.

Wu, Yiching (n.d.), "From the Entrails of the Old: Rethinking Cultural Revolution Politics and Ideology of Class." Unpublished ms.

Xiaoshu 笑蜀 (1999), *Liu Wencai zhenxiang* 刘文彩真相 (The true Liu Wencai). Xi'an: Shaanxi shifan daxue chubanshe.

Yang, Mayfair (1994), *Gifts, Favors, and Banquets: The Art of Social Relationships in China*. Ithaca, NY: Cornell University Press.

Yingquan 映泉 (2000), *Tianfu changye: haishi Liu Wencai* 天府长夜—还是刘文彩 (Long nights in Sichuan: Liu Wencai encore). Changsha: Hunan wenyi chubanshe.

HISTORY AND THE DEFINITION OF SPATIAL, CULTURAL AND TEMPORAL BOUNDARIES

CHAPTER NINE

REVOLUTION AS RESTORATION: MEANINGS OF "NATIONAL ESSENCE" AND "NATIONAL LEARNING" IN GUOCUI XUEBAO

Tze-ki Hon

In contemporary usage, the Chinese term *geming* (革命) means revolution. It refers to radical changes in political, social, and cultural systems with the support of the people.[1] An early example of this usage of *geming* is found in *The Revolutionary Army* (*Geming jun* 革命軍, 1903) by Zou Rong (鄒容 1885–1905). In this revolutionary pamphlet, Zou called for a break with the past, especially the recent past of the Manchu rule. He wrote: "Sweep away millennia of despotism in all its forms, throw out millennia of slavishness, annihilate the five million and more of the furry and horned Manchu race, cleanse ourselves of 260 years of harsh and unremitting pain.... How sublime is *geming*! How majestic!"[2] The goal of *geming*, according to Zou, was to liberate China from darkness and backwardness so that the country would advance from barbarism to civilization and from slavery to mastery.

In much of the twentieth century, Zou Rong's notion of *geming* dominated the political discourse in China. It informed many political movements that profoundly changed the country, such as the 1911 Revolution, the May Fourth New Culture Movement, the Communist Revolution, and the Great Proletarian Cultural Revolution. Distinct in their own right, these historical events were often linked together to project a historical teleology of "deepening the revolution." As the argument goes, twentieth-century China began with a political revolution to transform the Chinese empire into a nation-state; then it underwent a cultural revolution to replace Confucianism with Western science and democracy; finally it required a socialist revolution to drastically revamp the socioeconomic structure of the country. The moral of the story is that modernity in China can only be achieved by severing the country's ties with its past and

---

1 Chen 2000: 1–22.
2 Tsou 1968: 58, with modifications.

completely transforming its political, social, and cultural systems. Embedded in the revolutionary teleology is the idea of linear progression where time must move forward until the *telos* of time is reached.[3]

Yet, even in 1903 when Zou Rong was writing *The Revolutionary Army*, there was another meaning of *geming* that emphasized the continuity between past and present. For instance, Zhang Taiyan (章太炎 1869–1936), in his preface to *The Revolutionary Army*, pointed out that Zou Rong was misleading in using *geming* to refer to the anti-Manchu revolution: "When a government is changed by the people of the same race, this is known as *geming*. When an alien race is expelled, this is known as restoration (*guangfu* 光復). Because China was destroyed by the rebel barbarians, what is planned is restoration rather *geming*."[4] Three years later, shortly after arriving in Tokyo, Zhang reiterated the meaning of *geming* when he gave a speech to thousands of supporters of the Revolutionary Alliance (*Tongmenhui* 同盟會). Again, he told them the goal of *geming* was to restore the past: "What I mean by *geming* is not just a change of government, but restoration—restoring [the rule] of the Chinese race; restoring [the ancient system] of county and district; restoring the government of the Chinese. Thus I use the term *geming* to denote restoration."[5] In these bold statements, Zhang put forward a different understanding of the term *geming*. Racially, *geming* meant the restoration of the rule of the Han Chinese, who had lost control of their country since 1644. Politically, it meant the restoration of the decentralized *fengjian* (封建) system that had existed during the Western Zhou period (1045–771 BCE).

In the 1900s, Zhang Taiyan's *geming* as restoration was popular among a group of writers who contributed to *Guocui xuebao* (國粹學報 Journal of National Essence, 1905–1911). In addition to Zhang, the *Guocui* writers included Chen Qubing (陳去病 1874–1933), Deng Shi (鄧實 1877–1951), Huang Jie (黃節 1873–1935), Liu Shipei (劉師培 1884–1919), and Ma Xulun (馬敍倫 1885–1970). In current scholarship, due to the revolutionary historiography, these *Guocui* writers are often depicted as cultural conservatives because of their insistence on finding the Chinese "national essence" in ancient classics. They are grouped together with members of *Nanshe* (Southern Society, 1909–1923) and *Xueheng* (Critical Review, 1922–1933) to form what Laurence Schneider calls "a conservative movement" to

3 Chen 2000, Jin 2003.
4 Tsou 1968: 52, with modifications.
5 Zhang Taiyan 1906: 13.

preserve the Chinese national essence at all costs.[6] What is missing from this picture is that these writers were revolutionaries, although they believed in *geming* as restoration.

In what follows, I will examine how the *Guocui* writers promoted *geming* as restoration through developing the notions of *guocui* (國粹 national essence) and *guoxue* (國學 national learning). In the 1900s, these two terms were used by Qing officials to denote the collective identity of the Chinese when they faced a hostile international environment where the strong bullied the weak and the powerful crushed the powerless. And yet, while these two terms helped the Qing government garner support in its fight against imperialism and colonialism, they could also be used to challenge its authority and legitimacy. As will be shown, the *Guocui* writers deliberately reinterpreted and redeployed the two terms to subvert the Qing government. And it is this double nature of the terms—strengthening and yet subverting the central authority, unifying and yet distinguishing various groups of Chinese—that will be the focus of this chapter.

The Mobilization of the Educated Elite

Recent scholarship has shown that the late Qing "New Policies" (*xinzheng* 新政, 1901–1911) were in many respects ground-breaking in creating a Chinese nation-state. The goal of the New Policies was to establish modern systems of control and mobilization, including a parliament, local self-governance, a national school system, and a national transportation network. Together, these systems were to create a centralizing state that was powerful enough to mobilize the country to defend its sovereignty.[7] However, with so many changes taking place simultaneously, the New Policies were filled with tension and unbridled creativity. For instance, the New Policies ended many practices that had made the educated elite the de facto leaders of central and local governments.[8] On the other hand,

---

[6] Schneider 1976, Bernal 1976. I discuss the differences between the *Guocui xuebao* and the *Xueheng* in Hon 2003.

[7] Recent scholarship has shown the limits of Mary C. Wright's interpretation of the late Qing reforms as "the last stand of Chinese conservatism." For a sample of recent scholarship of the "New Policies," see Reynolds 1993, Esherick 1976, Fogel and Zarrow 1997, Zhao 2004: 1–59. For Mary C. Wright's original argument, see Wright 1968: introduction.

[8] Kuhn 1975. Here the distinction between the literati and the educated elite is crucial. As scholar-officials and scholar-gentry, the literati were closely tied to the civil examination system. Their political and social status was in many ways determined by their success in the examinations. In contrast, the "educated elite" refers broadly to educated

the New Policies also brought hope for creating what Joan Judge calls "the middle realm" to link the rulers and the ruled, thereby transforming the educated elite into the "spokesmen of the people."[9]

Viewing the "New Policies" in this light, it was not accidental that *Guocui xuebao* was founded in 1905—the same year that the Qing government replaced the civil service examinations with a national school system. For more than a millennium, the civil service examinations had been what Ping-ti Ho calls "the ladder of success" for the educated elite to become political, social, and cultural leaders.[10] Although plagued by problems of fairness and accountability, the examinations not only bestowed power on successful candidates to rule the country, but also established a body of knowledge that defined the membership of the learned community.[11] Known as "this culture of ours," the examinations created a common identity among the educated elite as the custodians of "the genealogy of the Way" (*daotong* 道統).[12] Seeing themselves as moral leaders, the educated elite believed that they had the responsibility of shaping political and social behavior by prescribing the moral and ritual codes. Because of the central role that the examinations had played in defining the membership and self-identity of the educated elite, the abolition of the examinations in 1905 triggered a profound soul-searching among the learned community.[13] Those who had passed the examinations needed to find ways to preserve their elite status in the new political and social environment. Those who were yet to pass the examinations needed to make sure that their efforts to prepare for the examinations would not be wasted.

If indeed the abolition of the examinations posed a challenge to the educated elite, the founding of a national school system offered them new opportunities. In *Zouding xuetang zhangcheng* (奏定學堂章程 Approved School Regulations, 1904), the Qing government spelled out unequivocally the structure and the goal of the national school system. The system consisted of four levels of schools—primary schools, junior high schools,

---

people who may or may not have been linked to the examinations. In fact, many of them became professionals working in publishing, popular art, and education. For a discussion of the changes in the social status of the educated elite in the early twentieth century, see Hon and Culp, 2007.

9 Judge 1996: 1–13.

10 Ho 1962, Chaffee 1995.

11 For a study of the social and political impact of the civil service examination system, see Elman 2000: especially 125–72, 239–370.

12 Bol 1992: 1–31.

13 Qing officials had a heated debate on the abolition of the examination system, see Franke 1963.

high schools, and universities.[14] Its goal was to create, in Pierre Bourdieu's terms, a "state nobility" who would spend years in school receiving training in order to serve the nation.[15] Although the founding of the school system was clearly part of the Qing court's plan to centralize power, the educated elite found ample opportunities to shape the school system. As political theorists, they joined the discussion of what constituted the "Chinese nation" and the "Chinese race." As educators, they wrote textbooks and gave advice on curriculum to win the hearts and minds of the young Chinese. As social leaders, they formed local educational associations, such as the powerful Jiangsu Provincial Education Association (*Jiangsu sheng jiaoyu hui* 江蘇省教育會), to gain influence in local and national politics.[16]

Published in the Anglo-American International Concession of Shanghai, *Guocui xuebao* was aimed at those who were perplexed and animated by the abolition of examinations. For instance, in the "Purposes of Publishing the Journal of National Essence" (Guocui xuebao fakanci 國粹學報發刊詞), Deng Shi, the chief editor of the journal, deliberately addressed "learners" (*shi* 士) in general without making a distinction between those who had passed the examinations and those who had not.[17] As a discursive strategy, Deng's broad and indiscriminate call to attention was a calculated move to eliminate the social distinction between successful and unsuccessful candidates of the examinations. Since Deng was a young scholar who had not pass the examinations, this move allowed him to take the moral high ground to discuss the mission of learning. He wrote:

> Since [the coming of Westerners] from the oceans, Western learning has been introduced to China, and Chinese learning barely survives in a single thread. Some uninformed scholars praise Western learning and denigrate Chinese learning. They consider the Chinese ancient texts as rubbish.... However, neither are the ancient texts outdated, nor is the ancient system outmoded. After taking a careful look at the ancient country and the ancient capital, [we find that] they are still full of vitality. Even hundreds of years later, they will still be admired and respected. Those who are born in this country they should have a fair assessment of their country. ("Fakanci," *Guocui xuebao* 1 [1905]: 1b)

---

14 Li and Wang 2000: 312–13, Liu Longxin 2001: 492.

15 Bourdieu 1996: 1–53.

16 For a discussion of the roles of the educated elite in the late Qing educational reforms, see Bastid 1988: 53–65, Borthwick 1983: 87–103, Schwintzer 1992: 55–270.

17 "Fakanci" 發刊詞, *Guocui xuebao* 1 (1905): 1a.

On the surface, Deng Shi seemed to repeat what the late Qing reformer Zhang Zhidong (張之洞 1837–1909) had said in *An Exhortation to Learning* (*Quanxue pian* 勸學篇, 1898). Like Zhang, Deng appeared to be calling on the educated elite to protect the country (*baoguo* 保國), the race (*baozhong* 保種), and Chinese learning (*baojiao* 保教) when China was threatened by Western powers and Western learning.[18] He seemed to position his journal as another attempt to preserve Chinese learning as "essence" (*ti* 體) and to limit Western learning to "function" (*yong* 用).[19] But unlike Zhang, Deng did not equate the protection of the country and the race with steadfast loyalty to the Qing government. Nor did he confine the protection of Confucian learning to preserving the "Three Bonds and Five Constants" (*sangang wuchang* 三綱五常)—the dominance of emperor over official, father over son, and husband over wife. Instead, he focused on building a strong Chinese nation based on developing a national learning and a national essence.[20] Here we see a fundamental difference between the *Guocui* writers and other anti-Manchu revolutionaries. Rather than directly calling on readers to take up arms against the Manchus as Zou Rong did in *The Revolutionary Army*, Deng skillfully reinterpreted Zhang's notions of "protecting the country," "protecting the race," and "protecting Chinese learning," and turned them into revolutionary slogans against the Manchu rulers. It was this strategy of subverting the late Qing reform discourse that distinguished the *Guocui* writers from other revolutionary thinkers.

The Revival of Ancient Learning

As Zheng Shiqu (鄭師渠) has pointed out, although the term *guocui* (國粹 national essence) was a loan word from the Japanese *kokusui*, the term had a special meaning for the *Guocui* writers.[21] Instead of referring to the preservation of a timeless national essence, *guocui* meant the revival of a particular kind of Chinese cultural heritage that would help integrate twentieth-century China into the modern world. Thus, when the *Guocui* writers spoke of reviving *guocui*, they did not mean to turn the clock back

---

18 See Zhang Zhidong 1898: section two *jiaozhong* 教忠, section three *minggang* 明綱, section four *baozhong* 保種.

19 For an analysis of Zhang Zhidong's *An Exhortation to Learning*, see Hon 2002.

20 "Fakanci," *Guocui xuebao* 1 (1905): 2b.

21 Zheng 1997: 111–39.

to a bygone era, but to move China forward into the twentieth century by restoring a particular type of Chinese cultural heritage.

For the *Guocui* writers, the restored Chinese cultural heritage was specific in time and content. Temporally speaking, it was a system of cultural practices developed prior to the formation of a bureaucratic empire under the First Emperor of Qin in 221 BCE. Viewing the Zhou dynasty (1050–256 BCE) as the "Golden Age," the *Guocui* writers considered what had happened in China after the Qin dynasty as the "Dark Age." Stretching from 221 BCE to the 1900s, the "Dark Age" of China included two elements. One was the centralization of power in the hands of absolute rulers. For the *Guocui* writers, two historical events exemplified the oppression of absolute rulers. First was the burning of books and burying alive of scholars during the reign of the First Emperor of Qin. Second was the suppression of the Hundred Schools by Emperor Wu of the Western Han.[22] In both cases, absolute rulers employed extreme measures to suppress and silence dissent voices. The other element of the "Dark Age" was the acceleration of the centralization of power when foreign races (particularly the Mongols and the Manchus) ruled China.[23] For the *Guocui* writers, foreign rulers expanded the system of absolute monarchy by using the examination system to corrupt men of letters.[24]

In contrast to the "Dark Age," the *Guocui* writers regarded the Zhou dynasty as the "Golden Age." The Zhou dynasty hitherto had been known as a time when the leaders of feudal states were politically powerful and men of letters were regarded as custodians of truth. To further develop this historical view, the *Guocui* writers saw the Zhou dynasty as a historical precedent where local autonomy and independent thinking were valued and protected. To drive home their point, the *Guocui* writers focused on the Hundred Schools of the Eastern Zhou period (770–256 BCE), or the *xianqin zhuzi* (先秦諸子 the thinkers of the pre-Qin period). For them, the goal of studying the Hundred Schools was not to question the authority of Confucianism, but to prove that there had been a wide variety of intellectual resources in ancient China, and the educated elite used to be the dominant players in politics.[25]

Of the *Guocui* writers, Huang Jie was the most eloquent in explaining the goal of reviving the Zhou learning. In his article "Preface to the Journal

---

22 Pan 1905, Deng 1905c.
23 Huang 1905: 1b.
24 Deng 1905d: 3b, see also Deng 1908a: 1a.
25 Deng 1905b: 2b–6b.

of National Essence" (Guocui xuebao xu 國粹學報敍, 1905), he stated that since the third century, Han Chinese had been repeatedly ruled by foreign dictators, and they were forced to alter their learning to serve absolute monarchs. To make his point, Huang described the government of imperial China as "the despotic government of foreigners" (*waizu zhuanzhi zhi guoti* 外族專制之國體) and its learning as "the despotic learning of foreigners" (*waizu zhuanzhi zhi xueshuo* 外族專制之學說).[26] Huang found that in additional to suffering from internal colonization, China, since 1840, was threatened by European powers. He criticized his contemporaries who, stunned by defeat at the hands of foreigners, wholeheartedly sought to learn from Europe and Japan. To encourage his contemporaries to look back to the past for guidance for the future, he wrote:

> We, the Bak tribe, are people of independent mind! Our learning—beginning from the Yellow Emperor, Yao, Shun, Yu, Tang, Wen, Wu, the Duke of Zhou, to Confucius—will last forever. Why would we give up our country? Why would we give up our learning?[27]

To justify this looking back in order to move ahead, Huang cited the example of the Italian Renaissance. He pointed out that during the Renaissance (*wenxue fuxing* 文學復興), Dante used his native tongue, Italian, to compose poems, thereby setting the stage for educational and cultural reforms in Europe. Huang foresaw the same in twentieth-century China. He envisaged the emergence of a Chinese Dante who would lead China into the modern world by reviving classical studies.[28]

Besides Huang Jie, other *Guocui* writers also invoked the Renaissance as a metaphor for reviving the Zhou cultural heritage. For instance, Deng Shi made elaborate comparisons between the revival of Zhou cultural practices and the rediscovery of Greek and Roman literature during the Renaissance. In his article "On Reviving the Ancient Learning" (*Guxue fuxing lun* 古學復興論, 1905), he coined the term "the rebirth of ancient studies" (*guxue fuxing* 古學復興) to refer to the recovery of Zhou cultural practices. To make his point, he drew a direct parallel between the Renaissance and the Chinese "rebirth of ancient studies." "Just as

---

[26] Huang 1905: 2a.

[27] Ibid., 1b. In referring to the Bak tribe who allegedly migrated from Mesopotamia to China in prehistoric times, Huang Jie equated the yellow race in China with the white race in Europe as advanced people with sophisticated skills in agriculture and pottery. For a discussion of the historical importance of Sino-Babylonianism (or *xilaishuo* 西來說), see Hon 2010.

[28] Huang 1905: 4b.

Europeans underwent their rebirth of ancient studies in the fifteenth century," he wrote, "so the Chinese experienced their rebirth of ancient studies in the twentieth century."[29] To make the comparison look compelling, he took pains to match events in Chinese history with those in European history, making sure that his readers would see a parallel between the Renaissance in the fifteenth century and the Chinese revival of ancient studies in the twentieth century.[30]

Like Deng Shi, Xu Shouwei (許守微) also stressed the theme of rebirth. In his article "On National Essence Being Not a Hindrance to Europeanization" (*Lun guocui wuzu yu ouhua* 論國粹無阻於歐化, 1905), he argued that there were many ways for the Chinese to become modern. Rather than indiscriminately adopting things from Europe or *Ouhua* (歐化 Europeanization), he suggested his readers to look back to the past to identify the Chinese national essence.[31] He admitted that Europe was clearly ahead of China in modernization, but China could catch up with Europe if she underwent a "revival of ancient learning" (*guxue zhi fuxing* 古學之復興). Like Deng Shi, Xu saw a parallel between the Renaissance and the Chinese "revival of ancient learning." With the Renaissance, Europeans laid the foundation for modern Europe by reviving ancient Greek learning, building libraries, and developing philosophical, legal, and historical studies. Five hundred years later, Xu suggested, the Chinese would do the same to join the modern world by reviving their ancient studies.[32]

From the writings of Huang Jie, Deng Shi, and Xu Shaowei, it is clear that the *Guocui* writers accepted that modernity began in Europe. Nevertheless, they argued that China's belated entry to the modern age was not caused by a lack in Chinese intelligence or ability; rather, it was caused by internal colonization, autocratic rule, and stifling of thoughts. Specifically, they blamed absolute monarchy, the Confucian orthodoxy, and the examination system as the stumbling blocks that prevented China from building a nation-state. Thus, for the *Guocui* writers, the modernization of China required both a removal and a recovery. On the one hand, the Chinese needed to remove the political, social, and cultural institutions that supported absolute monarchy. On the other hand, they needed to recover the Zhou cultural heritage, which would transform China into a modern nation-state.

---

29 Deng 1905b: 1b.
30 Ibid., 1b–2a.
31 Xu 1905: 1a–1b.
32 Xu 1905: 2a–2b.

The Public Realm of the Educated Elite

But what kind of Zhou cultural heritage did the *Guocui* writers want to recover? What was on the *Guocui* writers' minds when they said to recover the ancient learning? Why were the Hundred Schools of the Eastern Zhou period so important to them? To answer these questions, we need to turn to Liu Shipei, the Zhou specialist of *Guocui xuebao*. In a long article titled "The History of Learning at the End of the Zhou [Dynasty]" (*Zhoumo xueshu shi* 周末學術史, 1905), Liu gave an account of the various types of learning in the Eastern Zhou period that, he claimed, were the equivalent of modern academic disciplines such as psychology, ethics, logic, sociology, religion, law, mathematics, military science, education, natural science, philosophy, literature, law, and fine arts. In describing these different types of "modern" learning, Liu expanded the scope of Chinese ancient learning far beyond what was prescribed in the conventional fourfold division of scholarship—classics (*jing* 經), history (*shi* 史), philosophy (*zi* 子), and belles-lettres (*ji* 集). More important, he made clear that there was not only a wide variety of intellectual activities in the Eastern Zhou period, but also a close compatibility between those intellectual activities and modern forms of learning:

> In the past, in Europe the church had total control over schools. Since the fifteenth century the church lost its power, and the masses [*minshu* 民庶] were given the opportunity to learn. In comparing what had happened in Europe with what we find in Chinese documents, we gain a new perspective [on Zhou learning]. During the [Western] Zhou period, officials monopolized learning. But after the capital of the Zhou had been moved to the east, the imperial court lost its authority, giving talented individuals an opportunity to develop their thinking. What they said was all different, yet they all had their distinct contributions. These were the results of a learning developed from below.[33]

Like Huang Jie, Deng Shi, and Xu Shouwei, Liu saw a parallel between the Renaissance in fifteenth-century Europe and the "rebirth of ancient studies" in twentieth-century China. But unlike the other three, Liu clearly explained why the "rebirth of ancient studies" had to be a revival of Eastern Zhou learning. As suggested in the quotation, Liu regarded the Eastern Zhou period (770–256 BCE) as a time when the educated elite in China had the freedom to move from one feudal state to another to look

33 Liu Shipei 1905: 4b–5a.

for work. Under this fluid political environment, they could speak their mind to attract the attention of competing feudal lords. As a result, the "Hundred Schools" arose in a fierce competition to change the political system.

Eager to match events in Chinese history with those in European history, Liu compared the rise of the "Hundred Schools" in the Eastern Zhou with the sociopolitical changes in fifteenth-century Europe where the Catholic Church lost its monopoly in education with the rise of secular universities. The comparison might be far-fetched; nevertheless Liu made his point. He suggested that although separated by centuries, thinkers in the Eastern Zhou period were part of a drastic social and political change similar to what the Renaissance humanists later experienced. Like the humanists, the Hundred Schools thinkers challenged orthodoxy, discredited the central authority, spread culture and knowledge, and above all, expanded the public domain for exchanging ideas and information. More important, the comparison between the Hundred Schools thinkers with the Renaissance humanists pointed to what could happen in twentieth-century China. To expand the public realm of the educated elite, Liu believed there had to be a drastic reduction in government control over education, assembly, and the means of circulating ideas. Instead of the government controlling learning, it should be the scholars who decided what to learn and how to learn. In Liu's words, learning had to be "generated from below" (*xue xing yu xia* 學興於下) rather than controlled from above. At the same time, there must be more opportunities for the lower echelon of the educated class (or *minshu* 民庶) to receive higher learning and to participate in public debates.[34]

Supporting Liu's view, other *Guocui* writers such as Deng Shi, Ma Xulun, Lu Shaoming (陸紹明), and Zhang Taiyan also wrote on the competition of ideas during the Eastern Zhou period. Among them, Zhang Taiyan was the most important. In "A Brief Survey of the Learning of Various Thinkers [of the Eastern Zhou]" (*Zhuzi xue lüeshuo* 諸子學略說, 1906), he distinguished the "independence" (*duli* 獨立) of the Hundred Schools

---

[34] To prove his point, besides writing a historical account of pre-Qin learning, Liu published a series of articles in the *Guocui xuebao* to discuss the significance of the Hundred Schools, including "The Origins of Ancient Learning" (*Guxue qiyuan lun* 古學起源論, 1905), "On Ancient Learning Originated from the Government" (*Guxue chuyu guanshou lun* 古學出於官守論, 1906), "The True Confucian Teaching" (孔學真論, 1906), and "The Differences between the Confucians and the Legalists" (*Ruxue faxue fenqi lun* 儒學法學分歧論, 1907). In these articles, Liu showed that as the central authority lost its power, the educated elite increased their influence in the political, social, and cultural realms.

from the "harmony and compromise" (*tiaohe* 調和) of scholars in imperial China.[35] In the Hundred Schools, Zhang found a daring spirit to challenge the government, a determination to speak freely in public, and an unwavering commitment to thinking and acting independently. By contrast, in scholars of imperial China, he found pragmatic calculations to suit rulers' interests, a practical concern with wealth and power, and a willingness to adjust their scholarship for political purposes. For Zhang, the drastic differences between these two types of scholars were due to the establishment of state Confucian in the Han dynasty. Before then, scholars were free to express their views, establish their own schools of thought, and dispute with one another. After the establishment of Confucian orthodoxy in the Han, scholars were limited to explaining and substantiating the Confucian canon. As a result, the more scholars wanted to fit their scholarship in the Confucian orthodoxy, the further away they were "from their true concerns" (*shi qi ben zhen* 失其本真).[36]

Hence, by repeatedly comparing the Hundred Schools with the Renaissance, the *Guocui* writers expressed their demand for a new sociopolitical order in twentieth-century China. On the one hand, like the Catholic Church's loss of its monopoly in education in the face of the rise of secular universities in fifteenth-century Europe, the *Guocui* writers anticipated a transfer of power in cultural affairs from the central government to the learned community during their times. Particularly they had high hopes in the founding of the national school system. They believed they would play a leading role in running the schools, defining the curriculum, and shaping the hearts and minds of millions of young Chinese. On the other hand, like the Hundred Schools, they wanted to claim an autonomous realm free from the control of the government. For them, the history of absolute monarchy in China showed that rulers had been skillful in controlling and manipulating the cultural field to increase their power. To prevent that from happening again, the *Guocui* writers wanted to create an autonomous realm to express their views freely, to assemble on their own initiative, and to challenge the government if they so desired.

35 Zhang Taiyan 1906: 1b.
36 Zhang Taiyan 1906: 1a–1b.

Critiquing the Political Control of Culture

To demand an autonomous public realm for debate and dissent, the *Guocui* writers stressed the dichotomy between "national learning" (*guoxue* 國學) and "imperial learning" (*junxue* 君學). While the former was fully under the control of the educated elite, the latter was to serve the absolute rulers. During the first three years of the *Guocui xuebao*, from 1905 to 1908, this dichotomy between national learning and imperial learning was the main theme of many articles, expressing the wishes of the *Guocui* writers to take control of the cultural field.

Of the *Guocui* writers, Deng Shi wrote repeatedly on the meaning of "national learning." Usually appearing at the front of the journal under the heading "Editorial" (*sheshuo* 社說), his essays were full of historical references and classical imagery, but they were sharp and clear in delivering a political message. For instance, in his essay "On the True Meaning of National Learning" (*Guoxue zhenlun* 國學真論, 1907), he made a clear distinction between "national learning" and "imperial learning."

> With great pain and regret, I find that there has been no national learning in China. By national learning, I mean the kind of learning different from that sponsored by the emperor. Since the Qin and Han, the goal of learning in our Divine Continent had been to serve the emperor. There was neither a love of the nation nor a learning to serve the nation. Why was that so? [The reason was that scholars] only knew to serve the emperor. They did not know that they had to serve the nation.[37]

Similar to what Liang Qichao (梁啟超 1873–1929) had argued in his famous 1902 essay "The New Historiography" (*Xinshixue* 新史學), Deng made a categorical distinction between nation and dynasty.[38] For the two writers, dynasties might come and go, but the nation was always the basis upon which the Chinese built their collective identity. Whereas a dynasty was the private possession of the imperial family, the nation was the collective enterprise of all people living in the country. While a dynasty made its presence known through the absolute power of the imperial throne, the nation manifested itself in the loyalty and sacrifice of its citizens.

Although Liang Qichao and Deng Shi shared the need to distinguish nation from dynasty, they employed different strategies in making the

---

[37] Deng 1907a: 1a.

[38] Liang 1902. For a discussion of Liang's new historiography, see Tang Xiaobing 1996: 1–79.

distinction. While Liang focused on how history had been written to clarify the distinction between dynasty and nation, Deng concentrated on the forms of learning. For Liang, the fundamental difference between dynasty and nation was illustrated clearly in the contrast between dynastic history (*diwang lishi* 帝王歷史) and national history (*guoshi* 國史). For Deng, the distinction between dynasty and nation was best shown in the difference between national learning and imperial learning:

> In terms of political categories, recent scholars distinguishes nation from dynasty. But in terms of learning, they do not make the distinction between "national learning" and "imperial learning." As a result, they conflate these two types of learning, equating serving the emperor with serving the nation, and studying for the examinations with pursuing national learning. They do not know that national learning is a completely different type of learning.[39]

For Deng, national learning was fundamentally different from imperial learning because the latter was to lend support to the imperial system, and its goal was to offer legitimacy to absolute rulers. In contrast, national learning was a crystallization of critical thinking and the creative imagination of the educated elite, who articulated the collective identity of the Chinese people through a narration of the past.

To show the damage that "imperial learning" had done to the cultural field, Deng published the titles and abstracts of banned books and censored materials. In his 1907 article, "The Postscripts to the Combined Publication of Four Lists of Banned Books" (*Jinshu mulu sizhong heke ba* 禁書目錄四種合刻跋), he used the Qing literary inquisition of the Qianlong emperor as an example to show the extent to which the cultural field was harmed by imperial censorship. In order to suppress dissenting voices and to intimidate educated elites, Deng stated, Qianlong ordered the banning and burning of books on a scale exceeding that of the First Emperor of Qin. In the latter, Deng said, the banning and burning of books were confined to government collections, thereby leaving private libraries untouched and preserving precious materials for a later revival of learning in the Han period. In the former, however, the banning and burning of books included both public and private collections, creating a huge vacuum in scholarship. For this reason, Deng called on his readers to send him information about the banned and censored materials, in order to recover the lost scholarship.[40] To report on the rediscovery of

39 Deng 1907a: 1a.
40 Deng 1908b.

the lost scholarship, in 1909 he created a special section in the *Guocui xuebao* called "Introducing the Lost Books" (*Shaojie yishu* 紹介遺書), listing hundreds of books that were lost because of the Qing government's suppression of dissenting views.

Deng's other measure of promoting national learning was to compile the writings of the lone souls in history who had made tremendous sacrifices for the collective interest. Beginning in 1906 and continuing on for over two years, he serialized their writings in the *Guocui xuebao* under the title "The Collected Writings of Those with a Sense of Righteousness" (*zhengqi ji* 正氣集). In the preface, Deng explained the purpose of such an endeavor:

> What is the purpose of compiling "The Collected Writings of Those with a Sense of Righteousness"? The answer is to develop the national essence of the Divine Continent and to preserve the righteousness in heaven and earth. As we know, the essence of the ancient learning of our Divine Continent is morality. The pure form of morality that is most suitable to today's need is righteousness. Righteousness is the crystallization of the forces of heaven and earth, and the spirit of sun and stars. Indeed it is the basis upon which the Divine Continent has been built for five thousand years.[41]

Included in this "collected writings" were the prose, poems, and letters of scholar-generals such as Yue Fei (岳飛 1103–1142) and Wen Tianxiang (文天祥 1236–1283), who made ultimate sacrifices in fighting against the Mongols at the end of the Southern Song period. Also included were the writings of loyal officials such as Lu Xiufu (陸秀夫 1238–1279), who threw himself in the ocean with the child emperor of the Southern Song. With these examples, Deng showed that at critical moments in Chinese history, members of the educated elite had continuously stood up to sacrifice themselves for the public good. They were, in Deng's words, "the roosters that roused people on a windy and rainy day" and "the pine trees that remained green in winter."[42] For educated elites in the 1900s, as China was once again in a critical situation, they had to make sacrifices for the nation. Rather than practicing imperial learning to support the Qing dynasty, Deng urged them to serve the Chinese nation by using their cultural capital to topple the alien regime and to abolish absolute monarchy.

---

41 Deng 1906: 1a–1b.
42 Ibid.

The Significance of Geming *as Restoration*

As shown above, the terms "national essence" and "national learning" had special meanings for the *Guocui* writers. Rather than regressive and reactionary, as they were depicted by cultural iconoclasts of the New Culture Movement a decade later,[43] in the 1900s "national essence" and "national learning" were part of the critique of absolute monarchy and the Confucian orthodoxy. As tools to subvert the internal colonizers and to counter imperialism, the two terms evoked an image of modern nation-state where the central and local authorities shared power, educated elites led the discussion of governance, and citizens were armed with a collective memory of the past.

In addition, the ways the *Guocui* writers used the terms show the significance of *geming* as restoration. The premise of *geming* as restoration was that since ancient times China had been equipped with the resources to build a modern nation-state. Her late entry into modernity was due not to a lack of resources, but to long-standing institutional and intellectual barriers that had stopped her from fully developing her potential. Thus, for China, the task of joining the modern age had to be twofold—removing the barriers that kept China behind, and recovering the potential and energy that were latent since ancient times.

For the *Guocui* writers, the key issue in the rebirth of China was to identify a particular time in the past that was relevant and meaningful for the present. As shown above, the *Guocui* writers' choice of the Zhou period was deliberate. The Zhou period was selected not because it was part of the revered "Three Dynasties" (*sandai* 三代), but because it provided historical traces of a decentralized political system and a free public realm for intellectual debate. Seen in this light, the "Zhou period" was no longer the Zhou dynasty in history. It was a metaphor for what the *Guocui* writers considered as essential to constructing a Chinese nation in the twentieth century. Signifying political and intellectual pluralism, the "Zhou period" was lifted from its own time and transposed to the 1900s as a symbol of modernity.

This jump in time should not be seen as merely a strategic tactic to explain modern ideas through historical metaphors. Rather, the *Guocui* writers saw a direct parallel between the decentralized system in the "Zhou period" and the decentralized system that they wanted to create in twentieth-

43 See Mao Zishui's critique of Zhang Taiyan (Mao 1919).

century China. As shown in Liu Shipei's and Zhang Taiyan's interpretations of the Hundred Schools, when the *Guocui* writers read Zhou writings, they did not read them as historical documents, but as texts that revealed the conditions of modernity. Certainly, they were reading Zhou texts from a 1900s' perspective; at the same time, the texts responded because they portrayed an image of pluralism that was never possible under absolute monarchy and the Confucian orthodoxy.

Certainly this "fusion of horizons" had always been part of the educate elite's training for the civil service examinations; but in the 1900s classical training was re-deployed for a different purpose. Rather than serving the dynasty as in "imperial learning," classical learning now served the nation. In shifting the focus of classical learning, the *Guocui* writers made the anti-Manchu revolution relevant to the educated elite. By making "the Zhou" the symbol of political and social pluralism, they linked the dual goals of revolution—a racial revolution to topple the alien regime and a political revolution to abolish absolute monarchy—to the new identity of the educated elite. Speaking directly to the educated elite, who faced the challenges of the abolition of examinations, the *Guocui* writers reassured them that their classical learning would be useful in the post-examination era. In the past, classical learning had been essential to joining the officialdom and becoming local elites; now it would be an effective weapon to critique the imperial system and to build the Chinese nation. More important, through the symbol of "the Zhou," the *Guocui* writers promised the educated elite that they would have a bigger role to play in the new Chinese nation. As citizens, they would have the right to express their views freely, assemble on their own initiative, and challenge the government if they so desired. As political leaders, they would provide direction and leadership to the new nation where the central and local authorities worked in tandem. In both areas, the *Guocui* writers made the educated elite realize that they would have more to gain than to lose in supporting the anti-Manchu revolution. In this way, the *Guocui* writers transformed the educated elite by giving them the mission of "*geming* as revolution."

References Cited

Bastid, Marianne (1988), *Educational Reform in Early 20th-Century China.* Paul J. Bailey trans. Ann Arbor: Center for Chinese Studies, University of Michigan.

Bernal, Martin (1976), "Liu Shih-p'ei and National Essence," in Charlotte Furth, ed., *The Limits of Change: Essays on Conservative Alternatives in Republican China.* Cambridge: MA: Harvard University Press, 90–112.

Bol, Peter K. (1992), *"This Culture of Ours": Intellectual Transitions of T'ang and Sung China.* Stanford, CA: Stanford University Press.

Borthwick, Sally (1983), *Education and Social Change in China: The Beginnings of the Modern Era.* Stanford, CA: Hoover Institution Press.

Bourdieu, Pierre (1996), *The State Nobility: Elite Schools in the Field of Power.* Lauretta C. Clough trans. Stanford, CA: Stanford University Press.

Chaffee, John W. (1995), *The Thorny Gates of Learning in Sung China.* New edition. Albany: State University of New York Press.

Chen Jianhua 陳建華 (2000), *"Geming" de xiandai xing: Zhongguo geming huayu kaolun* 「革命」的現代型: 中國革命話語考論 (The modernity of "geming": A study of Chinese revolutionary discourse). Shanghai: Shanghai gujie chubanshe.

Cheng I-fan (1976), "Kung as an Ethos in Late Nineteenth-Century China: The Case of Wang Hisen-ch'ien (1842–1918)," in Paul Cohen and John Schrecker, eds., *Reform in Nineteenth-Century China.* Cambridge, MA: Harvard University Press, 170–80.

Deng Shi 鄧實 (1905a), "Guoxue yuanlun" 國學原論 (On national learning), in *Guocui xuebao* 1 (1905) *sheshuo*: 1a–6a.

—— (1905b), "Guoxue weilun" 國學微論 (A brief discussion of national learning), in *Guocui xuebao* 2 (1905) *sheshuo*: 1a–6b.

—— (1905c), "Guoxue tonglun" 國學通論 (A general discussion of national learning), in *Guocui xuebao* 3 (1905) *sheshuo*: 1a–9b.

—— (1905d), "Guxue fuxing lun" 古學復興論 (On reviving the ancient learning), in *Guocui xuebao* 9 (1905) *sheshuo*: 1a–4b.

—— (1906), "Zhengqi ji xu" 正氣集序 (Preface to the collected writings of those with a sense of righteousness), in *Guocui xuebao* 13 (1906) *zhuanlu*: 1a–1b.

—— (1907a), "Guoxue zhenlun" 國學真論 (On the true national learning), in *Guocui xuebao* 27 (1907) *sheshuo*: 1a–4a.

—— (1907b), "Guoxue wuyong bian" 國學無用辨 (A rebuttal to the argument that national learning is useless), in *Guocui xuebao* 30 (1907) *sheshuo*: 1a–2b.

—— (1908a), "*Guocui xuebao* di san nian zhu dian xu" 國粹學報第三年祝典序 (Preface to the third year anniversary of *Guocui xuebao*), in *Guocui xuebao* 38 (1908) *sheshuo*: 1a–2b.

—— (1908b), "Jinshu mulu sizhong heke ba" 禁書目錄四種合刻跋 (The postscripts to the combined publication of four lists of banned books), in *Guocui xuebao* 38 (1908) *wenpian*: 5a–6a.

—— (1908c), "Shenzhou guoguang ji xu" 神州國光集序 (Preface to the collection of the national glories of the divine continent), in *Guocui xuebao* 42 (1908) *wenpian*: 6a–8a.

Ding Weizhi 丁偉志 and Chen Song 陳崧 (1995), *Zhongxi tiyong zhijian: Wanqing zhongxi wenhua guan shulun* 中西体用之間: 晚清中西文化觀述論 (Between China and the West, substance and function: A study of the late Qing concepts of Chinese and Western culture). Beijing: Zhongguo shehui kexue chubanshe.

Elman, Benjamin A. (2000), *A Cultural History of Civil Examinations in Late Imperial China.* Berkeley: Univ. of California Press.

Esherick, Joseph W. (1976), *Reform and Revolution in China: The 1911 Revolution in Hunan and Hubei.* Berkeley: University of California Press.

Fogel, Joshua A. and Peter G. Zarrow (1997), *Imagining the People: Chinese Intellectuals and the Concept of Citizenship, 1890–1920.* Armork, NY: M.E. Sharpe.

Franke, Wolfgang (1963), *The Reform and Abolition of the Traditional Chinese Examination System*. Cambridge, MA: East Asian Research Center, Harvard University.

Guocui xuebao 國粹學報 (1905–1911). Taipei: Wenhai chubanshe reprint.

Ho, Ping-ti (1962), *The Ladder of Success in Imperial China: Aspects of Social Mobility, 1368–1911*. New York: Columbia University Press.

Hon, Tze-ki (2002), "Zhang Zhidong's Proposal of Reform: A New Reading of the *Quan xue pian*," in Rebecca Karl and Peter Zarrow, eds., *Rethinking the 1898 Reform Period: Political and Cultural Change in Late Qing China*. Cambridge, MA: Harvard University Asia Center, 77–98.

—— (2003), "National Essence, National Learning, and Culture: Historical Writings in *Guocui xuebao, Xueheng*, and *Guoxue jikan*," in *Historiography East and West* 1, 2 (2003): 242–86.

—— (2010), "From a Hierarchy in Time to a Hierarchy in Space: The Meanings of Sino-Babylonianism in Early Twentieth Century China," in *Modern China*, 36, 2 (2010): 139–69.

——, and Robert J. Culp (2007), "Introduction," in Tze-ki Hon and Robert J. Culp. eds., *The Politics of Historical Production in Late Qing and Republican China*. Leiden: Brill, 1–17.

Huang Jie (1905), "Guocui xuebao xu" 國粹學報敍 (Preface to *Guocui xuebao*), in *Guocui xuebao* 1 (1905): 1a–4a.

Jin Guantao 金觀濤 (2003), *Geming guannian zai Zhongguo de qiyuan he yanbian* 革命觀念在中國的起源和演變 (The origin and transformation of the concept of *geming* in China). Conference paper delivered at Chinese University of Hong Kong.

Judge, Joan (1996), *Print and Politics: "Shibao" and the Culture of Reform in Late Qing China*. Stanford, CA: Stanford University Press.

Kuhn, Philip A. (1975), "Local Self-Government under the Republic: Problems of Control, Autonomy and Mobilization," in Frederic Wakeman and Carolyn Grant, eds., *Conflict and Control in Late Imperial China*. Berkeley: University of California Press, 257–298.

Li Guojun 李國鈞 and Wang Bingzhao 王炳照 eds. (2000), *Zhongguo jiaoyu zhidu tongshi* 中國教育制度通史 (A history of the Chinese education system), vol. 6, part 2 (1840–1911). Ji'nan: Shandong jiaoyu chubanshe.

Liang Qichao 梁啟超 (1902), "Xin shixue" 新史學 (New historiography), in *Yinbingshi heji: wenji* 飲冰室合集: 文集 (The combined collections of writings from the Ice-Drinking Studio: Articles), vol. 9. Beijing: Zhonghua shuju, 1989 reprint, 1–31.

Liu Longxin 劉龍心 (2001), "Xueke tizhi yu jindai zhongguo shixue de jianli" 學科制度與中國史學的建立 (The curriculum system and the establishment of Chinese historiography), in Luo Zhitian 羅志田 ed., *Ershi shiji de Zhongguo: Xueshu yu shehui—shixue juan* 二十世紀的中國: 學術與社會—史學卷 (Scholarship and society in twentieth-century China: Section on historical studies). Ji'nan: Shandong renmin chubanshe, 449–585.

Liu Shipei 劉師培 [Liu Guanghan 劉光漢] (1905), "Zhoumo xueshu shi zongxu" 周末學術史總序 (Preface to the history of thought at the end of the Zhou period), in *Guocui xuebao* 1 (1905) *xuebian*: 4b–12b.

—— [Liu Guanghan] (1907a), "Ruxue faxue fenqi lun" 儒學法學分歧論 (On the differences between Confucianism and Legalism), in *Guocui xuebao* 29 (1907) *sheshuo*: 1a–4b; 32 (1907) *sheshuo*: 1a–5b; 35 (1907) *sheshuo*: 5a–7b.

Mao Zishui 毛子水 (1919), "Guogu yu kexue de jingshen" 國故與科學精神 (National past and scientific spirit), in *Xinchao* 5, 1 (1919): 731–45.

Pan Bo 潘博 (1905), "Guocui xuebao xu" 國粹學報敍 (Preface to the Journal of National Essence), in *Guocui xuebao* 1 (1905) *sheshuo*: 4a–4b.

Reynolds, Douglas R. (1993), *China, 1898–1912: The Xinzheng Revolution and Japan*. Cambridge, MA: Council on East Asian Studies, Harvard University.

Schneider, Laurence A. (1976), "National Essence and the New Intelligentsia," in Charlotte Furth, ed., *The Limits of Change: Essays on Conservative Alternatives in Republican China*. Cambridge: MA: Harvard University Press.

Schwintzer, Ernest Peter (1992), *Education to Save the Nation: Huang Yanpei and the Educational Reform Movement in Early Twentieth Century China*. PhD diss., University of Washington.

Tang Xiaobing (1996), *Global Space and the Nationalist Discourse of Modernity: The Historical Thinking of Liang Qichao*. Stanford, CA: Stanford University Press.

Tang Zhijun 湯志鈞 (1989), *Jindai jingxue yu zhengzhi* 近代經學與政治 (Classical studies and politics in modern China). Beijing: Zhonghua shuju.

Tsou Jung (1968), *The Revolutionary Army: A Chinese Nationalist Tract of 1903*. Introduction and trans. with notes by John Lust. The Hague: Mouton.

Wright, Mary C. ed. (1968), *China in Revolution: First Phase, 1900–1913*. New Haven, CT: Yale University Press.

Xu Shouwei 許守微 (1905), "Lun Guocui wuzu yu Ouhua" 論國粹無阻於歐化 (On national essence not being a barrier to Europeanization), in *Guocui xuebao* 7 (1905) *sheshuo*: 1a–4b.

Zhang Taiyan [pen name Zhang Jiang 章絳] (1906), "Zhuzi xue lüeshuo" 諸子學略説 (A brief discussion of the Hundred Schools), in *Guocui xuebao* 20 (1906) *xuepian*: 1a–6a; 21 (1906) *xuepian*: 1a–8a.

Zhang Zhidong 張之洞 (1898), "Quan xue pian" 勸學篇 (Exhortation to learning), in *Zhang wenxiang gong quanji* 張文襄公全集 (The collected works of Zhang Wenxiang). Taipei: Wenhai chubanshe, 1963 reprint.

Zhao, Suisheng (2004), *Nation-state by Construction: Dynamics of Modern Chinese Nationalism*. Stanford, CA: Stanford University Press.

Zheng Shiqu 鄭師渠 (1997), *Wanqing guocui pai—wenhua sixiang yanjiu* 晚清國粹派—文化思想研究 (The National Essence group of the late Qing: A study of culture and thought). Beijing: Beijing shifan daxue chubanshe.

CHAPTER TEN

TEMPORALITY OF KNOWLEDGE AND HISTORY WRITING IN EARLY TWENTIETH-CENTURY CHINA. LIU YIZHENG AND *A HISTORY OF CHINESE CULTURE*

Ya-pei Kuo

History imposes its law upon the faraway places that it conquers when it fosters the illusion that it is bringing them back to life.
—Michel de Certeau, "Making History"

Introduction: The Writing and Rewriting of History

Liang Qichao's "New Historiography" (*Xin shixue*) (1902) ushered in a new era in China's historical thinking. In this manifesto of modern historiography, Liang harshly criticized the traditional approach and proposed a set of new working precepts for future historians.

Liang lamented that although about sixty to seventy percent of books printed in China were historical works, including dynastic histories, chronologies, bibliographies, biographies, and local gazetteers, almost none of these books could be called real history. Each of them suffered from at least one of four problems: (1) they paid attention only to the events that took place in or around the royal court, mistakenly identifying the royal family with the entire country; (2) they dwelt on a few individuals and failed to see the people as a group, much less as a nation; (3) the accounts all prioritized ancient times; and (4) writers showed no interest in theoretical subjects such as the nature and movement of history. In addition, Liang complained, previous attempts at history lacked creativity and remained inaccessible to ordinary people. There was too much moral judgment and not enough causality. Meaning was established not by examining the course of history, but by focusing exclusively on individual events.[1] China, claimed Liang, needed a new history that would restore the nation to the center of its collective memories. The new history was to be a national history.

[1] Liang 1902a: 1628–31.

Liang's "New Historiography" was representative of its time. The first decade of the twentieth century witnessed a surge of interest in history writing. Prompted by the new national consciousness, intellectuals of different political stripes agreed on the need for a new narrative of the past. Zhang Taiyan, for example, confided to Liang Qichao in 1902 his private plan of writing a new general history of China.[2] Liu Shipei and Xia Zengyou contributed to the making of history in more practical ways. As they tried to fill the market niche for new-style textbooks, the two scholars of classical studies experimented with new styles of historical narrative in their composition of teaching materials.[3] Together with Liang's work, these writing projects shared the same aspiration of placing the nation at the center of history, and using history to naturalize the Chinese nation.[4] Zhang Taiyan, whose ambition went beyond producing a textbook, unabashedly asserted the utility of history for nation-building. The coherent narration of a national history was thought to inspire patriotism and solidify the China's new identity as a nation.

The significance of Liang's "New Historiography," however, went beyond promoting the writing of national history. Equally important was the notion of *jinhua* (進化 literally, "progression and change"), a subject that Liang devoted most of his essay to. Although the term has often been viewed as the Chinese translation of the word "evolution," Liang construed it almost as a synonym for "development," without the association with the modern idea of Progress. One example of such a use may be found in the title of his essay "On the *Jinhua* History of China's Political Despotism" (*Zhongguo zhuanzhi jinhua shi lun*) (1902).[5] Here *jinhua* lacks any positive connotations: it simply refers to the continuous development of an institution that was detrimental to the overall progress of the nation.

Jinhua was an irreversible and never-ending process; it was also the only appropriate topic for historical studies. According to Liang, history should not concern itself with the reversible and cyclical, because they were not phenomena of *jinhua*.[6] This irreversibility was due to the lineal coordinate system on which *jinhua* had to register itself. In this early attempt to give the modern concept of time, *shijian* (時間), a definition, Liang highlighted that time is the universal and all-inclusive framework of

2 Zhang 1902.
3 Hon and Culp 2007.
4 Tang 1996: 46–79.
5 Liang 1902b: 1648.
6 Ibid., 1632.

reference: "Everything exists in time and space."[7] Moreover, because of its linear shape—"Time passes and does not stop,"[8] any sequence of events registered on it, including *jinhua*, would be a directional movement towards the future.

But the future could not be made out. There were two reasons for this inscrutability. The first was the spiral nature of *jinhua*: at times, *jinhua* could appear to be retrogressive or cyclical. Historians needed to take into account the totality of human experience before they could make out the true direction of *jinhua*. The second reason, according to Liang, rested in the ongoing, incomplete nature of *jinhua*. Because history continued to unfold, the telos of *jinhua* remained unrevealed until the end was reached. There was no objective standard for determining what constituted progress and retrogress. Although, Liang averred, an underlying law did exist, grasping it was a matter of the historian's hermeneutic insight. Working to develop that insight was the historian's moral obligation.[9]

Trying to set the agenda of history writing for modern China, "New Historiography" based its main polemic points on a contrast between the modern and the traditional and the contrast was most apparent at the level of historical knowledge. To Liang, the ability to recognize history's true protagonist belonged uniquely to the modern historian. He was in this sense poles apart from his traditional predecessors, whose failure to give voice to the nation was merely symptomatic of their blindness to history's true nature. Similarly, in the notion of *jinhua*, Liang saw a new key to the meaning of history and history writing. It signified a modern consciousness that was unattainable to historians of pre-modern times. The rise of this new consciousness rendered histories of the past into nothing but collections of data and materials, something that might be used but which was sorely in need of correction and reworking. Liang's renunciation of traditional historiography was thus absolute and New Historiography had to be a total break from historical knowledge of the past.

Liang's totalistic renunciation of the old histories was couched in a particular sense of temporality, which rested the meaning of the modern on its rupture with its past. This temporality, however, stood in tension with the kind of histories that Liang aspired to produce. Both national history and the history of *jinhua* had to rely on data and material prepared

---

[7] Ibid.
[8] Ibid., 1645.
[9] Ibid., 1633–34.

by historians of previous generations. Certain faith in their professional competence would be a requisite. Moreover, as projects to naturalize the present through historicization, both national history and the history of *jinhua* presupposed history's internal coherence, and were prone to relegate conflicts and ruptures into a lower order of significance. After all, both the being of the nation and the principle of *jinhua* were supposed to have constant existence, cutting through all the epochal divides.

"New Historiography" prescribed a history that treated the past as simultaneously dead and alive. Liang never succeeded in producing such a history. How to mute the past while reanimating it? Later historians who aspired to put Liang's prescriptions into practice would continue to be confronted by the question.

Liu Yizheng and A History of Chinese Culture

Although Liu Yizheng (1880–1956) did not publish his *History of Chinese Culture* till the mid-1920s, the work was, wittingly or not, an answer to Liang Qichao's call for a new history.

Liu Yizheng was born in 1880 in the city of Dantu in Jiangsu province. Seven years Liang Qichao's junior, Liu too had witnessed the dramatic upheavals of the turn of the century. His classical education prepared him for the civil service examination. At sixteen, he passed the county-level examination and thus won state sponsorship for further studies. The degree also allowed Liu to work as a private tutor. Responding to the advice of his older friend Chen Shanyu, Liu decided not to pursue a career in public service. Instead he moved to Nanjing in 1901 and started to work under the renowned bibliographer Miao Quansun at the newly founded Jiang-Chu Compilation and Translation Bureau (Jiang-Chu bianyi ju). According to his later recollections, it was Miao who opened his eyes to the true nature of traditional scholarship. More importantly, Liu's short stay at the bureau marked the inception of his lifelong involvement with history writing.

At the Jiang-Chu Compilation and Translation Bureau, Liu spent most of his time writing history textbooks for new schools. The bureau's mission, established under the Qing government's reform initiatives after 1900, was to prepare teaching materials for the new, nationwide modern school system. This was not an easy task. The disparity between the old and new schools was more than institutional. The new curricular regime, largely imported from abroad, revolutionized China's disciplinary organization of knowledge. Old information had to be reorganized. Previous publications had to be "converted" into a new format and a new vocabulary. Teaching materials also needed to address a new educational mission. For subjects such as history, ethics, and geography, the new-style schools desperately needed books that would impart knowledge while inspiring patriotism. To fill the vacuum quickly, many education officials turned to Japan.

In 1902, Liu traveled with Miao Quansun to inspect Japan's school system. Recently reformed, the system impressed Liu with its innovative curriculum and school facilities. The main purpose of the trip, however, was to see how Japanese scholars had articulated Chinese history. Upon his return to China, close to the time when Liang Qichao published "New Historiography," Liu compiled *A Brief Historical Account of Different Eras* (*Lidai shi lüe*). The book was largely based on *A General History of China* (*Shina tsūshi*), written by Naka Michiyo,[10] and it was meant to be a textbook. It was Liu's first general history of China and one of the Jiang-Chu Bureau's few publications. In 1903, Liu left the bureau to participate in the founding of the Siyi Elementary School and the Jiangnan Commercial Middle School. At both schools, Liu was in charge of the history and literature division. During these years, Liu published two more textbooks: *A History of Chinese Commerce* (*Zhongguo shangye shi*) in 1905 and *A History of Chinese Education* (*Zhongguo jiaoyu shi*) in 1907.[11]

Liu joined the Nanjing Superior Normal College in 1915. When it was renamed Southeastern University in 1921, Liu became a professor in the Department of History. He met Wu Mi and Mei Guangdi there and was invited to participate in their journal, *The Critical Review* (*Xueheng zazhi*). From 1922 to 1933, Liu's contributions, largely on historical topics, appeared in almost every issue. Liu's *History of Chinese Culture* (*Zhongguo wenhua shi*), based on his lecture notes from Southeastern University, was serialized in *The Critical Review* from October 1925 to November 1929.

A History of Chinese Culture was one of the first works in Chinese that recounted the entire national history. In the book's preface, Liu presented the book as an effort to find the answers to three questions: What is Chinese culture? Where is it located? What distinguishes it from others, such as the Indo-European cultures? He hoped to answer these questions by asking others, including: How did China maintain such a vast territory? How did China incorporate various ethnic groups into its makeup? How

---

10 Liu Yizheng 1939: 39.

11 Liu Yizheng 1981.

did China continue its history for such a long period? Liu sought the final cause that shaped history, rather than the traces and effects of history, and his answer was a transhistorical entity—culture. *A History of Chinese Culture* narrates the Chinese nation as, fundamentally, a cultural entity, with its subjectivity constructed in cultural terms. The history of Chinese culture was the history of Chineseness.

This conception of culture or *wenhua* (文化) as the site of the nation's true being reflected a contemporaneous consensus. Revived from classical Chinese by Japanese scholars, *wenhua* (*bunka* in Japanese) was one of the compounds used to translate a specifically European concept to the East Asian world.[12] The 1910s and 1920s witnessed the rising importance of *wenhua* as a category in constructing and imagining the nation. In an effort to establish a degree of discursive control over the meaning of *wenhua*, in 1926 Hu Shi reintroduced *wenhua* and *wenming* as dual concepts and tersely defined civilization as "the totality of how a nation copes with its environment" and culture as "the lifestyle resulting from this coping."[13] *Wenhua*, Hu suggested, was a people's orientation as it presented itself in their outlook and life choices. In the famous treatise, "Research Methods for Chinese History" (*Zhongguo lishi yanjiu fa*) (1922), which originated as a lecture series on the history of Chinese culture at Nankai University, Tianjin, in 1921, Liang Qichao also designated culture as the proper subject of "general history" (*putong shi*). Contrasted with "specialized histories" (*zhuanmen shi*), which focused either on law, economics, society, war, literature, philosophy, the arts, or some other topic, general history synthesized the results of the narrower fields and offered a comprehensive description of culture. *Wenhua*, again, was general, diffuse, and all-inclusive; it was the totality of the nation's existence.[14]

---

12 Liu Lydia 1995: 32–34, 308, 312. However, it should be noted that when the two-character compound circulated back from Japan to China, the word *wenhua* failed to convey some of the connotations of the Japanese *bunka*, including its distinction from *wenming*/*bunmei* 文明, or "civilization." The strong ecumenism of Chinese intellectuals made them impervious to the implicit cultural particularism in the reinvented neologism. For a discussion of this ecumenism at the turn of the twentieth century, see Sun 2002: 3–35. As Lu Xun's 1908 essay "On the Extremity of Culture" (*Wenhua pianzhi lun*) illustrated, *wenhua* and *wenming* were often used interchangeably.

13 Hu 1930b: 3.

14 Tze-ki Hon has noted that *wenhua*, despite being reinvented as the equivalent of culture, still clung to its earlier meaning through its etymons. In this context, *wen*, which may be far too reductively translated as "the civil," stands in contradistinction with *wu* (the military), and *hua* (to transform) with *zhi* (to administer). Put together, *wenhua*, according to Hon, denotes the "gradual and civil process of transformation" (2004: 518–19).

The ascendance of culture in the 1910s and 1920s ushered in a new era in the conception of Chineseness. As Laurence A. Schneider has pointed out, the late Qing discourse of *guocui* (national essence) had paved the way for the construction of culture in China.[15] During the New Culture Movement, *guocui* was the target of harsh criticism. Hu Shi, for example, argued that what had been commonly regarded as "national essence" was often just "national debris."[16] The term quickly fell into oblivion in the 1920s. In its place, *wenhua* arose as the key term in discussions of national identity. The lexical shift was more than strategic: the displacement of the essentialist term coincided with a new approach to Chineseness.[17] More than a fixed quality, culture was to be construed as an organismic system, with various components intricately connected. It allowed a higher degree of complexity that had been lacking in the idea of national essence. Furthermore, signifying an autonomous sphere in itself, culture was also a possible bulwark against social, political, and economic changes.

In the "history of culture" one finds the aspiration to represent the nation in its entirety, simultaneously diachronic and synchronic. In such an account, the nation as a cultural entity was constantly in motion, asserting and creating its subjectivity throughout the movement of history. Liu Yizheng's *History of Chinese Culture* was the first completed project in modern China that consciously aimed to use culture to intercept history's diachrony and its relativizing effects. History, for him, was only the temporal space for culture's actualization; a complete narration of the national history was crucial for gaining a full idea of Chinese culture's potential and for demonstrating its consistency through its many metamorphoses. It thus helped correct the misconceptions about Chinese culture that were due to fragmented knowledge of history.[18]

To create a historical presentation of this transhistorical entity, Liu chose a narrative form that was organized thematically. The main text of the book consists of seventy-five chapters: each focuses on one cultural topic in history. The range of these topics is wide, including politics, government, philosophy, literature, and everyday life. Though all of the

---

[15] Schneider 1976: 57.

[16] Hu 1919a: 557–8.

[17] In spite of the essentialist connotation of the term *guocui*, not every individual that associated with the label understood Chineseness in an essentialist way. I thank Axel Schneider for bringing this point to my attention.

[18] Liu Yizheng 1932: 1.

chapters address aspects of Chinese culture, there is no apparent connection between these topics. Each chapter can be read as a self-contained essay, and taken together they offer a collection of insights into components of Chinese culture throughout history.

As an attempt to demonstrate Chinese culture's ability to overcome history, the narrative of *A History of Chinese Culture* is unique in its chronology. The overall arrangement of the topical chapters follows a general temporal order. However, There was no one sequence of computable time throughout the book. Instead, Liu's account used only traditional dynastic names for temporal registrations. The national time, in this sense, remained fragmented by the dynastic successions. Considering the ongoing debate on the creation of a national chronology, one cannot but find Liu's choice deliberately provocative. Not only does it place an extra demand on the reader—he is expected to be familiar with dynastic history before he begins—but by refusing to unify dynastic chronologies into one numerical system of reckoning, Liu avoids presenting the national time as a single, linear sequence, which would have the inevitable effect of distancing the historical from the present.

At the core of this new elaboration of culture one still finds an essentialist presentation of the nation. Underneath the assumed complexity of "culture" was the latent national character, *guomin xing* (國民性), whose genesis in antiquity defined the founding moment of the nation.[19] Specifically, Liu pointed to the ancient Tang and Yu dynasties, established respectively by the sage-kings Yao and Shun, as the moment of the nation's birth. Although by the time Liu started to write *A History of Chinese Culture* many historians had noted the dubious historicity of these dynasties and the sage-kings, proposing to view them as legends, Liu insisted on basing his narrative on traditional Confucian sources and identifying the Tang and Yu as the one of the most glorious moments in national history. He played up the technological advances in astronomy and hydraulics, breakthroughs in scientific knowledge that, he said, paved the way for sociopolitical achievements. Out of a meritocractic, harmonious, transparent, and just order emerged the nation's unique character, which he summed up in a pair of moral traits.

[19] Here Liu participated in the discourse of national character identified with the May Fourth movement but tried to upset the strongly negative connotations that movement had inscribed in the term. On the formation of that discourse, particularly the role of Lu Xun, see Liu Lydia 1995: 45–76.

The first national trait was moderation, or *zhong* (中 lit., middle). According to Liu, during Tang and Yu times, *Zhongguo* (中國 lit., middle country) became the name most commonly applied to the nation. The name not only described the temper of the people, it also articulated the nation's expectation for itself, its wish to differentiate itself from others. The second trait that Liu singled out was humility, or *rang* (讓 lit., to yield). He attributed this national trait to the rulers of the Tang and Yu, who recognized the competitive tendency in human nature and strived to counteract it by example. Later in the reigns of Yao and Shun, both searched for the country's most virtuous and capable persons and voluntarily transmitted political power to them. Their virtuous deeds made a deep impression on the minds of the Chinese people and *rang* was forever instilled into the national psyche. These traits gave the culture its resilience. They accounted for China's long history, large territory, and ethnic diversity, answering the questions that Liu had posed at the beginning of the book. Moderation made the Chinese accommodating towards different peoples and cultures, facilitating "the governance of an extensive land and the mixing of different ethnic groups."[20] Humility, as an antidote to belligerence and voracity, enabled the nation and the race to "endure without exhibiting weariness."[21] Their continuous presence determined the shape of Chinese history.

History after the ancient period, in Liu's account, consisted of two phases.[22] From the downfall of the Han up to the Ming dynasty (1368–1644) was China's medieval period, which was characterized by "the Indian culture's infiltration into China [through the spread of Buddhism] and its conflict with and assimilation into the original culture of China."[23] After this phase was China's modern period, which witnessed "the withering of both Chinese and Indian cultures and the importation of Western scholarship, thought, religion, and governmental principles to stimulate and interact with them."[24] In his preamble, Liu emphasized that this tripartite

[20] Liu Yizheng 1932: 37.
[21] Ibid., 55.
[22] A tripartite periodization was popular in China during the early twentieth century. Liang Qichao first enunciated this structure in 1901 (Liang 1901: 1626–27, Tang 1996: 43), and most general histories adopted it in their narratives (Hon 2004: 516). Liang explicitly cited Western historiography as his source of inspiration. In Liu's case, Japanese historiography, especially Naka Michiyo's work, might have played a significant role in his adoption of the tripartite divide into ancient, medieval, and modern periods.
[23] Ibid., 1.
[24] Ibid.

periodization was an artificial device to highlight moments of cultural transmutation, rather than interruption. China's continuous growth and expansion through its constant incorporation of foreign elements attested to the culture's creative and transformative strength, the common theme in all three periods.[25] Liu's narrative ends at the 1920s. Admitting that the remarkable transformations of recent years were uncanny, Liu nevertheless included his present as a part of the third period.[26] This made China's recent clash with the West simply another chapter in a long story of cultural conflicts and assimilations. And Liu refrained from making any statement about the future. Whether the nation could once again rise to the occasion and prove its cultural resilience remained an unanswered question in the book, but the author's hopes were clear enough.

A History of Chinese Culture was meant as a counternarrative to the May Fourth account of history and culture. In that account, modernity was a necessary break from tradition. An impotent traditional culture drained the nation of its creative energy and prevented it from moving forward into a new era. Culture represented the stagnation that history would overcome. This discourse of totalistic antitraditionalism set forth an image of Chinese culture that was no less tenacious than Liu's, though valued quite differently. Yet, in a dramatic dissent, Liu refused to relegate historical experience to the past. If the national crisis of the 1920s bore similarities to previous experiences, then the resolution of those earlier conflicts between China and alien groups were relevant to the present.[27] Debunking the myth of a unique present, Liu's *History* also questioned whether history needed to conform to the linearity of time. The structure of its narrative raised an implicit question: Why should the historian, he asked, let the forward movement of time prevent him from bringing the past into the play of the present?

The Politics of History Writing in the May Fourth Era

To bring the past into the present requires a new hermeneutics of history. In this regard, Liu meant to contend with the New Culture projects. The most prominent advocate of these projects was Hu Shi, a graduate of

[25] Ibid., 1–2.
[26] Ibid., 963–73.
[27] Ibid., 3–5.

the doctoral program of philosophy at Columbia University, John Dewey's best-known student in China, and one of the leading players in the New Culture Movement. An accomplished historian of Chinese philosophy and literature, Hu started to campaign in the late 1910s and early 1920s for "systematizing the national heritage," or *zhengli guogu* (整理國故). The main goal was to apply modern scientific methods to combing through and reexamining historical knowledge. Traditional knowledge, seen as utterly lacking in "logic, coherence, and systemization" needed a thorough reorganization and reevaluation.[28] A specific historical method was crucial to this endeavor: "Historical perspective allows one to return what belongs to the Han back to the Han, what belongs to the Wei and Jin to the Wei and Jin, what belongs to the Tang to the Tang.... Only after restoring everything back to the original [in history] can we then judge its significance."[29] History as a modern discipline would divide culture into isolated events and place them back to their respective contexts in time.

The modern conception of time as the scientific reckoning of sequential succession was the key to the method developed by Hu Shi and his followers. Its inexorable forward movement established the otherness of the past and turned it into an object, rather than the source, of modern knowledge. Mao Zishui, Hu's student at Beijing University, openly stated that to "systemize the national heritage, one should treat all heritage from the past as mere historical material."[30] Criticized for his obsession with the past, Hu, in as late as 1927, still insisted that the goal of the systemizing enterprise was to "transform the marvelous into the rotten, the magical into the ordinary."[31] The scientific approach to historical studies meant stripping the facts of the past of any meaning and reducing them to raw material awaiting interpretation.

Opposed to this trend toward objectivizing history, Liu upheld a more empathetic reading of the past. His narrative requested that his readers shine the most positive light on the past. For instance, when he wrote about the social hierarchy imposed by the sage-kings of the Tang and Yu,

28 Hu 1919a: 557.

29 Hu 1924: 10.

30 Mao 1919: 735.

31 Hu 1930a: 117. Hu Shi was famous for his admiration for the "scientific method" employed by the classical scholars of the Qing Dynasty. Yet, in a letter to Mao Zihui, he lamented that Qing scholars deployed the scientific method without being conscious of it (1919b: 328). Two years latter, in the foreword to a new journal dedicated to national learning (*guoxue*), Hu also criticized Qing scholarship for its narrow scope, fixation with details, and lack of comparative framework (1924: 7–9).

which he acknowledged was less than perfect, Liu still defended the system because he believed that it was intended to reward and foster moral behavior. The intention—which only a properly scientific historian could perceive—was to Liu more important than the harm done by social stratification: "Only those who refuse to empathize with the intention and unreasonably blame the system for dividing people into classes would call the hierarchy at that time unjust and a policy of obscurantism."[32] To fortify this hermeneutic strategy, Liu punctuated his narrative with reminders of how the present was a continuum of the past, and he constantly evoked a sense of rapport with the past. The invention of the agrarian calendar, for example, would not have been possible without many generations of continual effort. "The hardships and labors of the process are not recorded in histories. Yet, inferring from the conditions of the time, one should understand that the effort was tremendous. We, who were born thousands of years afterward, benefit from such inventions but remain ignorant [about the effort required]. It is indeed shameful!"[33] In passages like this one, Liu turned historical knowledge, the opposite of ignorance, into a form of gratitude. The historian and his readers are morally obligated to empathize with his subject. The rapport with the past is both the point of departure and the objective of the historical quest.

Empathy as a hermeneutic strategy may have set Liu's study apart from the other projects that emerged from the New Culture movement. But did it really enable the historian to bring the past into contact with the present? Mindful though he was of empathy, Liu nevertheless assumed the superiority of his own knowledge about history and placed himself on the epistemological high ground vis-à-vis the voices of the past. The suppression of China's traditional episteme is particularly blatant in Liu's maneuvering around two subjects. The first was the bottom-up view of history and the second was secularism in Chinese culture.

Liu's narrative of the ancient period was a deliberate reconstruction of civil relations in the Golden Age. He insisted that all the historical records of political and technological achievements should be read as manifestations of the collective effort of the people. Irrigation projects, for example, could not have been achieved without the involvement of the whole nation. The traditional historiography, Liu complained, wrongly attributed it to a single person and thus created a mythical image of the

[32] Liu Yizheng 1932: 47.
[33] Ibid., 49.

ancient leaders.[34] Liu's narrative consistently directed the reader's attention to the harmonious interactions between groups in ancient society. Quoting *The Book of Documents*, Liu pointed to the colloquial expressions used by ancient rulers addressing their subjects as evidence for easy communication across political boundaries during the Tang and Yu.[35] Even the choice of the name Zhongguo, he insinuated, was based on a consensus among "all citizens of the nation."[36]

Many of Liu's interpretative moves, which he used to extrapolate social information from political records, later became standard for social historians of early China. But for Liu they were part of a clearly planned strategy aimed at establishing a historical method. To write a national history had to mean producing a narrative that would balance the earlier histories so obsessed with politics. Relying on histories and historical records, Liu read them in a new light and extended their use beyond their original purpose. In his book's preamble, Liu explained that this hermeneutic maneuvering was a deliberate rectification of the misconceptions fostered by traditional histories. Agitated by how people such as Liang Qichao denounced the previous understanding of history as worthless "genealogies of imperial families," Liu tried to demonstrate in *History* that "the material for social and national history is everywhere":[37] one needed only know how to read the histories.

Parallel to Liu's social reading of political history was his secular reading of religious records. His portrait of ancient society was insistently secular. Heaven, or *tian* (天), was nature in the most neutral sense. When the ancient sages were creating the nation, they were most preoccupied with the knowledge of *tian* because, at the primordial stage of civilization, managing natural forces was crucial to people's basic subsistence. Astronomy and meteorology, that is, the knowledge of the movements of heavenly bodies, were important in assuring the food supply and therefore socioeconomic stability.[38] The sages' reverence for *tian* was a pillar of their sagehood because it signified their commitment to the people's wellbeing.[39] By the same token, the Mandate of Heaven, *tianming* (天命), was also explained away as a historical allusion to the failings of the bygone

34 Ibid., 65, 99.
35 Ibid., 76.
36 Ibid., 36.
37 Ibid., 7.
38 Ibid., 49.
39 Ibid., 79.

dynasties and thus a moral admonition directed towards those in power. Liu suggested that while the ancient texts did describe *tian* in a specifically religious manner, its *true* meaning consisted in its moral message.

To produce this secular representation of the Chinese experience in ancient times, again, entailed a methodical transformation of historical knowledge. In the chapter devoted to the ritual institution of the Zhou dynasty, Liu explained his decision to exclude from his narrative any information about sacrificial rites. While records of these rites were abundant, in the end they "revealed [little but] the polytheistic superstition of the Zhou people.... What they uttered [during these rites] was not appropriate for providing leadership to the people. I therefore do not present them here."[40] Liu embraced a particular conception of the proper subject for history, and he would dwell on the proper and suppress the improper. If improper topics could not be wholly suppressed, the historian would give them a transformative spin. Commenting on the difference among the Three Dynasties—Xia, Shang, and Zhou—Liu resorted to *The Book of Rites*. He gave a long quotation from the Classic: "The way (*dao* 道) of the Xia reveres destiny, patronizes ghosts, and reveres spirits.... That of the Yin [Shang] respects spirits, encourages the people to patronize them, and give a priority (*shang* 尚) first to ghosts and then rites.... That of the Zhou respects rites and emphasizes sacrifices, patronizes ghosts, and reveres spirits." Although the passage is evidently a comparison of the Three Dynasties' different attitudes towards the supernatural, Liu explained that *dao* and *shang* were two cue words for policy guidelines. This passage from *The Book of Rites*, Liu concluded, proved that the Three Dynasties' policies varied according to time and were by no means changeless.[41] Liu remained silent about the religious references in the passage and made no comments about how the religious and political were intertwined in ancient times.

From religious outlook to policy guideline, the distance between Liu's interpretation and the voice of the ancient text conveys how historical knowledge from the past was treated as a system of signs, whose true meaning lay beyond itself. As the historian deciphered texts that had not seemed too obscure until he got hold of them, he also dismissed the seeming coherence among them as illusory. Bringing the past into the present

40 Ibid., 194.
41 Ibid., 89.

meant writing over the past's self-knowledge.[42] This writing over was possible only because the historian of the national history lived in a new era of historical consciousness. He presumably had a more comprehensive knowledge of time and stood at higher vantage point in the history of writing history. In this regard, the historian of empathetic understanding did not differ much from his radical contemporaries. Upon the first publication of *A History of Chinese Culture* in book form in 1932, Hu Shi wrote a lukewarm review to point out many inadequacies of the book. Hu nevertheless ended his review on a positive note. The book's mindfulness of citing historical sources as evidence, according to Hu, set a positive example for historical method.[43] Hu Shi's compliment revealed the fact that in systematically and consciously turning traditional sources into nothing more than historical "material," Liu was very much in accord with the "systematization of the national heritage." Much as he resisted placing the past simply in the past, Liu Yizheng could not overcome the linear temporal movement of knowledge. His writing of history was inevitably a rewriting of history that altered the voices of the past rather than invigorating them.

References Cited

Certeau, Michel de (1988) "Making History," in *The Writing of History*, trans. Tom Conley. New York: Columbia University Press.

Gadamer, Hans-Georg (1975), *Truth and Method*. Ed. Garrett Bardden and John Cumming. New York: Seabury.

Hon, Tze-ki (2004), "Cultural Identity and Local Self-Government: A Study of Liu Yizheng's History of Chinese Culture," in *Modern China* 30, 4 (July 2004): 506–42.

——, and Robert J. Culp (2007), "Introduction," in Tze-ki Hon and Robert J. Culp eds., *The Politics of Historical Production in Late Qing and Republican China*. Leiden: Brill, 1–17.

[42] Hans-Georg Gadamer, in his critique of historicism, warned against the pitfall of naïve assimilation, when the historian tries to impose a historical horizon on tradition. For Gadamer, the recognition of the tension between the past and the present is the condition for the truth of history. "The hermeneutic task consists not in covering up this tension in a naïve assimilation but in developing it consciously" (1975: 273). Temporal distance sharpens the consciousness of one's own horizon and prepares the ground for its expansion. Conversely, naïve assimilation, by assuming the identity of the past and the present, forfeits the benefits of such distance. Liu's narrative of the history of Chinese culture is a case in point. By covering up the tension between the past and the present, Liu, to quote Gadamer, "has fundamentally given up the claim to find in the tradition a truth valid and intelligible for oneself" (270).

[43] Hu 1998: 772.

Hu Shi ([1919a] 1998), "Xin sichao de yiyi" 新思潮的意義 (The meaning of new thought), in *Hu Shi wencun* (Collection of Hu Shi's writings). Beijing: Beijing University Press, 551–58.

—— ([1919b] 1998), "Lun guoguxue (da Mao Zishui)" 論國故學 (答毛子水) (On the learning of national heritage (reply to Mao Zishui), in *Hu Shi wencun*, 327–28.

—— ([1924] 1998), "*Guoxue jikan* fakan xuanyan" 國學季刊發刊宣言 (Foreword to *National Learning Quarterly*), in *Hu Shi wencun er ji* (Collection two of Hu Shi's writings). Beijing: Beijing University Press, 5–17.

—— ([1930a] 1998), "Zhengli guogu yu 'dagui'—gei Hao Xu xiansheng xin" 整理國故與打鬼—給浩徐先生信 (The reorganization of the national past and "ghost hunting"—a letter to Mr. Xu Hao), in *Hu Shi wencun san ji* (Collection three of Hu Shi's writings). Beijing: Beijing University Press, 115–18.

—— ([1930b] 1998), "Women duiyu Xiyang jindai wenming de taidu" 我們對於西洋近代文明的態度 (Our attitude towards Western modern civilization), in *Hu Shi wencun san ji*, 3–22.

—— (1998), "Ping Liu Yizheng bian zhu *Zhongguo wenhua shi* 評柳詒徵編著中國文化史 (Comments on *A History of Chinese Culture*, compiled and written by Liy Yizheng), in *Hu Shi ji wei xueshu wenji* (Additional collection of Hu Shi's academic writings). Beijing: Beijing University Press, 768–72.

Liang Qichao ([1901] 2001), "Zhongguo shi xu lun," 中國史緒論 (Introduction to Chinese History), in *Yinbingshi wenji dianjiao*. Kunming: Yunnan jiaoyu chubanshe, 1620–27.

—— ([1902a] 2001), "Xin shixue," 新史學 (New Historiography), in *Yinbingshi wenji dianjiao*. Kunming: Yunnan jiaoyu chubanshe, 1628–47.

—— ([1902b] 2001), "Zhongguo zhuanzhi zhengzhi jinhua lun," 中國專制政治進化論 (On the Evolution of China's Authoritarian Politics), in *Yinbingshi wenji dianjiao*. Kunming: Yunnan jiaoyu chubanshe, 1648–70.

—— ([1922] 1980), "Zhongguo lishi yanjiu fa," 中國歷史研究法 (Methods for the Study of Chinese History), in *Liang Qichao shixue lunzhu sanzhong*. Hong Kong: Sanlian shudian, 41–177.

Liu, Lydia H. (1995), *Translingual Practice: Literature, National Culture, and Translated Modernity—China 1900–1937*. Stanford, CA: Stanford University Press.

Liu Yizheng ([1932] 2001), *Zhongguo wenhua shi* 中國文化史 (A History of Chinese Culture). Shanghai: Guji chubanshe.

—— (1981), "Wo de zishu" 我的自述 (My self-description), in *Wenxian* 7 (July 1981): 124–25.

—— ([1939] 2002), "Ji zaonian shi" 記早年事 (On things of early age), in *Qutang xueji* (Notes on learning in Qutang). Shanghai: Shanghai shudian, 21–41.

Mao Zishui (1919), "Guoku he kexue de jingshen" 國故和科學的精神 (National heritage and scientific spirit), in *Xinchao* 1, 5: 731–43.

Schneider, Laurence A. (1976), "National Essence and the New Intelligentsia," in Charlotte Furth ed., *The Limits of Change*. Cambridge, MA: Harvard University Press, 57–89.

Sun, Lung-kee (2002), *The Chinese National Character: From Nationhood to Individuality*. Armonk, NY: M. E. Sharpe.

Tang, Xiaobing (1996), *Global Space and The Nationalist Discourse of Modernity: The Historical Thinking of Liang Qichao*. Stanford, CA: Stanford University Press.

Zhang Taiyan 張章太炎 (1902), "Zhang Taiyan laijian" 張章太炎來件 (Correspondence from Zhang Taiyan), in *Xinmin congbao* 13 (1902): 57–60.

INDEX